Advances in Comparative and Environmental Physiology 14

Volumes already published

Vol. 1: NaCl Transport in Epithelia. Edited by *R. Greger* (1988)

Vol. 2: From the Contents: Hibernation – Vapor Absorption. Nutrient Transport – Stable Pollutants (1988)

Vol. 3: Molecular and Cellular Basis of Social Behavior in Vertebrates. Edited by *J. Balthazart* (1989)

Vol. 4: Animal Adaptation to Cold. Edited by *L. C. H. Wang* (1989)

Vol. 5: From the Contents: Osmoregulation – Chemoreception – Temperature – Spectroscopy – Metalloproteins – Prostaglandins (Eicosanoids) – Maternal-Fetal Relationships (1989)

Vol. 6: Vertebrate Gas Exchange: From Environment to Cell. Edited by *R. G. Boutilier* (1990)

Vol. 7: From the Contents: Protein Turnover – Xenobiotic Metabolism Lipoproteins (1991)

Vol. 8: From the Contents: Dormancy in Aquatic Invertebrates-Root Effect – Red Blood Cell Functions in Fish Diving Mammals (1991)

Vol. 9: Volume and Osmolality Control in Animal Cells. Edited by *R. Gilles, E. K. Hoffmann,* and *L. Bolis* (1991)

Vol. 10: Comparative Aspects of Mechanoreceptor Systems. Edited by *F. Ito* (1992)

Vol. 11: Mechanics of Animal Locomotion. Edited by *R. McN. Alexander* (1992)

Vol. 12: Muscle Contraction and Cell Motility: Molecular and Cellular Aspects. Edited by *H. Sugi* (1992)

Vol. 13: Blood and Tissue Oxygen Carriers. Edited by *Ch. P. Mangum*

Vol. 14: Interaction of Cell Volume and Cell Function. Edited by *F. Lang* and *D. Häussinger*

Volumes in preparation

Biomechanics of Feeding in Vertebrates.
Edited by *V. L. Bels, M. Chardon,* and *P. Vandewalle*

Ion Transport in Vertebrate Colon. Edited by *W. Clauss*

Electrogenic Chloride Transporters in Biological Membranes.
Edited by *G. A. Gerencser*

Effects of High Pressure on Biological Systems. Edited by *A. G. Macdonald*

Mechanisms of Systematic Regulation. Vol. 1: Respiration and Circulation.
Vol. 2: Acid-Base Regulation, Ion Transfer and Metabolism. Edited by *N. Heisler*

Advances in

Comparative and Environmental Physiology 14

Interaction of Cell Volume and Cell Function

Guest Editors: F. Lang and D. Häussinger

With Contributions by

L. Bianchini · W. Gerok · J. Graf · S. Grinstein · W. B. Guggino
A. P. Halestrap · D. Häussinger · E. K. Hoffmann · H. Kettenmann
H. K. Kimelberg · A. Kurtz · I. H. Lambert · F. Lang
E. R. O'Connor · W. C. O'Neill · M. Ritter · H. Scholz
L. O. Simonsen · K. Turnheim · H. Völkl

With 81 Figures

Springer-Verlag
Berlin Heidelberg New York
London Paris Tokyo Hong Kong
Barcelona Budapest

Guest Editors:

Prof. Dr. *Florian Lang*
Physiologisches Institut
Universität Tübingen
Gmelinstraße 5
W-7400 Tübingen, FRG

Prof. Dr. *Dieter Häussinger*
Abteilung für Innere Medizin
Universität Freiburg
Hugstetterstraße 55
W-7800 Freiburg, FRG

ISBN-13:978-3-642-77126-2 e-ISBN-13:978-3-642-77124-8
DOI: 10.1007/978-3-642-77124-8

ISSN 0938-2673

Softcover reprint of the hardcover 1st edition 1993

Typesetting: Best-set, Hong Kong, and K+V Fotosatz GmbH, Beerfelden
31/3145-5 4 3 2 1 0 – Printed on acid-free paper

Foreword to the Series

The aim of the series is to provide comprehensive, integrated reviews giving sound, critical, and provocative summaries of our present knowledge in environmental and comparative physiology, from the molecular to the organismic level.

Living organisms have evolved a widespread range of basic solutions to cope with the different problems, both organismal and environmental, with which they are faced. A clear understanding of these solutions is of course of fundamental interest for all biologists, zoologically or medically oriented. They can be best comprehended in the framework of the environmental and/or comparative approaches. These approaches demand either wide surveys of animal forms or a knowledge of the specific adaptive features of the species considered. This diversity of requirements, both at the conceptual and technological level, together with the fact that physiology and biochemistry have long been mainly devoted to the service of medicine, can account for the fact that these approaches emerged only slowly amongst the other new, more rapidly growing disciplines of the biological sciences.

The field has now gained the international status it deserves and the organization of a series devoted to it appeared timely to me in view of its actual rapid development and of the interest it arouses for a growing number of biologists, physiologists, and biochemists, independently of their basic, major orientation.

Liège, Belgium, Spring 1988 *Raymond Gilles*

List of Editors

Preface

One of the prerequisites for survival is the ability of cells to maintain their volume. Thus, during the course of evolution, cells have learned a variety of strategies to achieve cell volume homeostasis. This volume regulatory machinery involves regulation of both cellular metabolism and cellular transport and is exploited by hormones and transmitters to regulate cellular function.

It is the aim of this book to illustrate the complex interplay of cell volume regulatory mechanisms and cellular function in a variety of tissues. Our knowledge is still far from conclusive, however, and the present collection of reviews is thought to foster further experimental efforts to unravel the role of cell volume in the integrated function of cells.

Tübingen and Freiburg, November 1992

Florian Lang
Dieter Häussinger

Contents

Chapter 7

Chapter 8

Chapter 9

Chapter 1

Cell Volume Regulatory Mechanisms – An Overview

F. Lang[1,2], *M. Ritter*[1], *H. Völkl*[1], and *D. Häussinger*[3]

Contents

[1] Department for Physiology, University of Innsbruck, Fritz-Pregl-Straße 3, 6010 Innsbruck, Austria
[2] Present address: Physiologisches Institut I, Gmelinstrape S, W-7400 Tübingen, FRG
[3] Department of Internal Medicine, University of Freiburg, Hugstetterstraße 55, W-7800 Freiburg, FRG

Advances in Comparative and Environmental Physiology, Vol. 14

1 Introduction

Most cell membranes are highly permeable to water. Movement of water across the cell membranes is driven by an osmotic and hydrostatic pressure gradient. Animal cells are not able to withstand significant hydrostatic pressure gradients. Thus, any osmotic gradient across the cell membrane will lead to the respective movements of water and alterations of cell volume.

An osmotic gradient could arise from a change in extracellular or intracellular osmolarity. With only few exceptions (e.g. kidney medulla) the extracellular osmolarity undergoes little alteration in most mammalian tissues. However, the intracellular generation or consumption of osmotically active substances as well as the transport of osmotically active substances across the cell membrane impose an ever-changing burden on cell volume homeostatic mechanisms.

Thus, it is not surprising that despite the constant osmolarity of extracellular fluid most cells have maintained the ability to regulate their volume. This ability is usually disclosed by exposure of cells to anisotonic media. Most cells, if suddenly exposed to hypotonic media, initially swell more or less like perfect osmometers but within minutes retain (almost) their original cell volume. This behavior has been labeled regulatory cell volume decrease (RVD). If the cells are suddenly exposed to hypertonic media, they initially shrink but within minutes retain (almost) their original cell volume. This behavior has been labeled regulatory cell volume increase (RVI).

The regulation of cell volume is primarily achieved by transport of electrolytes across the cell membrane. However, in addition, cell volume regulatory mechanisms modify the transport of nonelectrolytes across the cell membrane as well as cellular metabolism, which is amplified in a separate chapter of this book (Häussinger et al. this Vol.) and will thus not be dealt with here.

In this chapter, a synopsis of the diverse cell volume regulatory mechanisms will be followed by a discussion of some factors interfering with cell volume regulation. Only cell volume regulation in mammalian tissues will be dealt with. Intracellular mechanisms linking cell volume to the regulatory mechanisms will be detailed in a separate chapter (Hoffmann et al., this Vol.). For comparative aspects of cell volume regulation, the reader is referred to reviews focusing on this topic (Gilles 1987, 1988; Goldstein and Kleinzeller 1987). Furthermore, excellent reviews have appeared on various aspects of cell volume regulation (Macknight and Leaf 1977; Macknight 1987, 1988; Ballanyi and Grafe 1988; Beck et al. 1988a; Deutsch and Lee 1988; Gilles 1988; Graf et al. 1988; Hoffmann et al. 1988; Lauf 1988; Reuss 1988; Völkl et al. 1988; Chamberlin and Strange 1989; Hoffmann and Simonsen 1989; Schultz 1989a; Beck and Guder 1990; Geck 1990; Kleinzeller and Ziyedeh 1990; McConnell and Goldstein 1990; Ussing 1990), and the interaction of cell volume regulation with cell function in various tissues is detailed in the other chapters of this book.

2 Mechanisms of Regulatory Cell Volume Decrease (RVD)

RVD is achieved by the reduction of intracellular osmotic activity, accomplished by cellular loss of electrolytes and nonelectrolytes and by metabolic disposal of osmotically active substances. The cellular loss of electrolytes, which is the most efficient and rapid means to decrease intracellular osmolarity, is in most cells due to activation of ion channels and/or KCl symport. In *Amphiuma* and dog red cells, K^+/H^+ exchange and Na^+/Ca^{2+} exchange, respectively, have been invoked to mediate RVD (see Table 1).

2.1 Ion Channels

In a variety of different cells (see Table 1), cell swelling leads to the activation of ion channels, which allows the volume regulatory loss of cellular K^+ and anions. As illustrated in Fig. 1, activation of the ion channels leads to a decrease in cell membrane resistance. In Table 2, a variety of different ion channels are listed that are activated by cell swelling.

The movement of ions across the channels is dictated by the respective electrochemical gradients (see Table 2), which depend on both the chemical ion gradient across the cell membrane and the cell membrane potential. In almost every cell the electrochemical gradient for potassium is usually outwardly directed and activation of K^+ channels leads to the extrusion of cellular potassium. The exit of potassium tends to hyperpolarize the cell membrane, thus increasing the driving force for anion exit. Depending on the respective electrochemical gradient, the activation of anion channels allows the exit of cellular anions favoring depolarization of the cell membrane. The activation of unselective ion channels, which do not discriminate between K^+ and Na^+, usually depolarizes the cell membrane, since, due to the difference in driving forces, the entry of sodium surpasses the exit of potassium through these channels.

The effect of cell swelling on the cell membrane potential depends on the ion channels activated and on the cell membrane potential prior to cell swelling, as illustrated by the following two examples:

The cell membrane potential of isolated hepatocytes has been reported to be low (Graf et al. 1988), i.e., close to the chloride equilibrium potential and far below the potassium equilibrium potential. Accordingly, due to the high electrochemical driving force for potassium, the activation of K^+ channels in these cells leads to rapid cellular loss of potassium and to marked hyperpolarization of the cell membrane, which increases the driving force for chloride exit (Howard and Wondergem 1987; Graf et al. 1988). Activation of chloride channels would have little effect on cell membrane potential in these cells and would have little effect on ion movements across

Table 1. Cell volume regulatory mechanisms

Regulatory cell volume decrease (RVD)

Parallel activation of K^+ and Cl^- channels:
- Ehrlich ascites tumor cells (Hoffmann et al. 1984, 1986, 1988; Hoffmann 1985a,b, 1987; Lang et al. 1987; Hudson and Schultz 1988)
- lymphocytes (Cheung et al. 1982; Grinstein et al. 1982a,b,c, 1983a, 1984; Sarkadi et al. 1984a,b, 1985; Deutsch and Lee 1988)
- platelets (Livne et al. 1987)
- endothelial cells (O'Neill 1992)
- fibroblasts (Rugolo et al. 1989)
- astrocytes (Kimelberg and Kettenmann 1990; Kimelberg et al. 1990b, 1992)
- choroid plexus (Christensen 1987; Christensen et al. 1989)
- ciliary epithelium (Farahbahksh and Fain 1987, Yantorno et al. 1989)
- eye lens (Beebe et al. 1990)
- intestine (Lau et al. 1984; Schultz et al. 1985; Germann et al. 1986; Hazama and Okada 1988; Montero and Ilundain 1989; MacLeod and Hamilton 1990; Wiener and Turnheim 1990)
- Madin-Darby canine kidney (MDCK) cells (Simmons 1984; Roy and Sauvé 1987; Völkl et al. 1988; Paulmichl et al. 1989)
- frog skin (Ussing 1986; Costa et al. 1987)
- frog urinary bladder (Davis and Finn 1985, 1987)
- convoluted proximal renal tubules (Welling et al. 1985; Kirk et al. 1987a,b; Sackin 1987, 1989; Tauc et al. 1990; Welling and O'Neill 1990; Schild et al. 1991)
- liver (Kristensen and Folke 1984; Kristensen 1986; Häussinger et al. 1990d)
- trachea (McCann et al. 1989; Butt al. 1990)

Activation of electrogenic bicarbonate exit:
- straight proximal renal tubules of the mouse kidney (Völkl and Lang 1988a,b; Völkl et al. 1988)
- MDCK cells (Weiss and Lang 1992)

Activation of KCl cotransport:
- erythrocytes from toadfish (Lauf 1982), birds (Kregenow 1981), sheep (Dunham and Ellory 1981; Ellory et al. 1985; Lauf 1985) and man (O'Neill 1987)
- *Necturus* gallbladder (Larson and Spring 1984)
- calcium-depleted Ehrlich ascites tumor cells (Kramhoft et al. 1986)
- dog red cells with low intracellular sodium concentration (Parker 1983a)

Activation of Na^+ entry:
- frog skin (Costa et al. 1987)
- urinary bladder (Finn and Reuss 1975; Lewis et al. 1985)

Parallel operation of K^+/H^+ exchange and Cl^-/HCO_3^- exchange:
- *Amphiuma* red cells (Cala 1983; Cala et al. 1986)
- frog skin (Ussing 1990)
- cornea (Bonanno 1991)

K^+/H^+ ATPase:
- gastric epithelium (Sernka 1990)

Na^+/Ca^{++} exchange and Ca^{2+} ATPase:
- dog red cells with high intracellular sodium concentration (Parker et al. 1975; Parker 1978, 1979)

Loss of cellular amino acids (Hoffmann and Hendil 1976; Fugelli and Rohrs 1980; Thurston et al. 1981; Hoffmann and Lambert 1983; King and Goldstein 1983; Vislie 1983; Fincham et al. 1987; Law and Turner 1987; Roy and Sauve 1987; Pasantes-Morales and Schousboe 1988; Bevan et al. 1990; Kimelberg et al. 1990a; Sanchez Olea et al. 1991)

Table 1. Continued

Regulatory cell volume increase (RVI)

Parallel activation of Na^+/H^+ exchange and Cl^-/HCO_3^- exchange:
- *Amphiuma* (Cala 1980, 1983, 1985; Kregenow et al. 1985; Siebens and Kregenow 1985), rabbit (Jennings et al. 1986) and dog (Parker 1983b; Parker and Castranova 1984) red cells
- lymphocytes (Grinstein et al. 1983b, 1984, 1985a,b)
- osteoclasts (Green et al. 1988)
- endothelial cells (O'Neill and Klein 1992)
- parotis (Soltoff et al. 1989)
- pancreas (Muallem and Loessberg 1990)
- gallbladder (Fisher et al. 1981; Ericson and Spring 1982; Reuss 1984; Spring 1985)
- liver (Häussinger et al. 1990d)
- medullary thick ascending limb of the kidney (Blumenfeld et al. 1989)
- medullary collecting duct (Sun and Hebert 1989)
- proximal tubule (Lohr et al. 1989)
- MDCK cells (Ritter et al. 1991b)

NaCl-KCl symport or NaCl symport:
- avian (Schmidt and McManus 1977a,b,c; Bakker-Grunwald 1981; Kregenow 1981), ferret (Mercer and Hoffman 1985), rat (Duhm and Göbel 1984) and human (Adragna and Tosteson 1984; O'Neill and Mikkelsen 1987) red cells
- Ehrlich ascites tumor cells (Geck et al. 1980; Hoffmann et al 1983, 1988; Geck and Heinz 1986)
- 3T3 cells (Bakker-Grunwald et al. 1982)
- HeLa cells (Tivey et al. 1985)
- frog skin (Ussing 1985, 1986)
- astrocytes (Kimelberg and Frangakis 1986; Tas et al. 1987)
- retinal pigment epithelium (Adorante and Miller 1990)
- C6 glioma cells (Chassande et al. 1988)
- intestine (MacLeod and Hamilton 1990)
- trachea (Musch and Field 1989)
- rabbit medullary thick ascending limb cells (Eveloff and Calamia 1986; Blumenfeld et al. 1989)
- OK cells (Montrose et al. 1988)
- MDCK cells (Ritter et al. 1991b)
- parotis (Soltoff et al. 1989)

Inhibition of K^+ and/or Cl^- channels:
- MDCK cells (Ritter et al. 1991a)
- frog skin (Costa et al. 1987; Leibowish et al. 1988)
- urinary bladder (Lewis et al. 1985; Donaldson et al. 1989)
- gallbladder (Stoddard and Reuss 1989)
- intestine (Montero and Ilundain 1989)
- proximal tubule (Lapointe et al. 1990)
- heart cells (Frelin et al. 1986; Drewnowska and Baumgarten 1991)

Na^+/K^+ ATPase:
- intestine (Montero and Ilundain 1989)
- liver (Häussinger et al. 1990d)
- thick ascending limb (Hebert 1986)

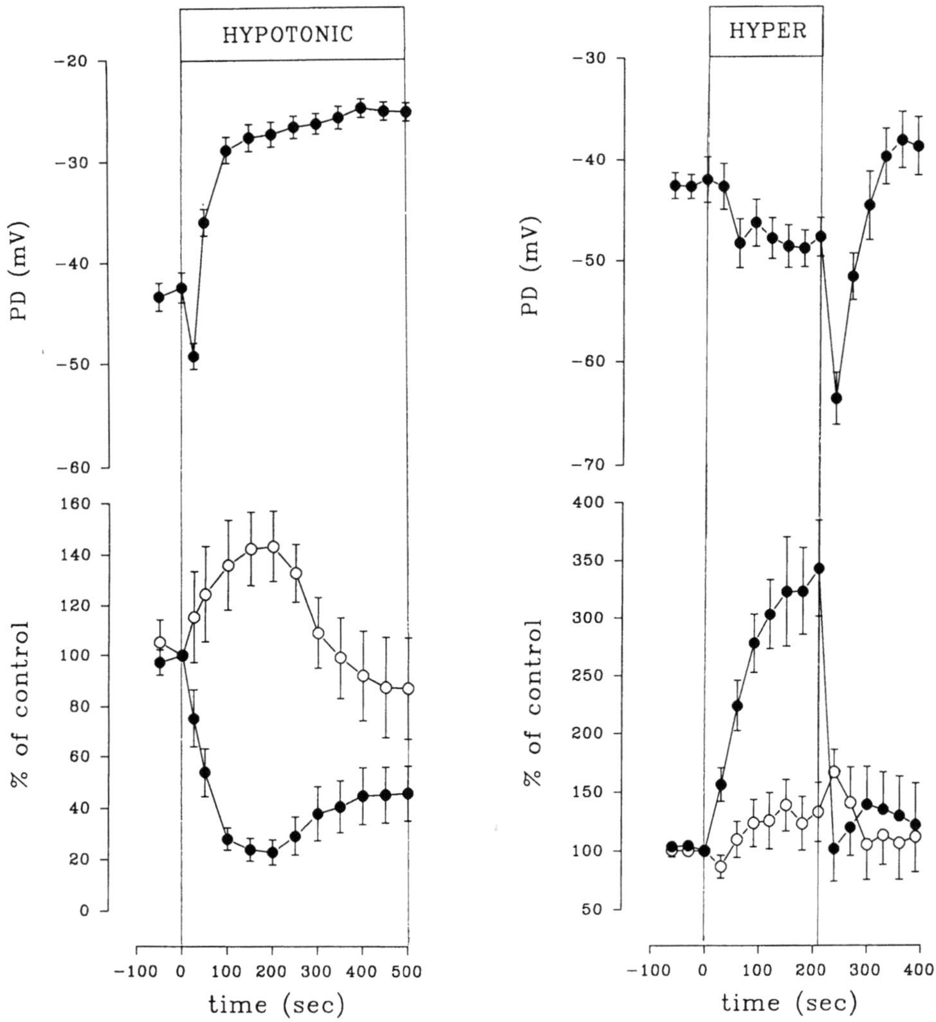

Fig. 1. Influence of cell swelling (*left*) and cell shrinkage (*right*) on the potential difference across the cell membrane (*PD*, *upper panel*), the cell membrane resistances (*lower panels*, *filled symbols*) and the gap junctional resistances (*lower panels*, *open symbols*) of MDCK cells. PD has been determined by conventional microelectrodes, the resistances by cellular cable analysis. (Ritter et al. 1991a,b)

the cell membrane. Clearly, the activation of K^+ channels and the continued patency of anion channels, but not the additional activation of anion channels, are prerequisites for electrolyte extrusion in these cells. Other cells reported to hyperpolarize following cell swelling are found in the early proximal renal tubules (Lopes and Guggino 1987) and the small intestine (Lau et al. 1984).

Table 2. Intracellular composition and driving forces for ion transport systems in proximal renal tubules and thick ascending limbs bathed in isotonic extracellular fluid (Lang 1988). Approximate values for ion activities in intracellular (a_i) and extracellular (a_e) fluid, equilibrium potentials (E_o) and differences in E_o and cell membrane potential ($E_m = -70\,mV$)

	a_i (mmol/l)	a_e (mmol/l)	E·Z·F R·T	E_o (mV)	$E_m - E_o$ (mV)
Sodium conductance	12	115	2.3	+60	−130
Potassium conductance	80	4	−3.0	−80	+10
Chloride conductance	15	90	−1.8	−50	−20
Bicarbonate conductance	10	20	0.7	−20	−50
Calcium conductance	0.0001	0.8	9.0	+120	−190
Na^+/H^+ exchange			3.0	–	–
$Na^+ (HCO_3^-)_3$ cotransport			4.3	−58	−15
Na^+/Ca^{2+} exchange (3:1)			−2.2	−60	−10
NaClKCl cotransport			2.8	–	–
KCl symport			1.2	–	–
Cl^-/HCO_3^- exchange			1.1	–	–

Values for E_m and intracellular ion activities are from (Edelman et al. 1978; Frömter 1979, 1982, 1984; Lee et al. 1980; Biagi et al. 1981a,b; Cemerikic et al. 1982; Cassola et al. 1983; Greger et al. 1983, 1984; Cardinal et al. 1984; Wang et al. 1984; Yoshitomi and Frömter 1984, 1985; Greger 1985; Yoshitomi and Burckhardt 1985; Völkl et al. 1986)

In Ehrlich ascites tumor cells (Lang et al. 1987; Lambert et al. 1989) or even more so in straight proximal renal tubule cells (Völkl et al. 1986), the cell membrane potential is rather high and closer to the potassium equilibrium potential than to the chloride equilibrium potential. In these cells the activation of conductive anion exit depolarizes the cell membrane (Lang et al. 1987; Völkl and Lang 1988a). Preferable activation of K^+ channels in these cells would be less efficient in stimulating cellular electrolyte loss than preferable activation of anion channels. Accordingly, the activation of anion channels and the continued patency of K^+ channels, but not the additional activation of K^+ channels, are prerequisites for electrolyte extrusion in these cells. Other cells reported to depolarize following cell swelling include lymphocytes (Grinstein et al. 1982a; Sarkadi et al. 1984a), astrocytes (Kimelberg and O'Connor 1988), neuroblastoma cells (Falke and Misler 1989), OK-cells (Ubl et al. 1988b) and intestinal cells (Hazama and Okada 1988).

As illustrated in Fig. 1, cell swelling leads to a variable transient hyperpolarization, followed by an invariable sustained depolarization of the cell membrane in Madin Darby canine kidney (MDCK) cells (Paulmichl et al. 1989; Ritter et al. 1991a). The transient hyperpolarization is the result of the transient activation of K^+ channels; the sustained depolarization is the result of the activation of the anion channels (Weiss and Lang 1992).

The anion extruded by the cells is considered to be chloride. However, a close inspection of the intracellular chloride concentrations prior to cell swelling (Table 1) and the driving forces during cell swelling reveals that in many cells chloride may contribute to but cannot quantitatively account for the required loss of cellular anions (Rome et al. 1988). Accordingly, in proximal renal tubules RVD had been observed to be dependent on the presence of exogenous bicarbonate and CO_2 (Lopes and Guggino 1987; Völkl and Lang 1988b) and it has been speculated that extrusion of cellular bicarbonate may contribute to RVD in these cells. The intracellular bicarbonate concentration is similarly low as the intracellular chloride concentration. However, the cellular loss of bicarbonate could be replenished from CO_2 and the formed H^+ ions could be bound to intracellular buffers or consumed during cellular metabolism. The net result of parallel electrogenic bicarbonate and potassium loss would thus be the exchange of osmotically active K^+ for osmotically inactive H^+.

Recently, we have observed that the anion channels activated by cell swelling in MDCK cells are more permeable to bicarbonate than to chloride (Weiss and Lang 1992), further suggesting a role of bicarbonate in cell volume regulation (Fig. 2). However, unlike proximal renal tubules, MDCK cells are able to regulate their volume in the absence of exogenous bicarbonate and CO_2 (Paulmichl et al. 1989). Inhibition of K^+ channels, on the other hand, almost abolishes the RVD, indicating that it does depend on ion flux through the channels. Thus, in those cells, the exit of chloride or the intracellular formation of CO_2 and bicarbonate may be sufficient to support cell volume regulation. Alternatively, the bicarbonate and chloride conductive anion channel is conductive to other anions such as phosphate. In Ehrlich ascites tumor cells, the anion pathway has similarly been described as highly unselective (Hoffmann et al. 1988).

The unselective ion channels activated during cell swelling presumably do not mediate any net exit of ions but rather serve to allow the entry of calcium. The calcium entering the cells subsequently activates calcium-sensitive K^+ channels, thus indirectly accounting for cellular loss of potassium (Guharay and Sachs 1984; Christensen 1987; Christensen and Zeuthen 1987; Kullberg 1987; Palmer 1987; Sackin and Palmer and Sackin 1988; Ubl et al. 1988a,b; Falke and Misler 1989; Taniguchi and Guggino 1989; Dubé et al. 1990; Hunter 1990; Filipovic and Sackin 1991).

2.2 KCl Symport

In some cells, the cellular loss of KCl has been demonstrated to be independent of the cell membrane potential. Thus, in those cells, the loss of KCl is thought to be achieved by activation of electroneutral KCl symport. Inasmuch as this carrier selectively accepts chloride, it depends on the

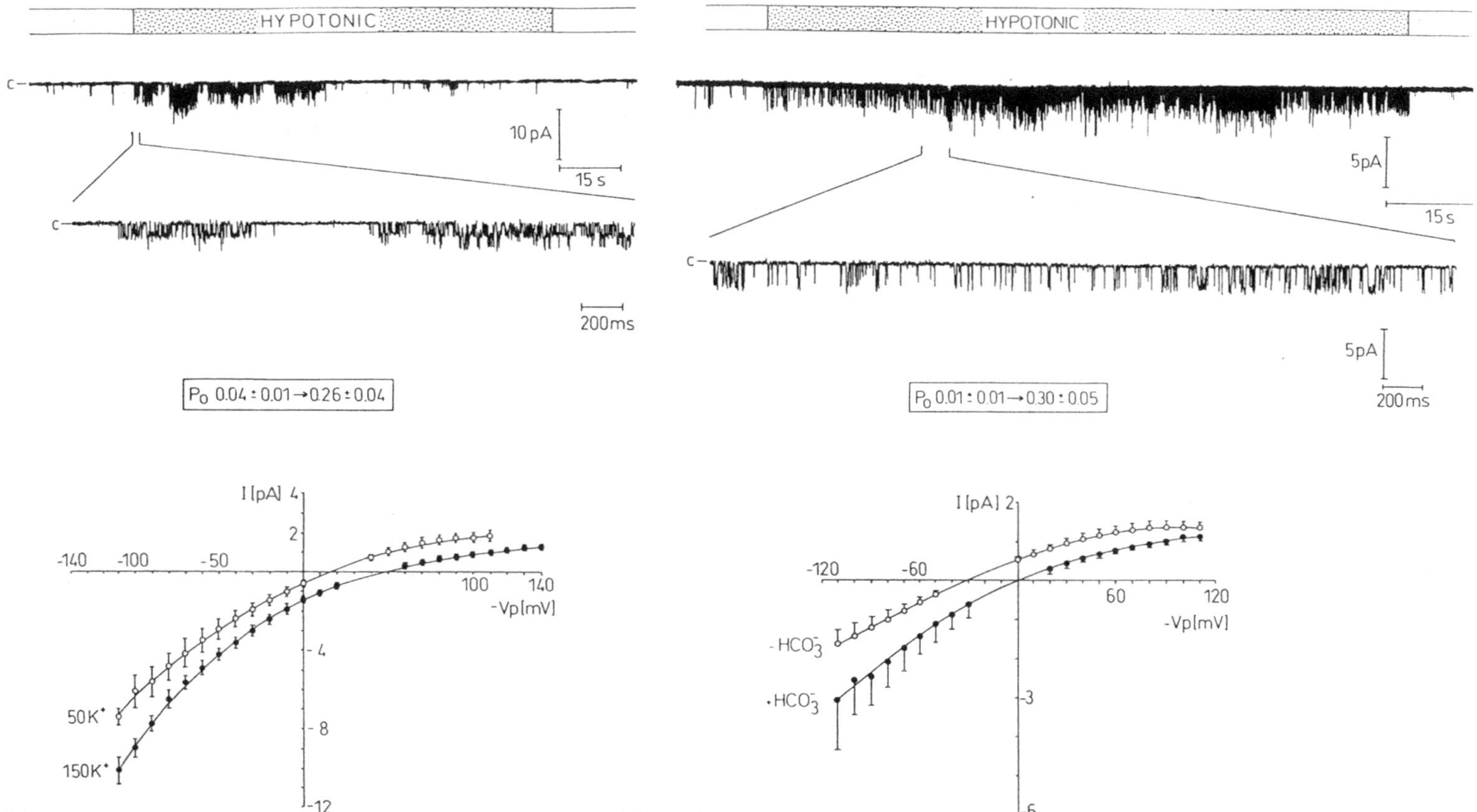

Fig. 2. Ion channels activated by swelling of MDCK cells. *Left* Transient activation of K^+ channels. *Right* Sustained activation of unselective anion channels. *Above* Original tracings in cell-attached configuration; *below* current (I)-voltage (V)-relationships (arithmetic means ± SEM) in excised patches. Cell swelling increases the open probability (P_o) of the K^+ channels from 0.04 ± 0.01 to 0.26 ± 0.04 (*left*) and of the anion channels from 0.01 ± 0.01 to 0.30 ± 0.05 (*right*). The K^+ channels are highly selective for K^+, as evident from the shift of reversal potential in excised patches (pipette solution 150 KCl) upon a replacement of 150 mmol/l KCl (*filled symbols*) by 100 mmol/l NaCl and 50 mmol/l KCl (*open symbols*). The anion channels are more conductive to bicarbonate than to chloride: replacement of 45 mmol/l choline chloride (*open symbols*) by 5 KCl and 40 choline bicarbonate (*filled symbols*) leads to a marked shift of the reversal potential (pipette solution 145 choline chloride). (After Weiss and Lang 1992)

combined driving force for K^+ and Cl^-, which is usually small in cells with low intracellular chloride activity (see Table 2).

2.3 K^+/H^+ Exchange

In amphiuma red cells Cala (Cala 1983) presented evidence in favor of regulatory cell volume decrease by parallel activation of K^+/H^+ exchange and Cl^-/HCO_3^- exchange. The H^+ and HCO_3^- entering the cells form carbonic acid, which is dehydrated to CO_2 by the carbonic anhydrase. CO_2 easily equilibrates across the cell membrane. Thus, the net effect of the parallel activation of these transport systems is the cellular loss of KCl.

2.4 Na^+/Ca^{2+} Exchange

A completely different mode of RVD has been described in dog red cells (Parker et al. 1975; Parker 1978). In these cells, intracellular sodium is high and intracellular potassium low. Accordingly, these cells extrude sodium during RVD. The extrusion of sodium is driven by calcium entry via Na^+/Ca^{2+} exchange. Calcium is probably eliminated by the Ca^{2+} ATPase. If rendered potassium-rich, these cells, like others, lose potassium and chloride by activation of KCl cotransport during regulatory cell volume decrease (Parker 1983a).

2.5 Na^+/K^+ ATPase

The transport systems reported above operate without a direct supply of energy, however, they depend on the ion gradients built up by the Na^+/K^+ ATPase (Kregenow 1971; Cheung et al. 1982; Gagnon et al. 1982; Parker 1983a; Grinstein et al. 1984; Larson and Spring 1984; Siebens and Kregenow 1985). In proximal renal tubules, RVD is not inhibited following acute inhibition of Na^+/K^+ ATPase with ouabain (Völkl and Lang 1988b). However, prolonged exposure of the tubules to ouabain does impair RVD in these cells, since the dissipation of the ion gradients interferes with the volume regulatory mechanisms (Völkl et al. 1988). Inhibition of Na^+/K^+ ATPase has been shown to impair VRD and stimulation of Na^+/K^+ ATPase activity during VRD is assumed to occur in a number of tissues (Ussing 1965; Lipton 1972; Cala 1977; Pollack et al. 1981; Ericson and Spring 1982; Grinstein et al. 1984; Lewis et al. 1985; Siebens and Kregenow 1985; Farahbakhsh and Fain 1987; Venosa 1991). On the other hand, cell swelling has been reported to inhibit Na^+/K^+ ATPase activity in chicken enterocytes (Montero and Ilundáin 1989).

2.6 Transport of Nonelectrolytes

In some tissues RVD has been shown to be paralleled by loss of cellular amino acids (Hoffmann and Hendil 1976; Hoffmann and Lambert 1983; Law and Turner 1987; Roy and Sauvé 1987; Sánchez Olea et al. 1991), of polyols, such as sorbitol and inositol (Garcia-Perez and Burg 1990, 1991; Garty et al. 1991; Moriyama et al. 1990; Siebens and Spring 1989; Yamauchi et al. 1991) and methylamines, such as betaine and glycerophosphorylcholine (Siebens and Spring 1989; Garcia-Perez and Burg 1990, 1991; Moriyama et al. 1990; Garty et al. 1991; Yamauchi et al. 1991). The polyols and the methylamines serve as intracellular osmolytes during exposure to high extracellular osmolarity (see Sect. 3.5). Very high concentrations of these osmolytes are found in kidney medulla during antidiuresis, as detailed in a separate chapter (Graf et al. this Vol.). Rapid reduction of extracellular osmolarity in kidney medulla during transition from antidiuresis to diuresis requires the efficient elimination of these intracellular osmolytes. This is achieved by activation of the respective transport systems (permeases).

2.7 Metabolism

In liver cells, the metabolism of glycogen and proteins is greatly influenced by the cell volume (Häussinger and Lang 1991a,b,c; Häussinger et al., this Vol.). Cell swelling inhibits the degradation and stimulates the formation of glycogen (Graf et al. 1988; Lang et al. 1989; Baquet et al. 1990) and proteins (Häussinger and Lang 1990, 1991a,b,c; Häussinger et al. 1990a,b,c). The disposal of cellular carbohydrate metabolites and amino acids into the osmotically far less active macromolecules is expected to reduce intracellular osmolarity. On the other hand, as outlined elsewhere (Graf et al., this Vol.), cellular formation and accumulation of polyols and methylamines are inhibited during cell swelling in nonmammalian tissues (Forster and Goldstein 1979; Somero 1986), in mammalian brain (Pollock and Arieff 1980) and in mammalian renal medulla (Ullrich and Jarausch 1956; Beck et al. 1985, 1988a; Bagnasco et al. 1986; Balaban and Burg 1987; Law and Turner 1987; Wirthensohn et al. 1987a,b; Beck and Guder 1990).

2.8 Other Mechanisms

Van Rossum et al. (1987) produced liver cell swelling by exposure to ouabain and cooling to 4 °C. Rewarming of these cells was followed by a reduction in cell volume despite the continued presence of ouabain. Further experiments revealed the inhibitory action of furosemide and agents which interfere with the cytoskeleton on this type of RVD. The authors concluded that the RVD was due to formation and exocytosis of intracellular vesicles which were loaded with cellular electrolytes by furosemide-sensitive Na^+, K^+, $2Cl^-$ cotransport (van Rossum and Russo 1981, 1984; van Rossum et al. 1987).

For nonperfused proximal renal tubules, chronically exposed to ouabain, a model for apparent cell volume regulation has been presented by Welling and Welling (1988). In their calculation they assume that two types of pores exist at the basolateral cell membrane with different reflection coefficients. If the parameters are chosen appropriately, the time course of cell volume changes predicted by the model exactly matches the time course of observed volume changes. The model requires substantial pressure gradients across the cell membrane during cell swelling which may indeed arise in this particular model from the extension of the tubular basement membrane as well as in myocardium from the interstitial collagen network (Pine et al. 1981). The model is probably irrelevant for systems without a limitation to cell swelling by the extracellular matrix.

3 Mechanisms of Regulatory Volume Increase (RVI)

RVI is achieved by an increase in intracellular osmotic activity, accomplished by the net uptake of ions and nonelectrolytes across the cell membrane and by metabolic generation of osmotically active substances. The net uptake of ions, which is the most rapid means to increase intracellular osmolarity, results from inhibition of ion channels, on the one hand, and from stimulation of Na^+/H^+ exchange and Na^+, K^+, $2Cl^-$ cotransport, on the other.

3.1 Ion Channels

As illustrated in Fig. 1, shrinkage of subconfluent Madin Darby canine kidney (MDCK) cells leads to a marked increase in cell membrane resistance (Ritter et al. 1991b). The major cell membrane conductances of these cells are a K^+ conductance (Paulmichl et al. 1985) and a Cl^- conductance (Lang et al. 1986b). Shrinkage of the MDCK cells leads to a moderate increase in the cell membrane K^+ selectivity and hyperpolarization of the cell membrane (Ritter et al. 1991b), indicating that the decrease in Cl^- conductance prevails. The effect of cell shrinkage on ion channels is not just the reversal of cell swelling: the ion channels activated by cell swelling are virtually closed in isotonic extracellular fluid (Weiss and Lang 1992) and their conductance cannot be reduced further. The volume regulatory efficacy of ion channel closure obviously depends on the loss of ions through these channels, i.e., on the prevailing ion gradients. The activation of ion-accumulating transport systems leads to the intracellular accumulation of chloride and is thus expected to markedly enhance the chemical gradient for chloride. The resulting chloride movement across the Cl^- channels would depolarize the cell membrane and consequently enhance the electrochemical gradient for potassium ions. Inhibition of the ion channels thus prevents a futile operation of the chloride-accumulating transport systems. The inhibition of the Na^+, K^+, $2Cl^-$ cotransport, on the other hand, should, in the absence of other chloride-accumulating systems, dissipate the chloride gradient across the cell membrane, thus minimizing the chloride movement across the Cl^- channels. Thus, the contribution of Cl-channel closure to RVI would not show up during inhibition of chloride-accumulating transport systems. Accordingly, RVI in suspended MDCK cells is virtually abolished in the presence of both Na^+, K^+, $2Cl^-$ cotransport and Na^+/H^+ exchange (Ritter et al. 1991b). However, we cannot be certain about the behavior of cell membrane conductances in suspended MDCK cells.

3.2 Na^+/H^+ Exchange

In a variety of cells (see Table 1), cell shrinkage leads to the parallel activation of Na^+/H^+ exchange and Cl^-/HCO_3^- exchange. The sodium thus taken up is replaced by potassium by the Na^+/K^+ ATPase. Accordingly, the parellel operation of these pumps eventually leads to cellular uptake of KCl. The advantage of the Na^+/H^+ exchanger is its relatively high extra driving force (see Table 2) and thus its independence of moderate alterations in intra- and extracellular ion composition. The disadvantage of recruiting the Na^+/H^+ exchanger for RVI is the interference with intracellular pH regulation. The predominant activation of the Na^+/H^+ exchanger has indeed been shown to result in intracellular alkalinization in some cells, especially when bicarbonate is removed from the extracellular fluid, thus impairing the Cl^-/HCO_3^- exchanger. It may be this intracellular alkalinization which eventually limits the operation of the Na^+/H^+ exchanger.

3.3 Na^+, K^+, $2Cl^+$ Cotransport

As can be seen in Table 1, RVI may be accomplished by activation of Na^+, K^+, $2Cl^-$ cotransport in a variety of cells. The Na^+, K^+, $2Cl^-$ cotransport may be effective alone or in cooperation with the Na^+/H^+ exchanger, as in MDCK cells (Ritter et al. 1991b) or in *ras* oncogene expressing NIH 3T3 fibroblasts (Lang et al. 1992). The main disadvantage of the Na^+, K^+, $2Cl^-$ cotransport as a volume regulatory system is its small driving force (Table 2) which can be easily dissipated by moderate alterations of intra- or extracellular ion activities. Furthermore, the carrier may be inhibited by high intracellular chloride activity. Accordingly, in some cells RVI cannot be elicited if extracellular osmolarity is increased but in the same cells RVI is observed if the cells are depleted of chloride by prior cell swelling and subsequent RVD (Hoffmann et al. 1983; Grinstein et al. 1985a; Montrose et al. 1988; O'Neill 1992; O'Neill and Klein 1992).

3.4 Other Transport Systems

As mentioned above (see Sect. 2.6), the cellular uptake and release of amino acids (Hoffmann and Hendil 1976; Hoffman and Lambert 1983; Law and Turner 1987; Roy and Sauvé 1987; Häussinger and Lang 1990; Sánchez Olea et al. 1991), of polyols, such as sorbitol and inositol (Garcia-Perez and Burg 1990, 1991; Garty et al. 1991; Moriyama et al. 1990; Siebens and Spring 1989, Yamauchi et al. 1991) and methylamines, such as betaine and glycerophosphorylcholine (Siebens and Spring 1989; Garcia-Perez and Burg 1990, 1991; Moriyama et al. 1990; Garty et al. 1991; Yamauchi et al. 1991) are dependent on cell volume. The transport of these substances is stimulated by cell swelling and inhibited by cell shrinkage.

Efficient RVI could be observed only after addition of alanine, glucose, citrate, lactate, acetate, or butyrate to proximal renal tubules (Rome et al. 1989), after addition of butyrate to cortical collecting tubules (Natke 1990) and only after addition of ADH to mouse medullary thick ascending limbs (Hebert 1986) and medullary collecting ducts (Sun and Hebert 1989). Butyrate probably enters the cell as nonionized acid and dissociates within the cell to butyrate and H^+, which is extruded in exchange for Na^+ by the Na^+/H^+ exchanger. The net effect of extracellular butyrate addition is the intracellular accumulation of Na^+ butyrate. The resulting increase in intracellular osmolarity is expected to trigger mechanisms of regulatory cell volume decrease. These mechanisms are presumably turned off during cell shrinkage leading to the apparent RVI. Accordingly, butyrate transport may not be considered a volume regulatory mechanism but may shift the volume regulatory machinery into a more favorable range for RVI. Similarly, ADH may act by stimulating volume-independent electrolyte entry into the cells.

3.5 Metabolism

RVI is paralleled by profound alterations of cell metabolism (Häussinger and Lang 1991a,b,c). In liver cells, cell shrinkage favors degradation of glycogen and proteins (see Häussinger et al., this Vol.). The generation of cellular carbohydrates and amino acids from the osmotically far less active macromolecules is expected to enhance intracellular osmolarity as long as they do not become equally distributed in intra- and extracellular space.

On the other hand, high ambient osmolarity stimulates the intracellular formation of polyols such as sorbitol and inositol and amines such as betaine and glycerophosphorylcholine in various nonmammalian tissues (Forster and Goldstein 1979; Somero 1986), in mammalian brain (Pollock and Arieff

1980) and in mammalian renal medulla (Ullrich and Jarausch 1956; Beck et al. 1985, 1988a; Bagnasco et al. 1986; Balaban and Burg 1987; Law and Turner 1987; Wirthensohn et al. 1987a,b; Beck and Guder 1990; Law and Burg 1991). Excessive concentrations of these organic osmolytes are found in kidney medulla (see Graf et al., this Vol.).

4 Factors Activating or Modifying Cell Volume Regulatory Mechanisms

In the following sections a few examples will be given of how cell volume may be compromised and cell volume regulatory mechanisms come into play.

4.1 Extracellular Osmolarity

In most mammalian tissues extracellular osmolarity is finely tuned and thus the cells are rarely exposed to rapidly altering extracellular osmolarity. There are, however, some notable exceptions. The most profound alterations of extracellular osmolarity are encountered in renal medulla, where osmolarity can range from less than 300 mOsmol/l to more than 2000 mOsmol/l, depending on the animal species and the diuretic state of the animal. As detailed in another chapter (see Graf et al., this Vol.) and in a variety of excellent recent reviews (Beck et al. 1988a; Beck and Guder 1989; Garcia-Perez and Burg 1990, 1991; Montrose-Rafizadeh and Guggino 1990), the extreme environment requires powerful mechanisms to alter intracellular osmolarity in the renal medullary cells. Furthermore, blood cells passing the kidney medulla are suddenly exposed to tremendous alterations of extracellular osmolarity. Apart from kidney medulla, rapid alterations of osmolarity may occur at the luminal side of several epithelia, such as ureter, urinary bladder, gland ducts and intestinal tract. Since the basolateral side is usually bathed in isotonic fluid and/or the hydraulic permeability of the apical cell membrane is low, the alterations of luminal osmolarity are usually not a heavy burden for the volume homeostasis of the respective epithelial cells. During intestinal absorption, some alterations of liver cell volume may occur (Haberich 1968), which are again not a serious challenge for cell volume regulatory mechanisms.

4.2 Extracellular Ion Composition

Even at constant extracellular osmolarity, alterations of extracellular ion composition could modify cell volume. An increase in extracellular potassium concentration reduces conductive potassium efflux and thus leads to accumulation of intracellular potassium. The reduced potassium efflux leads to a depolarization and thus impairs chloride or bicarbonate extrusion across the cell membrane (Lang et al. 1988). As a result, potassium, chloride and bicarbonate are accumulated within the cell and induce cellular swelling (Dellasega and Grantham 1973; Pine et al. 1979; Hermansson and Spring 1986; Kirk et al. 1987a; Ballanyi et al. 1988; Lang et al. 1988).

An increase in extracellular bicarbonate concentration is similarly expected to impair the extrusion of bicarbonate. The reduced efflux of negatively charged bicarbonate should hyperpolarize the cell membrane and thus decrease the driving force for potassium efflux. As a result cellular bicarbonate accumulation should be paralleled by potassium accumulation and slight cellular swelling, observations actually made in renal tissue during acute metabolic alkalosis (Beck et al. 1988b).

Cell volume can be manipulated by isotonic replacement of chloride with organic anions such as gluconate, propionate, lactate, or acetate (Lewis et al. 1985; Macknight 1985, 1988; Jakubovicz et al. 1987; Macknight and Scott 1989; Walz and Mukerji 1990). Gluconate leads to cell shrinkage due to loss of intracellular chloride, whereas acetate, lactate and propionate lead to cell swelling, since they enter the cell as nonionized acid and dissociate within the cell. The resulting intracellular acidosis activates the Na^+/H^+ exchanger. Eventually, the cell accumulates the anion, sodium and potassium. A similar effect is exerted by increasing CO_2 (Kempski et al. 1988).

4.3 Transport Across the Cell Membrane

In epithelial tissues, cell volume regulation is constantly challenged by transcellular transport. Alterations of epithelial cell volume may arise from an imbalance of ion transport across the apical and basolateral cell membrane. In small intestine (Schultz 1981, 1989b), proximal renal tubule (Lang et al. 1983, 1986a; Beck and Potts 1990) and liver (Bakker-Grunwald 1983; Kristensen and Folke 1984; Kristensen 1986), sodium-coupled transport for organic substrates, such as glucose and amino acids, represents a main threat for cell volume homeostasis. The transport systems operate according to the luminal availability of substrates and the respective driving forces; they are obviously not turned off, even if sodium entry exceeds the capacity of the cell to extrude sodium (Lang et al. 1984). Stimulation of sodium-coupled transport across the luminal cell membrane leads to a delayed activation of basolateral potassium conductance in both the small intestine (Schultz 1981) and renal proximal tubules (Lang et al. 1986a). The activation of basolateral K^+ channels in intestine has been attributed to cell volume regulation, and it has indeed been shown that cell swelling in this tissue by perfusion with hypotonic perfusates leads to an increase in basolateral potassium conductance (Lau et al. 1984). Even though other mechanisms may contribute to the action of sodium-coupled transport on potassium conductance (Lang et al. 1986a; Lang and Rehwald 1992), sodium-coupled transport is a challenge for cell volume control, which certainly modifies the transport properties of the proximal tubule cell.

Inhibition of Na^+/K^+ ATPase impairs the extrusion of sodium in exchange for potassium across the cell membrane, leading to accumulation of sodium and a decrease in intracellular potassium activity, and to depolarization and accumulation of anions such as chloride (Macknight and Leaf 1977; Lang et al. 1983). The accumulation of NaCl should eventually result in cell swelling. However, the time required for appreciable cell swelling may be quite variable, depending on the rate of sodium entry. In the absence of substrates for sodium-coupled transport in proximal renal tubules, for instance, cell membrane potential (Lang et al. 1984, Völkl et al. 1986) and cell volume (Macknight and Leaf 1977) could be stable for hours in the presence of ouabain. In the presence of substrates for sodium-coupled transport, the loss of potassium leads to a rapid reduction in potassium conductance at the basolateral cell membrane. The resulting depolarization reduces the driving force for sodium-coupled transport and impairs bicarbonate extrusion across the basolateral cell membrane, which, in turn, leads to intracellular alkalinization, inhibition of the Na^+/H^+ exchanger and thus to reduced entry of sodium via this route (Lang et al. 1983). In the diluting segment of the amphibian kidney and medullary thick ascending limb of the mammalian kidney (Hebert 1986), ouabain leads to cell swelling only if the luminal NaCl-KCl cotransporter is operative, i.e., cell volume is unaffected if ouabain is given together with furosemide (Guggino et al. 1985). In the gallbladder the initial swelling is reversed by volume regulatory mechanisms (Koch Jensen et al. 1984). In liver, cell volume is maintained even if the Na^+/K^+ ATPase is inhibited by ouabain, presumably due to extrusion of electrolytes and water by exocytosis (van Rossum and Russo 1981, 1984; van Rossum et al. 1981, 1987; Russo et al. 1985). In nonperfused renal tubules swelling is limited by the rigidity of the basement membrane (Linshaw and Grantham 1980; Linshaw 1980).

4.4 Metabolism

The degradation of glycogen to glucose phosphates and proteins to amino acids should increase the cellular osmolarity as long as the metabolites are retained within the cell. The degradation of glucose and amino acids eventually decreases cellular osmolarity. Thus, alterations of cellular metabolism are expected to modify cell volume, if they are not paralleled by respective volume regulatory mechanisms. However, the effect of cell metabolism on cell volume has not been elucidated in any detail.

4.5 Hormones

Activation of transport systems by hormones, mediators or other stimuli should result in respective alterations of cell volume. As detailed elsewhere (see Häussinger et al., this Vol.), a number of hormones do in fact modify liver cell volume: insulin, e.g., stimulates Na^+, K^+, $2Cl^-$ cotransport and Na^+/H^+ exchange and thus leads to cell swelling; glucagon activates conductive loss of K^+ and Cl^- and thus leads to cell shrinkage (Häussinger and Lang 1991c). It is the alteration of cell volume, which triggers part of the metabolic effects of these hormones. Activation by the hormones of transport systems in epithelial cells may similarly be followed by the respective alterations of cell volume (see Graf et al., this Vol.), for instance, the stimulation of Na^+ channels may swell the cells in tight epithelia (Schultz 1989b). The volume regulatory response may then participate in the activation of the K^+ channels. Conversely, the activation of chloride channels by cAMP in MDCK cells (Lang et al. 1986b) has been shown to shrink these cells (Mills and Skiest 1985; Mills and Lubin 1986; Mills 1987). The volume regulatory response probably leads to activation of Na^+, K^+, $2Cl^-$ cotransport. Since the chloride channels are at the apical cell membrane and the Na^+, K^+, $2Cl^-$ cotransporter at the basolateral cell membrane, the activation of both leads to transepithelial transport of chloride (Brown and Simmons 1981; Simmons 1982; Lifschitz 1986; Lang et al. 1990).

Activation of excitable cells may similarly modify cell volume (for details, see Kimelberg et al. 1992). For instance, in rat sympathetic ganglia GABA activates Cl^- channels (Ballanyi and Grafe 1985). The depolarization due to chloride exit drives potassium exit via K^+ channels. The cellular loss of KCl shrinks the cell. This is probably the signal for stimulation of Na^+, K^+, $2Cl^-$ cotransport. The carrier not only restores cell volume but in addition maintains intracellular chloride. Thus, depolarization is sustained in the presence of GABA. If GABA is applied together with furosemide, depolarization is only transient due to dissipation of intracellular chloride activity. As outlined by Bianchini and Grinstein (this Vol.), mitogenic signals similar to insulin stimulate Na^+/H^+ exchange and Na^+, K^+, $2Cl^-$ cotransport in a variety of cells. In addition to their effects on intracellular ion composition, the transporters may serve to enhance cell volume.

4.6 *Ras* Oncogenes

Recently, we could observe that in growth factor-depleted media cells expressing the *ras* oncogene are some 40% larger than cells not expressing the oncogene (Meyer et al. 1991; Lang et al. 1992). Both *ras* oncogene-expressing cells and normal cells displayed a regulatory cell volume decrease and an increase in hypotonic and hypertonic extracellular fluid, respectively

(Lang et al. 1992). Thus, it appears that the expression of *ras* oncogenes leads to a shift of the set point for cell volume regulation to larger volumes (Fig. 3). The growth factor-independent proliferation of the *ras* oncogene-expressing cells can be blocked by amiloride and furosemide (unpubl. observ.). Thus, the activation of the Na^+/H^+ exchanger and Na^+, K^+, $2Cl^-$ cotransporter may be crucial for the proliferation of these cells. Possibly, mitogenic signals, by phosphorylating the carriers, similarly shift the set point of the Na^+/H^+ exchanger and Na^+, K^+, $2Cl^-$ cotransport, thus leading to the well-known activation of these carriers (see Bianchini and Grinstein, this Vol.).

5 Conclusions

One of the most obvious requirements for cell survival is the ability to maintain the constancy of cell volume under a variety of conditions. Accordingly, cell volume regulatory mechanisms have been found in virtually

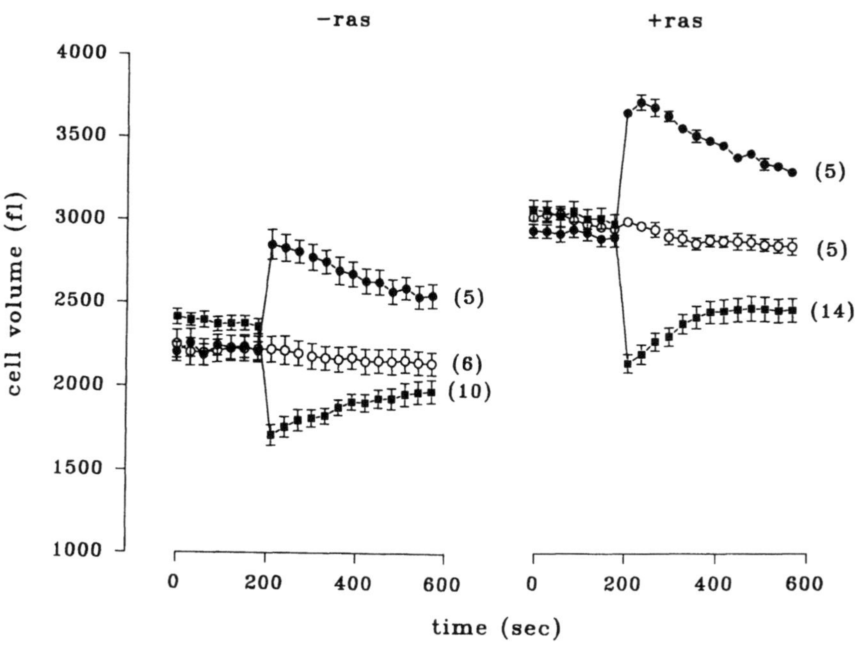

Fig. 3. Set point shift of cell volume regulation in *ras* oncogene-expressing cells. Cell volume has been determined by Coulter counter in cells expressing the *ras* oncogene upon treatment with dexamethasone (+ras) as well as in cells not expressing the oncogene (−ras). At 180 s, extracellular osmolarity was increased by addition of 50 mmol/l NaCl (*filled squares*), decreased by removal of 35 mmol/l NaCl (*filled circles*) or left unchanged (*open symbols*). Arithmetic means ± SEM, number of experiments in *parentheses*. (After Lang et al. 1992)

every cell studied. Cell volume regulatory mechanisms include ion and nonelectrolyte transport systems at the cell membrane and cellular metabolism of macromolecules and small osmolytes such as amino acids, polyols and methylamines. Hormones, growth factors and *ras* oncogenes may alter cell volume by activation of cellular ion uptake or release, possibly through shifting of their cell volume regulatory set point. The alterations of cell volume could trigger the other cell volume regulatory mechanisms, eventually leading to the effects of the hormones on transport and metabolism. Accordingly, the machinery of cell volume regulation is an integral part of the intracellular transducing mechanisms.

References

Adorante JS, Miller SS (1990) Potassium-dependent volume regulation in retinal pigment epithelium is mediated by Na, K, Cl cotransport. J Gen Physiol 96: 1153–1176

Adragna NC, Tosteson DC (1984) Effect of volume changes on ouabain-insensitive net outward cation movements in human red cells. J Membr Biol 78: 43–52

Bagnasco S, Balaban R, Fales HM, Yang Y-M, Burg M (1986) Predominant osmotically active organic solutes in rat and rabbit renal medullas. J Biol Chem 261: 5872–5877

Bakker-Grunwald T (1981) Hormone-induced diuretic-sensitive potassium transport in turkey erythrocytes is anion dependent. Biochim Biophys Acta 641: 427–431

Bakker-Grunwald T (1983) Potassium permeability and volume control in isolated rat hepatocytes. Biochim Biophys Acta 731: 239–242

Bakker-Grunwald T, Ogden P, Lamb JF (1982) Effects of ouabain and osmolarity on bumetanide-sensitive potassium transport in simian virus-transformed 3T3 cells. Biochim Biophys Acta 687: 333–336

Balaban RS, Burg MB (1987) Osmotically active organic solutes in the renal inner medulla. Kidney Int 31: 562–564

Ballanyi K, Grafe P (1985) An intracellular analysis of Γ-aminobutyric-acid-associated ion movements in rat sympathetic neurones. J Physiol 365: 41–58

Ballanyi K, Grafe P (1988) Cell volume regulation in the nervous system. Renal Physiol Biochem 11: 142–157

Ballanyi K, Serve G, Grafe P, Schlue W-R (1988) Electrophysiological measurements of K^+-induced volume change in leech glial cells (Abstract). Pflügers Arch Eur J Physiol 411 (Suppl 1): R161, 310

Baquet A, Hue L, Meijer AJ, van Woerkom GM, Plomp PJAM (1990) Swelling of rat hepatocytes stimulates glycogen synthesis. J Biol Chem 265: 955–959

Beck F-X, Guder W (1990) Cell osmoregulation in the counter current system of the renal medulla. Comp Physiol 4: 132–158

Beck F-X, Potts DJ (1990) Cell swelling, co-transport activation and potassium conductance in isolated perfused rabbit kidney proximal tubules. J Physiol 425: 369–378

Beck F, Dörge, A, Rick R, Thurau K (1985) Osmoregulation of renal papillary cells. Pflügers Arch Eur J Physiol 405: S28–S32

Beck, F-X, Dörge A, Thurau K (1988a) Cellular osmoregulation in renal medulla. Renal Physiol Biochem 11: 174–186

Beck F-X, Dörge A, Rick R, Schramm M, Thurau K (1988b) The distribution of potassium, sodium and chloride across the apical membrane of renal tubular cells: effect of acute metabolic alkalosis. Pflügers Arch Eur J Physiol 411: 259–267

Beebe DC, Parmelee JT, Belcher KS (1990) Volume regulation in lens epithelial cells and differentiating lens fiber cells. J Cell Physiol 143: 455–459

Bevan C, Theiss C, Kinne RK (1990) The role of Ca^{2+} in sorbitol release from rat inner medullary collecting duct cells under hypoosmotic stress. Biochem Biophys Res Commun 170: 563–568

Biagi B, Kubota T, Sohtell M, Giebisch G (1981a) Intracellular potentials in rabbit proximal tubules perfused in vitro. Am J Physiol 240: F200–F210

Biagi B, Sohtell M, Giebisch G (1981b) Intracellular potassium activity in the rabbit proximal straight tubule. Am J Physiol 241: F677–F686

Blumenfeld JD, Hebert SC, Heilig, CW, Balschi JA, Stromski ME, Gullans SR (1989) Organic osmolytes in inner medulla of Brattleboro rat: effects of ADH and dehydration. Am J Physiol 256: F916–F922

Bonanno JA (1991) K^+-H^+ exchange, a fundamental cell acidifier in corneal epithelium. Am J Physiol 260: C618–C625

Brown CDA, Simmons NL (1981) Catecholamine-stimulation of Cl-secretion in MDCK cell epithelium. Biochim Biophys Acta 649: 427–435

Butt AG, Clapp WL, Frizzell RA (1990) Potassium conductance in tracheal epithelium activated by secretion and cell swelling. Am J Physiol 258: C630–C638

Cala PM (1977) Volume regulation by flounder red blood cells in anisotonic media. J Gen Physiol 69: 537–552

Cala PM (1980) Volume regulation by Amphiuma red blood cells. The membrane potential and its implications regarding the nature of the ion-flux pathways. J Gen Physiol 76: 683–708

Cala PM (1983) Cell volume regulation by *Amphiuma* red blood cells. The role of Ca^{2+} as a modulator of alkali metal/H^+ exchange. J Gen Physiol 82: 761–784

Cala PM (1985) Volume regulation by *Amphiuma* red blood cells: characteristics of volume-sensitive K/H and Na/H exchange. Mol Physiol 8: 199–214

Cala PM, Mandel LJ, Murphy E (1986) Volume regulation by *Amphiuma* red blood cells: cytosolic free Ca and alkali metal-H exchange. Am J Physiol 250: C423–C429

Cardinal J, Lapointe J-Y, Laprade R (1984) Luminal and peritubular ionic substitutions and intracellular potential of the rabbit proximal convoluted tubule. Am J Physiol 247: F352–F364

Cassola AC, Mollenhauer M, Frömter E (1983) The intracellular chloride activity of rat kidney proximal tubular cells. Pflügers Arch Eur J Physiol 399: 259–265

Cemerikic D, Wilcox CS, Giebisch G (1982) Intracellular potential and K^+ activity in rat kidney proximal tubular cells in acidosis and K^+ depletion. J Membr Biol 69: 159–165

Chamberlin ME, Strange K (1989) Anisosmotic cell volume regulation: a comparative view. Am J Physiol 257: C159–C173

Chassande O, Frelin C, Farahifar D, Jean T, Lazdunski M (1988) The $Na^+/K^+/Cl^-$ cotransport in C6 glioma cells. Properties and role in volume regulation. Eur J Biochem 171: 425–433

Cheung RK, Grinstein S, Dosch H-M, Gelfand EW (1982) Volume regulation by human lymphocytes: characterization of the ionic basis for regulatory volume decrease. J Cell Physiol 112: 189–196

Christensen O (1987) Mediation of cell volume regulation by Ca^{2+} influx through stretch-activated channels. Nature 330: 66–68

Christensen O, Zeuthen T (1987) Maxi K^+ channels in leaky epithelia are regulated by intracellular Ca^{2+}, pH and membrane potential. Pflügers Arch Eur J Physiol 408: 249–259

Christensen O, Simon M, Randlev T (1989) Anion channels in a leaky epithelium. A patch clamp study of choroid plexus. Pflügers Arch Eur J Physiol 415: 37–46

Costa PM, Fernandes PL, Ferreira HG, Ferreira KT, Giraldez F (1987) Effect of cell volume changes on membrane ionic permeabilities and sodium transport in frog skin (*Rana ridibunda*). J Physiol 393: 1–17

Davis CW, Finn AL (1985) Cell volume regulation in frog urinary bladder. Fed Proc 44: 2520–2525

Davis CW, Finn AL (1987) Interactions of sodium transport, cell volume, and calcium in frog urinary bladder. J Gen Physiol 89: 687–702

Dellasega M, Grantham JJ (1973) Regulation of renal tubule cell volume in hypotonic media. Am J Physiol 224: 1288–1294

Deutsch C, Lee SC (1988) Cell volume regulation in lymphocytes. Renal Physiol Biochem 11: 260–276
Donaldson PJ, Chen LK, Lewis SA (1989) Effects of serosal anion composition on the permeability properties of rabbit urinary bladder. Am J Physiol 256: F1125–F1134
Drewnowska K, Baumgarten CM (1991) Regulation of cellular volume in rabbit ventricular myocytes: bumetanide, chlorothiazide, and ouabain. Am J Physiol 260: C122–C131
Dubé L, Parent L, Sauvé R (1990) Hypotonic shock activates a maxi K^+ channel in primary cultured proximal tubule cells. Am J Physiol 259: F348–F356
Duhm J, Göbel BO (1984) Na^+-K^+ transport and volume of rat erythrocytes under dietary K^+ deficiency. Am J Physiol 246: C20–C29
Dunham PB, Ellory JC (1981) Passive potassium transport in low potassium sheep red cells: dependence upon cell volume and chloride. J Physiol 318: 511–530
Edelman A, Curci S, Samarzija I, Frömter E (1978) Determination of intracellular K^+ activity in rat kidney proximal tubular cells. Pflügers Arch Eur J Physiol 378: 37–45
Ellory JC, Hall AC, Stewart GW (1985) Volume-sensitive cation fluxes in mammalian red cells. Mol Physiol 8: 235–246
Ericson A-C, Spring KR (1982) Volume regulation by Necturus gallbladder apical Na^+-H^+ and Cl^--HCO_3^- exchange. Am J Physiol 243: C146-C150
Eveloff JL, Calamia J (1986) Effect of osmolarity on cation fluxes in medullary thick ascending limb cells. Am J Physiol 250: F176–F180
Falke LC, Misler S (1989) Activity of ion channels during volume regulation by clonal N1E115 neuroblastoma cells. Proc Natl Acad Sci USA 86: 3919–3923
Farahbakhsh NA, Fain, GL (1987) Volume regulation of nonpigmented cells from ciliary epithelium. Invest Ophthalmol Visual Sci 28: 934–944
Filipovic D, Sackin H (1991) A calcium-permeable stretch-activated cation channel in renal proximal tubule. Am J Physiol 260: F119–F129
Fincham DA, Wolowyk MW, Young JD (1987) Volume-sensitive taurine transport in fish erythrocytes. J Membr Biol 96: 45–56
Finn AL, Reuss L (1975) Effects of changes in the ionic composition of the serosal solution on the electrical properties of the toad urinary bladder epithelium. J Physiol 250: 541–558
Fisher RS, Persson BE, Spring KR (1981) Epithelial cell volume regulation: bicarbonate dependence. Science 214: 1357–1359
Forster RP, Goldstein L (1979) Amino acids and cell volume regulation. Yale J Biol Med 52: 497–515
Frelin C, Chassande O, Lazdunski M (1986) Biochemical characterization of the $Na^+/K^+/Cl^-$ co-transport in chick cardiac cells. Biochem Biophys Res Commun 134: 326–331
Frömter E (1979) Solute transport across epithelia: what can we learn from micropuncture studies on kidney tubules? J Physiol 288: 1–31
Frömter E (1982) Electrophysiological analysis of rat renal sugar and amino acid transport. I. Basic phenomena. Pflügers Arch Eur J Physiol 393: 179–189
Frömter E (1984) Viewing the kidney through microelectrodes. Am J Physiol 247: F695–F705
Fugelli K, Rohrs H (1980) The effect of Na^+ and osmolality on the influx and steady state distribution of taurine and gamma-aminobutyric acid in flounder (*Platichthys flesus*) erythrocytes. Comp Biochem Physiol 67A: 545–551
Gagnon J, Ouimet D, Nguyen H, Laprade R, Le Grimellec C, Carriere S, Cardinal J (1982) Cell volume regulation in the proximal convoluted tubule. Am J Physiol 243: F408–F415
Garcia-Perez A, Burg MB (1990) Importance of organic osmolytes for osmoregulation by renal medullary cells. Hypertension 16: 595–602
Garcia-Perez A, Burg MB (1991) Role of organic osmolytes in adaptation of renal cells to high osmolality. J Membr Biol 119: 1–13
Garty H, Furlong, TJ, Ellis DE, Spring KR (1991) Sorbitol permease: an apical membrane transporter in cultured renal papillary epithelial cells. Am J Physiol 260: F650–F656

Geck P (1990) Volume regulation in Ehrlich ascites tumor cells. Comp Physiol 4: 26–58
Geck P, Heinz E (1986) The Na-K-2Cl cotransport system. J Membr Biol 91: 97–105
Geck P, Pietrzyk C, Burckhardt B-C, Pfeiffer B, Heinz E (1980) Electrically silent cotransport of Na^+, K^+ and Cl^- in Ehrlich cells. Biochim Biophys Acta 600: 432–447
Germann WJ, Ernst SA, Dawson DC (1986) Resting and osmotically induced basolateral K conductances in turtle colon. J Gen Physiol 88: 253–274
Gilles R (1987) Volume regulation in cells of euryhaline invertebrates. Curr Top Membr Transp 30: 205–247
Gilles R (1988) Comparative aspects of cell osmoregulation and volume control. Renal Physiol Biochem 11: 277–288
Goldstein L, Kleinzeller A (1987) Cell volume regulation in lower vertebrates. Curr Top Membr Transp 30: 181–204
Graf J, Haddad P, Häussinger D, Lang F (1988) Cell volume regulation in liver. Renal Physiol Biochem 11: 202–222
Green J, Yamaguchi DT, Kleeman CR, Muallem S (1988) Selective modification of the kinetic properties of Na^+/H^+ exchanger by cell shrinkage and swelling. J Biol Chem 263: 5012–5015
Greger R (1985) Ion transport mechanisms in thick ascending limb of Henle's loop of mammalian nephron. Physiol Rev 65: 760–797
Greger R, Oberleithner H, Schlatter E, Cassola AC, Weidtke C (1983) Chloride activity in cells of isolated perfused cortical thick ascending limbs of rabbit kidney. Pflügers Arch Eur J Physiol 399: 29–34
Greger R, Weidtke C, Schlatter E, Wittner M, Gebler B (1984) Potassium activity in cells of isolated perfused cortical thick ascending limbs of rabbit kidney. Pflügers Arch Eur J Physiol 401: 52–57
Grinstein S, Clarke CA, Rothstein A (1982a) Increased anion permeability during volume regulation in human lymphocytes. Philos Trans R Soc Lond B299: 509–518
Grinstein S, Dupre A, Rothstein A (1982b) Volume regulation by human lymphocytes. Role of calcium. J Gen Physiol 79: 849–868
Grinstein S, Clarke CA, Dupre A, Rothstein A (1982c) Volume-induced increase of anion permeability in human lymphocytes. J Gen Physiol 80: 801–823
Grinstein S, Clarke CA, Rothstein A, Gelfand EW (1983a) Volume-induced anion conductance in human B lymphocytes is cation independent. Am J Physiol 245: C160–C163
Grinstein S, Clarke CA, Rothstein A (1983b) Activation of Na^+/H^+ exchange in lymphocytes by osmotically induced volume changes and by cytoplasmic acidification. J Gen Physiol 82: 619–638
Grinstein S, Rothstein A, Sarkadi B, Gelfand EW (1984) Responses of lymphocytes to anisotonic media: volume-regulating behavior. Am J Physiol 246: C204–C215
Grinstein S, Goetz JD, Cohen S, Furuya W, Rothstein A, Gelfand EW (1985a) Mechanism of regulatory volume increase in osmotically shrunken lymphocytes. Mol Physiol 8: 185–198
Grinstein S, Rothstein A, Cohen S (1985b) Mechanism of osmotic activation of Na^+/H^+ exchange in rat thymic lymphocytes. J Gen Physiol 85: 765–787
Guggino W, Oberleithner H, Giebisch G (1985) Relationship between cell volume and ion transport in the early distal tubule of the *Amphiuma* kidney. J Gen Physiol 86: 31–58
Guharay F, Sachs F (1984) Stretch-activated single ion channel currents in tissue-cultured embryonic chick skeletal muscle. J Physiol 352: 685–701
Haberich FJ (1968) Osmoreception in the portal circulation. Fed Proc 27: 1137–1141
Häussinger D, Lang F (1990) Exposure of perfused liver to hypotonic conditions modifies cellular nitrogen metabolism. J Cell Biochem 43: 355–361
Häussinger D, Lang F (1991a) The mutual interaction between cell volume and cell function: a new principle of metabolic regulation. Biochem Cell Biol 69: 1–4
Häussinger D, Lang F (1991b) Regulation of liver cell function and cell volume: a new principle for metabolic control. Biochim Biophys Acta 1071: 331–350

Häussinger D, Lang F (1991c) Cell volume-a "second messenger" in the regulation of metabolism by amino acids and hormones. Cell Physiol Biochem 1: 121–130
Häussinger D, Lang F, Bauers K, Gerok W (1990a) Interactions between glutamine metabolism and cell volume regulation in perfused rat liver. Eur J Biochem 188: 689–695
Häussinger D, Hallbrucker C, vom Dahl S, Lang F, Gerok W (1990b) Cell swelling inhibits proteolysis in perfused rat liver. Biochem J 272: 239–242
Häussinger D, Lang F, Bauers K, Gerok W (1990c) Control of hepatic nitrogen metabolism and glutathione release by cell volume regulatory mechanisms. Eur J Biochem 193: 891–898
Häussinger D, Stehle T, Lang F (1990d) Volume regulation in liver: further characterization by inhibitors and ionic substitutions. Hepatology 11: 243–254
Hazama A, Okada Y (1988) Ca^{2+} sensitivity of volume regulatory K^+ and Cl^- channels in cultured human epithelial cells. J Physiol 402: 687–702
Hebert SC (1986) Hypertonic cell volume regulation in mouse thick limbs. II. Na^+/H^+ and Cl^-/HCO_3^- exchange in basolateral membranes. Am J Physiol 250: C920–C931
Hermansson K, Spring KR (1986) Potassium induced changes in cell volume of gallbladder epithelium. Pflügers Arch Eur J Physiol 407: S90–S99
Hoffmann EK (1985a) Role of separate K^+ and Cl^- channels and of Na^+/Cl^- cotransport in volume regulation in Ehrlich cells. Fed Proc 44: 2513–2519
Hoffmann EK (1985b) Regulatory volume decrease in Ehrlich ascites tumor cells: role of inorganic ions and amino compounds. Mol Physiol 8: 167–184
Hoffman EK (1987) Volume regulation in cultured cells. Curr Top Membr Transp 30: 125–180
Hoffmann EK, Hendil KB (1976) The role of amino acids and taurine in isosmotic intracellular regulation in Ehrlich ascites mouse tumour cells. J Comp Physiol 108: 279–286
Hoffmann EK, Lambert IH (1983) Amino acid transport and cell volume regulation in Ehrlich ascites tumour cells. J Physiol 338: 613–625
Hoffmann, EK, Simonsen LO (1989) Membrane mechanisms in volume and pH regulation in vertebrate cells. Physiol Rev 69: 315–382
Hoffmann EK, Sjoholm C, Simonsen LO (1983) Na^+, Cl^- cotransport in Ehrlich ascites tumor cells activated during volume regulation (regulatory volume increase). J Membr Biol 76: 269–280
Hoffmann EK, Simonsen LO, Lambert IH (1984) Volume-induced increase of K^+ and Cl^- permeabilities in Ehrlich ascites tumor cells. Role of internal Ca^{2+}. J Membr Biol 78: 211–222
Hoffmann EK, Lambert IH, Simonsen LO (1986) Separate, Ca^{2+}-activated K^+ and Cl^- transport pathways in Ehrlich ascites tumor cells. J Membr Biol 91: 227–244
Hoffmann EK, Lambert IH, Simonsen LO (1988) Mechanisms in volume regulation in Ehrlich ascites tumor cells. Renal Physiol Biochem 11: 221–247
Howard LD, Wondergem R (1987) Effects of anisosmotic medium on cell volume, transmembrane potential and intracellular K^+ activity in mouse hepatocytes. J Membr Biol 100: 53–61
Hudson RL, Schultz SG (1988) Sodium-coupled glycine uptake by Ehrlich ascites tumor cells results in an increase in cell volume and plasma membrane channel activities. Proc Natl Acad Sci USA 85: 279–283
Hunter M (1990) Stretch-activated channels in the basolateral membrane of single proximal cells of frog kidney. Pflügers Arch Eur J Physiol 416: 448–453
Jakubovicz DE, Grinstein S, Klip A (1987) Cell swelling following recovery from acidification in C6 glioma cells: an in vitro model of postischemic brain edema. Brain Res 435: 138–146
Jennings ML, Douglas SM, McAndrew PE (1986) Amiloride-sensitive sodium-hydrogen exchange in osmotically shrunken rabbit red blood cells. Am J Physiol 251: C32–C40
Kempski O, Staub F, Jansen M, Schodel F, Baethmann A (1988) Glial swelling during extracellular acidosis in vitro. Stroke 19: 385–392

Kimelberg HK, Frangakis MV (1986) Volume regulation in primary astrocyte cultures. Adv Biosci 61: 177–186
Kimelberg HK, Kettenmann H (1990) Swelling induced changes in electrophysiological properties of cultured astrocytes and oligodendrocytes. I. Effects on membrane potentials, input impedance and cell-cell coupling. Brain Res 529: 255–261
Kimelberg HK, O'Connor ER (1988) Swelling-induced depolarization of astrocyte potentials. Glia 1: 219–224
Kimelberg HK, Goderie SK, Higman S, Pang S, Waniewski RA (1990a) Swelling-induced release of glutamate, aspartate, and taurine from astrocyte cultures. J Neurosci 10: 1583–1591
Kimelberg HK, Anderson E, Kettenmann H (1990b) Swelling induced changes in electrophysiological properties of cultured astrocytes and oligodendrocytes. II. Whole cell currents. Brain Res 529: 262–268
King PA, Goldstein L (1983) Organic osmolytes and cell volume regulation in fish. Mol Physiol 4: 53–66
Kirk KL, DiBona DR, Schafer JA (1987a) Regulatory volume decrease in perfused proximal nephron: evidence for a dumping of cell K^+. Am J Physiol 252: F933–F942
Kirk KL, Schafer JA, DiBona DR (1987b) Cell volume regulation in rabbit proximal straight tubule perfused in vitro. Am J Physiol 252: F922–F932
Kleinzeller A, Ziyedeh FN (1990) Volume regulation in epithelia: emphasis on the role of osmolytes and the cytoskeleton. Comp Physiol 4: 59–86
Koch Jensen P, Fisher RS, Spring KR (1984) Feedback inhibition of NaCl entry in *Necturus* gallbladder epithelial cells. J Membr Biol 82: 95–104
Kramhoft B, Lambert IH, Hoffmann EK, Jorgensen F (1986) Activation of Cl-dependent K transport in Ehrlich ascites tumor cells. Am J Physiol 251: C369–C379
Kregenow FM (1971) The response of duck erythrocytes to nonhemolytic hypotonic media. Evidence for a volume-controlling mechanism. J Gen Physiol 58: 372–395
Kregenow FM (1981) Osmoregulatory salt transporting mechanisms: control of cell volume in anisotonic media. Annu Rev Physiol 43: 493–505
Kregenow FM, Caryk T, Siebens AW (1985) Further studies of the volume-regulatory response of *Amphiuma* red cells in hypertonic media. Evidence for amiloride-sensitive Na/H exchange. J Gen Physiol 86: 565–584
Kristensen LO (1986) Associations between transports of alanine and cations across cell membrane in rat hepatocytes. Am J Physiol 251: G575–G584
Kristensen LO, Folke, M (1984) Volume-regulatory K^+ efflux during concentrative uptake of alanine in isolated rat hepatocytes. Biochem J 221: 265–268
Kullberg R (1987) Stretch-activated ion channels in bacteria and animal cell membranes. TINS 10: 38–39
Lambert IH, Hoffman EK, Jorgensen F (1989) Membrane potential, anion and cation conductances in Ehrlich ascites tumor cells. J Membr Biol 111: 113–132
Lang F (1988) NaCl transport in the kidney. In: Greger R (ed) Advances in comparative and environmental physiology, vol 1. Springer, Berlin Heidelberg New York, pp 153–188
Lang F, Rehwald W (1992) Potassium channels in renal epithelial transport regulation. Physiol Rev 20: 1–39
Lang F, Messner G, Wang W, Oberleithner H (1983) Interaction of intracellular electrolytes and tubular transport. Klin Wochenschr 61: 1029–1037
Lang F, Messner G, Wang W, Paulmichl M, Oberleithner H, Deetjen P (1984) The influence of intracellular sodium activity on the transport of glucose in proximal tubule of frog kidney. Pflügers Arch Eur J Physiol 401: 14–21
Lang F, Messner G, Rehwald W (1986a) Electrophysiology of sodium-coupled transport in proximal renal tubules. Am J Physiol 250: F953–F962
Lang F, Defregger M, Paulmichl M (1986b) Apparent chloride conductance of subconfluent Madin Darby canine kidney cells. Pflügers Arch Eur J Physiol 407: 158–162

Lang F, Paulmichl M, Völkl H, Gstrein E, Friedrich F (1987) Electrophysiology of cell volume regulation. In: Kovacevic Z, Guder WG (eds) Molecular physiology: biochemical aspects of kidney function. de Gruyter, Berlin, pp 133–139

Lang F, Oberleithner H, Kolb HA, Paulmichl M, Völkl H, Wang W (1988) Interaction of intracellular pH and cell membrane potential. In: Häussinger D (ed) pH homeostasis. Academic Press, New York, pp 27–42

Lang F, Stehle T, Häussinger D (1989) Water, K^+, H^+, lactate and glucose fluxes during cell volume regulation in perfused rat liver. Pflügers Arch Eur J Physiol 413: 209–216

Lang F, Friedrich F, Paulmichl M, Schobersberger W, Jungwirth A, Ritter M, Steidl M, Weiss H, Wöll E, Tschernko E, Paulmichl R, Hallbrucker C (1990) Ion channels in Madin-Darby canine kidney cells. Renal Physiol Biochem 13: 82–93

Lang F, Ritter M, Wöll E, Weiss H, Häussinger D, Maly K, Grunicke H (1992) Altered cell volume regulation in *ras* oncogene expressing NIH fibroblasts. Pflügers Arch Eur J Physiol (in press)

Lapointe JY, Garneau L, Bell PD, Cardinal J (1990) Membrane crosstalk in the proximal tubule during alterations in transepithelial sodium transport. Am J Physiol 258: F339–F345

Larson M, Spring KR (1984) Volume regulation by *Necturus* gallbladder: basolateral KCl exit. J Membr Biol 81: 219–232

Lau KR, Hudson RL, Schultz SG (1984) Cell swelling increases a barium-inhibitable potassium conductance in the basolateral membrane of *Necturus* small intestine. Proc Natl Acad Sci USA 81: 3591–3594

Lauf PK (1982) Evidence for chloride dependent potassium and water transport induced by hyposmotic stress in erythrocytes of the marine teleost, *Opsanus tau*. J Comp Physiol 146: 9–16

Lauf PK (1985) On the relationship between volume- and thiol-stimulated K^+ Cl^- fluxes in red cell membranes. Mol Physiol 8: 215–234

Lauf PK (1988) K:Cl cotransport: emerging molecular aspects of a ouabain-resistant, volume-responsive transport system in red blood cells. Renal Physiol Biochem 11: 248–259

Law RO, Burg MB (1991) The role of osmolytes in the regulation of mammalian cell volume. Adv Comp Environ Physiol 9: 189–225

Law RO, Turner DPJ (1987) Are ninhydrin-positive substances volume-regulatory osmolytes in rat renal papillary cells? J Physiol 386: 45–61

Lee CO, Taylor A, Windhager EE (1980) Cytosolic calcium ion activity in epithelial cells of *Necturus* kidney. Nature 287: 859–861

Leibowich S, DeLong J, Civan MM (1988) Apical Na^+ permeability of frog skin during serosal Cl^- replacement. J Membr Biol 102: 121–130

Lewis SA, Butt AG, Bowler MJ, Leader JP, Macknight ADC (1985) Effects of anions on cellular volume and transepithelial Na^+ transport across toad urinary bladder. J Membr Biol 83: 119–137

Lifschitz MD (1986) Prostaglandins may mediate chloride concentration gradient across domes formed by MDCK1 cells. Am J Physiol 250: F525–F531

Linshaw MA (1980) Effect of metabolic inhibitors on renal tubule cell volume. Am J Physiol 239: F571–F577

Linshaw MA, Grantham JJ (1980) Effect of collagenase and ouabain on renal cell volume in hypotonic media. Am J Physiol 238: F491–F498

Lipton P (1972) Effect of changes in osmolarity on sodium transport across the isolated toad bladder. Am J Physiol 222: 821–828

Livne A, Grinstein S, Rothstein A (1987) Volume-regulating behavior of human platelets. J Cell Physiol 131: 354–363

Lohr JW, Sullivan LP, Cragoe EJ, Grantham JJ (1989) Volume regulation determinants in isolated proximal tubules in hypertonic medium. Am J Physiol 256: F622–F631

Lopes AG, Guggino WB (1987) Volume regulation in the early proximal tubule of the *Necturus* kidney. J Membr Biol 97: 117–125

Macknight ADC (1985) The role of anions in cellular volume regulation. Pflügers Arch Eur J Physiol 405: S12–S16

Macknight ADC (1987) Volume maintenance in isosmotic conditions. Curr Top Membr Transp 30: 3–44
Macknight ADC (1988) Principles of cell volume regulation. Renal Physiol Biochem 11: 114–141
Macknight ADC, Leaf A (1977) Regulation of cellular volume. Physiol Rev 57: 510–573
Macknight ADC, Scott RJ (1989) Effects of impermeant medium ions on the composition of rabbit renal cortical slices. Renal Physiol Biochem 12: 118–136
MacLeod RJ, Hamilton JR (1990) Regulatory volume increase in isolated mammalian jejunal villus is due to bumetanide-sensitive NaKCl2 cotransport. Am J Physiol 258: G665–G674
McCann JD, Li M, Welsh MJ (1989) Identification and regulation of whole-cell chloride currents in airway epithelium. J Gen Physiol 94: 1015–1036
McConnell F, Goldstein L (1990) Volume regulation in elasmobranch red blood cells. Comp Physiol 4: 114–131
Mercer RW, Hoffman JF (1985) Bumetanide-sensitive Na/K cotransport in ferret red blood cells. (Abstract). Biophys J 47: 157a
Meyer M, Maly K, Überall F, Hoflacher J, Grunicke H (1991) Stimulation of K^+ transport systems by *Ha-ras*. J Biol Chem 266: 8230–8235
Mills JW (1987) The cell cytoskeleton: possible role in volume control. Curr Top Membr Transp 30: 75–101
Mills JW, Lubin M (1986) Effect of adenosine 3′,5′-cyclic monophosphate on volume and cytoskeleton of MDCK cells. Am J Physiol 250: C319–C324
Mills JW, Skiest DL (1985) Role of cyclic AMP and the cytoskeleton in volume control in MDCK cells. Mol Physiol 8: 247–262
Montero MC, Ilundáin A (1989) Effects of anisosmotic buffers on K^+ transport in isolated chicken enterocytes. Biochim Biophys Acta 979: 269–271
Montrose MH, Knoblauch C, Murer H (1988) Separate control of regulatory volume increase and Na^+-H^+ exchange by cultured renal cells. Am J Physiol 255: C76–C85
Montrose-Rafizadeh C, Guggino WB (1990) Cell volume regulation in the nephron. Annu Rev Physiol 52: 761–772
Moriyama T, Carcia-Perez A, Burg MB (1990) Factors affecting the ratio of different organic osmolytes in renal medullary cells. Am J Physiol 259: F847–F858
Muallem S, Loessberg PA (1990) Intracellular pH-regulatory mechanisms in pancreatic acinar cells. II. Regulation of H^+ and HCO_3^- transporters by Ca^{2+}-mobilizing agonists. J Biol Chem 265: 12813–12819
Musch MW, Field M (1989) K-independent Na-Cl cotransport in bovine tracheal epithelial cells. Am J Physiol 256: C658–C665
Natke E (1990) Cell volume regulation of rabbit cortical collecting tubule in anisotonic media. Am J Physiol 258: F1657–F1665
O'Neill WC (1987) Volume-sensitive Cl-dependent K transport in human erythrocytes. Am J Physiol 253: C883–C888
O'Neill WC, Klein JD (1992) Regulation of vascular endothelial cell volume by Na-K-2Cl cotransport. Am J Physiol (in press)
O'Neill WC, Mikkelsen RB (1987) Furosemide-sensitive Na^+ and K^+ transport and human erythrocyte volume. Biochim Biophys Acta 896: 196–202
Palmer LG, Sackin H (1988) Regulation of renal ion channels. FASEB J 2: 3061–3065
Parker JC (1978) Sodium and calcium movements in dog red blood cells. J Gen Physiol 71: 1–17
Parker JC (1979) Active and passive Ca movements in dog red blood cells and resealed ghosts. Am J Physiol 237: C10–C16
Parker JC (1983a) Hemolytic action of potassium salts on dog red blood cells. Am J Physiol 244: C313–C317
Parker JC (1983b) Volume-responsive sodium movements in dog red blood cells. Am J Physiol 244: C324–C330
Parker JC, Castranova V (1984) Volume-responsive sodium and proton movements in dog red blood cells. J Gen Physiol 84: 379–401

Parker JC, Gitelman HJ, Glosson PS, Leonard DL (1975) Role of calcium in volume regulation by dog red blood cells. J Gen Physiol 65: 84–96
Pasantes-Morales H, Schousboe A (1988) Volume regulation in astrocytes: a role for taurine as osmoeffector. J Neurosci Res 20: 505–509
Paulmichl M, Gstraunthaler G, Lang F (1985) Electrical properties of Madin-Darby canine kidney cells. Effects of extracellular potassium and bicarbonate. Pflügers Arch Eur J Physiol 405: 102–107
Paulmichl M, Friedrich F, Maly K, Lang F (1989) The effect of hypoosmolarity on the electrical properties of Madin-Darby canine kidney cells. Pflügers Arch Eur J Physiol 413: 456–462
Pine MB, Rhodes D, Thorp K, Tsai Y (1979) Anion exchange and volume regulation during metabolic blockade of renal cortical slices. J Physiol 297: 387–403
Pine MB, Brooks WW, Nosta JJ, Abelmann WH (1981) Hydrostatic forces limit swelling of rat ventricular myocardium. Am J Physiol 241: H740–H747
Pollack LR, Tate EM, Cook JS (1981) Turnover and regulation of Na, K-ATPase in HeLa cells. Am J Physiol 67: 537–552
Pollock AS, Arieff AI (1980) Abnormalities of cell volume regulation and their functional consequences. Am J Physiol 239: F195–F205
Reuss L (1984) Independence of apical membrane Na^+ and Cl^- entry in *Necturus* gallbladder epithelium. J Gen Physiol 84: 423–445
Reuss L (1988) Cell volume regulation in non-renal epithelia. Renal Physiol Biochem 11: 187–201
Ritter M, Paulmichl M, Lang F (1991a) Further characterization of volume regulatory decrease in cultured renal epitheloid (MDCK) cells. Pflügers Arch Eur J Physiol 418: 35–39
Ritter M, Steidl M, Lang F (1991b) Inhibition of ion conductances by osmotic shrinkage of Madin Darby canine kidney (MDCK-) cells. Am J Physiol 261: C602–C607
Rome L, Grantham J, Savin V, Lechene C (1988) Volume regulation in S2 segments: analysis of intracellular osmolal and anion gaps. Proceedings of the Annual Meeting of the American Society of Nephrology, San Antonio: 339A, 107, (Abstr)
Rome L, Grantham J, Savin V, Lohr J, Lechene C (1989) Proximal tubule volume regulation in hyperosmotic media: intracellular K^+, Na^+, and Cl^-. Am J Physiol 257: C1093–C1100
Roy G, Sauvé R (1987) Effect of anisotonic media on volume, ion and amino-acid content and membrane potential of kidney cells (MDCK) in culture. J Membr Biol 100: 83–96
Rugolo M, Mastocola T, Flamigni A, Lenaz G (1989) Chloride transport in human fibroblasts is activated by hypotonic shock. Biochem Biophys Res Commun 160: 1330–1338
Russo MA, Ernst SA, Kapoor SC, van Rossum GDV (1985) Morphological and physiological studies of rat kidney cortex slices undergoing isosmotic swelling and its reversal: a possible mechanism for ouabain-resistant control of cell volume. J Membr Biol 85: 1–24
Sackin H (1987) Stretch-activated potassium channels in renal proximal tubule. Am J Physiol 253: F1253–F1262
Sackin H (1989) A stretch-activated K^+ channel sensitive to cell volume. Proc Natl Acad Sci USA 86: 1731–1735
Sackin H, Palmer LG (1987) Basolateral potassium channels in renal proximal tubule. Am J Physiol 253: F476–F487
Sànchez Olea R, Pasantes-Morales H, Lázaro A, Cereijido M (1991) Osmolarity-sensitive release of free amino acids from cultured kidney cells (MDCK). J Membr Biol 121: 1–9
Sarkadi B, Mack E, Rothstein A (1984a) Ionic events during the volume response of human peripheral blood lymphocytes to hyotonic media. I. Distinctions between volume-activated Cl^- and K^+ conductance pathways. J Gen Physiol 83: 497–512
Sarkadi B, Mack E, Rothstein A (1984b) Ionic events during the volume response of human peripheral blood lymphocytes to hypotonic media. II. Volume- and time-dependent activation and inactivation of ion transport pathways. J Gen Physiol 83: 513–527

Sarkadi B, Cheung R, Mack E, Grinstein S, Gelfand EW, Rothstein A (1985) Cation and anion transport pathways in volume regulatory response of human lymphocytes to hyposmotic media. Am J Physiol 248: C480–C487

Schild L, Aronson PS, Giebisch G (1991) Basolateral transport pathways for K^+ and Cl^- in rabbit proximal tubule: effects on cell volume. Am J Physiol 260: F101–F109

Schmidt WF III, McManus TJ (1977a) Ouabain-insensitive salt and water movements in duck red cells. I. Kinetics of cation transport under hypotonic conditions. J Gen Physiol 70: 59–79

Schmidt WF III, McManus TJ (1977b) Ouabain-insensitive salt and water movements in duck red cells. II. Norepinephrine stimulation of sodium plus potassium cotransport. J Gen Physiol 70: 81–97

Schmidt WF III, McManus TJ (1977c) Ouabain-insensitive salt and water movements in duck red cells. III. The role of chloride in the volume response. J Gen Physiol 70: 99–121

Schultz SG (1981) Homocellular regulatory mechanisms in sodium-transporting epithelia: avoidance of extinction by "flush-through". Am J Physiol 241: F579–F590

Schultz SG (1989a) Volume preservation: then and now. News Physiol Sci 4: 169–172

Schultz SG (1989b) Intracellular sodium activities and basolateral membrane potassium conductances of sodium absorbing epithelia. Curr Top Membr Transp 34: 21–44

Schultz SG, Hudson RL, Lapointe JY (1985) Electrophysiological studies of sodium cotransport in epithelia: towards a cellular model. In: Semenza G, Kinne R (eds) Membrane transport driven by ion gradients. Ann N Y Acad Sci 456: 127–135

Sernka TJ (1990) Direct hyposmotic stimulation of gastric acid secretion. Membr Biochem 9: 1–7

Siebens AW, Kregenow FM (1985) Volume-regulatory responses of *Amphiuma* red cells in anisotonic media. The effect of amiloride. J Gen Physiol 86: 527–564

Siebens AW, Spring KR (1989) A novel sorbitol transport mechanism in cultured renal papillary epithelial cells. Am J Physiol 257: F937–F946

Simmons NL (1982) Cultured monolayers of MDCK cells: a novel model system for the study of epithelial development and function. Gen Pharmacol 13: 287–291

Simmons NL (1984) Epithelial cell volume regulation in hypotonic fluids: studies using a model tissue culture renal epithelial cell system. Q J Exp Physiol 69: 83–95

Soltoff SP, McMillian MK, Cantley LC, Cragoe EJ Jr, Cantley LC, Talamo BR (1989) Effects of muscarinic, alpha-adrenergic, and substance P agonists and ionomycin on ion transport mechanisms in the rat parotid acinar cell. The dependence of ion transport on intracellular calcium. J Gen Physiol 93: 285–319

Somero GN (1986) Protons, osmolytes, and fitness of internal milieu for protein function. Am J Physiol 251: R197–R213

Spring KR (1985) Determinants of epithelial cell volume. Fed Proc 44: 2526–2529

Stoddard JS, Reuss L (1989) Electrophysiological effects of mucosal Cl^- removal in *Necturus* gallbladder epithelium. Am J Physiol 257: C568–C578

Sun A, Hebert SC (1989) Rapid hypertonic cell volume regulation in the perfused inner medullary collecting duct. Kidney Int 36: 831–842

Taniguchi J, Guggino WB (1989) Membrane stretch: a physiological stimulator of Ca^{2+}-activated K^+ channels in thick ascending limb. Am J Physiol 257: F347–F352

Tas PWL, Massa PT, Kress HG, Koschel K (1987) Characterization of an $Na^+/K^+/Cl^-$ co-transport in primary cultures of rat astrocytes. Biochim Biophys Acta 903: 411–416

Tauc M, Le Maout S, Poujeol P (1990) Fluorescent video-microscopy study of regulatory volume decrease in primary culture of rabbit proximal convoluted tubule. Biochim Biophys Acta 1052: 278–284

Thurston JH, Hauhart RE, Naccarato EF (1981) Taurine: possible role in osmotic regulation of mammalian heart. Science 214: 1373–1374

Tivey DR, Simmons NL, Aiton JF (1985) Role of passive potassium fluxes in cell volume regulation in cultured HeLa cells. J Membr Biol 87: 93–105

Ubl J, Murer H, Kolb HA (1988a) Hypotonic shock evokes opening of Ca^{2+} activated K channels in opossum kidney cells. Pflügers Arch Eur J Physiol 412: 551–553

Ubl J, Murer H, Kolb H-A (1988b) Ion channels activated by osmotic and mechanical stress in membranes of opossum kidney cells. J Membr Biol 104: 223–232
Ullrich KJ, Jarausch KH (1956) Untersuchungen zum Problem der Harnkonzentrierung und Harnverdünnung. Pflügers Arch Eur J Physiol 262: 537–550
Ussing HH (1965) Relationship between osmotic reactions and active sodium transport in frog skin epithelium. Acta Physiol Scand 63: 141–155
Ussing HH (1985) Volume regulation and basolateral cotransport of sodium, potassium, and chloride ions in frog skin epithelium. Pflügers Arch Eur J Physiol 405: S2–S7
Ussing HH (1986) Epithelial cell volume regulation illustrated by experiments in frog skin. Renal Physiol 9: 38–46
Ussing HH (1990) Volume regulation of frog skin epithelium. Comp Physiol 4: 87–113
van Rossum GDV, Russo MA (1981) Ouabain-resistant mechanism of volume control and the ultrastructural organization of liver slices recovering from swelling in vitro. J Membr Biol 59: 191–209
van Rossum GDV, Russo MA (1984) Requirement of Cl^- and Na^+ for the ouabain-resistant control of cell volume in slices of rat liver. J Membr Biol 77: 63–76
van Rossum GDV, Russo MA, Schisselbauer JC (1987) Role of cytoplasmic vesicles in volume maintenance. Curr Top Membr Transp 30: 45–74
van Rossum GDV, Ernst SA, Russo MA (1981) Relative effects of furosemide and ethacrynic acid on ion transport and energy metabolism in slices of rat kidney-cortex. Naunyn-Schmiedeberg's Arch Pharmacol 317: 90–96
Venosa RA (1991) Hypo-osmotic stimulation of active Na^+ transport in frog muscle: apparent upregulation of Na^+ pumps. J Membr Biol 120: 97–104
Vislie T (1983) Cell volume regulation in fish heart ventricles with special reference to taurine. Comp Biochem Physiol 76A: 507–514
Völkl H, Lang F (1988a) Electrophysiology of cell volume regulation in proximal tubules of the mouse kidney. Pflügers Arch Eur J Physiol 411: 514–519
Völkl H, Lang F (1988b) Ionic requirement for regulatory cell volume decrease in renal straight proximal tubules. Pflügers Arch Eur J Physiol 412: 1–6
Völkl H, Geibel J, Greger R, Lang F (1986) Effects of ouabain and temperature on cell membrane potentials in isolated perfused straight proximal tubules of the mouse kidney. Pflügers Arch Eur J Physiol 407: 252–257
Völkl H, Paulmichl M, Lang F (1988) Cell volume regulation in renal cortical cells. Renal Physiol 11: 158–173
Walz W, Mukerji S (1990) Simulation of aspects of ischemia in cell culture: changes in lactate compartmentation. Glia 3: 522–528
Wang W, Messner G, Oberleithner H, Lang F, Deetjen P (1984) The effect of ouabain on intracellular activities of K^+, Na^+, Cl^-, H^+ and Ca^{2+} in proximal tubules of frog kidneys. Pflügers Arch Eur J Physiol 401: 6–13
Weiss H, Lang F (1992) Ion channels activated by swelling of Madin Darby canine kidney (MDCK-) cells. J Membr Biol (in press)
Welling DJ, Welling LW (1988) Model of renal cell volume regulation without active transport: role of a heteroporous membrane. Am J Physiol 255: F529–F538
Welling PA, Linshaw MA, Sullivan LP (1985) Effect of barium on cell volume regulation in rabbit proximal straight tubules. Am J Physiol 249: F20–F27
Welling PA, O'Neill RG (1990) Cell swelling activates basolateral membrane Cl and K conductances in rabbit proximal tubule. Am J Physiol 258: F951–F962
Wiener H, Turnheim K (1990) Calcium-activated potassium channels in basolateral membranes of colon epithelial cells; reconstitution and functional properties. Wien Klin Wochenschr 26: 622–628
Wirthensohn G, Lefrank S, Guder WG, Beck FX (1987a) Studies on the role of glycerophosphorylcholine and sorbitol in renal osmoregulation. In: Kovacevic Z, Guder WG (eds) Molecular nephrology; biochemical aspects of kidney function. de Gruyter, Berlin, pp 323–327
Wirthensohn G, Beck F-X, Guder WG (1987b) Role and regulation of glycerophosphorylcholine in rat renal papilla. Pflügers Arch Eur J Physiol 409: 411–415

Yamauchi A, Kwon HM, Uchida S, Preston AS, Handler JS (1991) Myo-inositol and betaine transporters regulated by tonicity are basolateral in MDCK cells. Am J Physiol 261: F197–F202

Yantorno RE, Coca-Prados M, Krupin T, Civan MM (1989) Volume regulation of cultured transformed, non-pigmented epithelial cells from human ciliary body. Exp Eye Res 49: 423–437

Yoshitomi K, Frömter E (1984) Cell pH of rat renal proximal tubule in vivo and the conductive nature of peritubular HCO_3^--(OH^-) exit. Pflügers Arch Eur J Physiol 402: 300–305

Yoshitomi K, Frömter E (1985) How big is the electrochemical potential difference of Na^+ across rat renal proximal tubular cell membranes in vivo? Pflügers Arch Eur J Physiol 405: S121–S126

Yoshitomi K, Burckhardt B-Ch, Frömter E (1985) Rheogenic sodium-bicarbonate cotransport in the peritubular cell membrane of rat renal proximal tubule. Pflügers Arch Eur J Physiol 405: 360–366

Chapter 2

Cell Volume and Hepatic Metabolism

D. Häussinger[1], *W. Gerok*[1], and *F. Lang*[2]

Contents

1 Introduction

Cell volume homeostasis is crucial for overall cellular function. A great variety of volume regulatory mechanisms has been disclosed and found under appropriate conditions in almost every cell under study (see Chap. 1, this Vol.). These mechanisms are not only activated in order to compensate for cell volume deviations as induced by anisoosmotic exposure or during

[1] Medizinische Universitätsklinik Freiburg, Hugstetterstraße 55, W-7800 Freiburg, FRG
[2] Physiologisches Institut, Universität Innsbruck, Austria

Advances in Comparative and Environmental Physiology, Vol. 14

cumulative substrate uptake into the cell, but they can also be activated in the resting state by hormones so that new steady states of cell volume are generated. Accordingly, cell volume homeostasis does not mean the absolute constancy of cell volume, but rather the integration of events allowing cell volume changes to play a regulatory role in cellular function. Indeed, rapid and persistent cell volume changes may occur physiologically on a short-term time scale. Specifically in the liver, cellular volume can change considerably within minutes following physiological alterations of the hormonal and substrate milieu. Recently, it became clear that such alterations of cell volume markedly influence a variety of metabolic pathways. It appears that cell swelling and cell shrinkage lead to certain opposite patterns of cellular metabolic function. Apparently, hormones and amino acids can trigger these patterns simply by altering cell volume and it was suggested that cell volume alterations may represent a new principle of metabolic control (Häussinger and Lang 1991a,b).

It is the aim of this review to illustrate the interplay between liver cell volume and liver cell metabolism.

2 Cell Volume Control in Liver

Anisoosmotic exposure of cells is widely used for studies on cell volume regulation. This approach may primarily be seen as an experimental tool to modify liver cell volume, although during intestinal absorption portal venous blood may become slightly hypo- or hypertonic and liver swelling has been shown during intestinal absorption of water (Haberich et al. 1966). Physiologically more important, however, are liver cell volume changes during cumulative substrate uptake into hepatocytes and under the influence of hormones. The principles of anisoosmotic cell volume regulation are outlined in detail in Chapter 1; accordingly, only some liver-specific aspects are briefly described here.

2.1 Anisoosmotic Cell Volume Modulation

If liver cells are suddenly exposed to hypotonic (hypertonic) media, they initially swell (shrink) like more or less perfect osmometers but within minutes retain (almost) their original cell volume. This behaviour has been labelled regulatory cell volume decrease (regulatory cell volume increase). Both, regulatory cell volume increase (RVI) and regulatory cell volume decrease (RVD), however, do not completely restore the initial cell volume, i.e. the cells are left in either a slightly shrunken or swollen state. The extent of this volume deviation apparently acts as a signal modifying cellular function persistently (see Sects. 3–6).

In perfused rat liver, RVD is largely achieved by a release of cellular potassium, chloride and bicarbonate (Graf et al. 1988; Haddad and Graf 1989; Haddad et al. 1989; Lang et al. 1989; Häussinger et al. 1990d). The release of potassium is inhibited by barium, quinidine and SITS (Haddad et al. 1989; Häussinger et al. 1990d). Accordingly, cell swelling apparently leads to activation of potassium channels in parallel to anion channels. In isolated hepatocytes, RVD has similarly been inhibited by barium and quinidine (Corasanti et al. 1990b). The activation of the K^+ channels leads to the respective hyperpolarization of the cell membrane (Graf et al. 1988). Hepatocyte swelling apparently opens stretch-activated nonselective cation channels, which allow passage of Ca^{2+} into the cell (Bear 1990). The increase in intracellular Ca^{2+} then may activate Ca^{2+}-sensitive K^+ channels (Bear and Petersen 1987). However, Ca^{2+} activation of K^+ channels in liver may not be the only mechanism allowing regulatory cell volume decrease; indeed, K^+ channels have also been described in isolated hepatocytes, which do not require an increase in intracellular calcium activity (Henderson et al. 1989). It is conceivable that these K^+ channels are directly activated by cell membrane stretch, as it is observed in rat proximal tubule (Sackin 1987).

RVI in liver is at least in part achieved by parallel activation of Na^+/H^+ exchange and Cl^-/HCO_3^- exchange. In perfused liver, a sudden increase in extracellular perfusate osmolarity stimulates amiloride- and ouabain-sensitive K^+ uptake (Haddad and Graf 1989; Lang et al. 1989; Häussinger et al. 1990d), which eventually leads to RVI. Loop diuretics do not appreciably modify the cell volume regulatory K^+ uptake. Accordingly, RVI in liver is apparently accomplished by activation of Na^+/H^+ exchange with subsequent extrusion of sodium in exchange with potassium. Na-K-2Cl cotransport does not appear to appreciably participate in hepatic RVI (Häussinger et al. 1990d), even though the carrier exists in the hepatic cell membrane and its activation by insulin is followed by an increase in cell volume (Häussinger et al. 1991; Hallbrucker et al. 1991b,c). RVI is similar in perfused livers, if perfusate osmolarity is increased from isotonic to hypertonic or from hypotonic to isotonic levels. Isolated hepatocytes, on the other hand, are unable to regulate their volume, if they are exposed to hypertonic extracellular fluid (Corasanti et al. 1990b) but do undergo RVI, if extracellular osmolarity is reduced before to elicit RVD and then increased to isotonic levels. This "secondary" RVI is sensitive to amiloride and paralleled by intracellular alkalosis and is – similar to RVI in perfused livers – thought to be accomplished by activation of Na^+/H^+ exchange.

2.2 Cumulative Substrate Transport

In liver, one of the most important challenges for cell volume homeostasis is the cumulative uptake of osmotically active substances, such as amino acids (Kristensen 1980; Bakker-Grunwald 1983; Kristensen and Folke 1984;

Häussinger et al. 1990a; Wettstein et al. 1990; Hallbrucker et al. 1991a). Four major amino acid transport systems have been distinguished on the basis of substrate specificities and ion dependence (for reviews, see Guidotti et al. 1978; Shotwell et al. 1983; Kilberg et al. 1985) and were termed systems A, ASC, N and L. Systems A and ASC are both Na^+-dependent, exhibit overlapping substrate specificities and mediate the electrogenic transport mainly of small aliphatic amino acids, whereas the Na^+-independent system L accomplishes the electroneutral transport of branched chain and aromatic amino acids. System N is Na^+-dependent and electrogenic and is specific for glutamine, histidine and asparagine. System A activity is increased during starvation, feeding of a high protein diet and by hormones such as glucagon and insulin, whereas system N is much less responsive to starvation (Hayes and McGivan 1982). The Na^+-dependent transporters build up intra/extracellular amino acid concentration gradients of up to 20. Na^+ entering the hepatocyte together with the amino acid is extruded in exchange for K^+ by the electrogenic Na^+/K^+ ATPase.

The accumulation of amino acids and K^+ into the cells leads to hepatocyte swelling, which in turn triggers volume regulatory K^+ efflux (Kristensen 1980; Bakker-Grunwald 1983; Kristensen and Folke 1984, 1986; Häussinger et al. 1990a; Wettstein et al. 1990). The volume-regulatory mechanisms which are activated during isoosmotic (e.g. cumulative substrate transport-induced) cell swelling are probably identical to those involved in anisoosmotic cell volume regulation. As illustrated in Fig. 1, addition of glutamine to isolated perfused rat liver creates within about 12 min an intra/extracellular glutamine concentration gradient of about 12. During the first 2 min of glutamine accumulation, liver mass increases rapidly due to cell swelling, as validated by measurements of intracellular water space (vom Dahl et al. 1991d). Hepatic net K^+ uptake during this phase is probably due to extrusion of Na^+ by Na^+/K^+ ATPase. Thereafter, no further increase in liver mass is observed, despite continuing accumulation of glutamine inside the cell. This is achieved by a volume-regulatory K^+ efflux from the liver which occurs until the steady-state intracellular glutamine concentration of about 35 mM is reached (Fig. 1). During the rapid washout of glutamine following its withdrawal from the influent perfusate, the swollen hepatocytes shrink rapidly as evident from the respective decrease in liver mass and a secondary volume-regulatory net K^+ uptake which lasts for more than 10 min, restores cell volume to the starting level (Fig. 1). Cell swelling and volume-regulatory K^+ efflux were shown to occur upon addition of glutamine, alanine, proline, serine, glycine, aminoisobutyrate, phenylalanine, hydroxyproline and a complete amino acid mixture in physiological concentrations, but not upon addition of glucose or leucine (Wettstein et al. 1990), i.e. compounds which are not concentrated by liver cells. No cell swelling and volume-regulatory K^+ responses are observed upon addition of glutamate, although this amino acid is taken up by liver cells via Na^+-independent and Na^+-dependent mechanisms. Glutamate-induced volume changes prob-

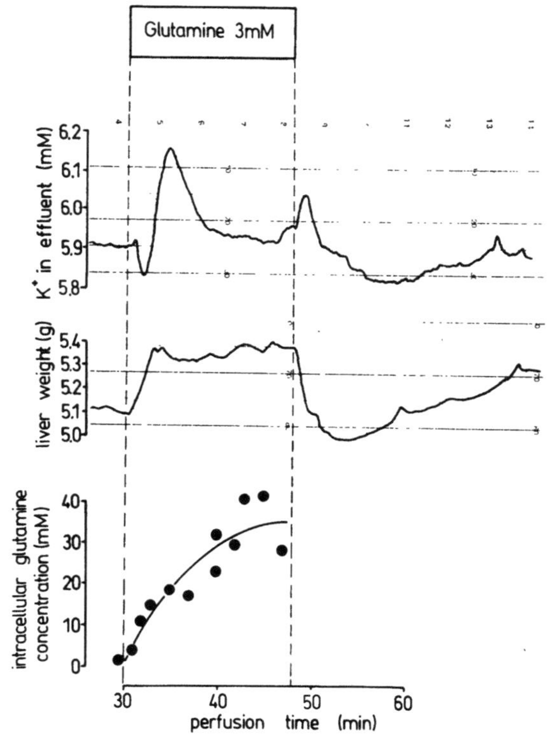

Fig. 1. Effect of glutamine (3 mM) addition to influent perfusate of isolated, single-pass perfused rat liver on intracellular glutamine accumulation, cell volume and volume-regulatory K^+ fluxes. Liver cell volume changes were assessed as mass changes in the perfused organ. As shown recently (vom Dahl et al. 1991d) glutamine-induced liver weight changes are exclusively due to alterations of cellular volume. Addition of glutamine to portal perfusate leads to rapid cell swelling due to cumulative, Na^+-dependent uptake of glutamine into liver cells. The initial net K^+ uptake is explained by exchange of cotransported Na^+ against K^+ by Na^+/K^+ ATPase. Glutamine-induced cell swelling during the first 2 min of glutamine infusion activates volume-regulatory K^+ (plus Cl^- and HCO_3^-) efflux. This volume-regulatory response prevents further cell swelling despite continuing glutamine accumulation inside the cell until a steady-state intracellular glutamine concentration of about 35 mM is reached. However, the liver cell remains in a swollen state as long as glutamine is infused. This degree of cell swelling modifies cellular function (see Sect. 3). For further details, see text. (Häussinger et al. 1990a)

ably can not be picked up, because the uptake of this amino acid is restricted to only a very small perivenous hepatocyte population, which comprises less than 10% of all hepatocytes (for review, see Häussinger 1990).

It recently became clear that amino acid-induced cell swelling and volume-regulatory responses occur upon exposure to amino acids in the physiological concentration range. Glutamine-induced cell swelling is half-maximal at a concentration of 0.6–0.8 mM, i.e. the concentration normally found in portal venous blood, and is maximal at 2 mM (Wettstein et al.

1990). Accordingly, physiological fluctuations of the portal amino acid concentration are accompanied by parallel alterations of liver cell volume. The degree of amino acid-induced cell swelling seems to be largely related to the steady-state intra/extracellular amino acid concentration gradient. This gradient and, accordingly, the degree of cell swelling are modified by hormones and the nutritional state in a complex way. Their action not only involves effects on the expression of plasma membrane transport systems and the electrochemical Na^+ gradient as a driving force for Na^+-coupled transport, but also alterations of intracellular amino acid metabolism. For example in livers from rats fed insulin, which is known to stimulate amino acid transport via system A (Le Cam and Freychet 1978), enhances the glycine-induced cell swelling (Hallbrucker et al. 1991b). Similarly, starvation increases the swelling potency of glycine about four-fold (Hallbrucker et al. 1991a) probably due to adaptive up-regulation of the glycine-transporting system A (Hayes and McGivan 1982).

The osmotic gradient built up during amino acid accumulation and therefore the degree of cell swelling are sometimes dissipated by draining amino acids into metabolic pathways, which yield products that equally distribute on both sides on the plasma membrane. Because alanine transport in liver is controlling its metabolism (Groen et al. 1982), inhibition of alanine metabolism by transaminase inhibitors markedly enhances the alanine-induced volume-regulatory K^+ efflux (Wettstein et al. 1990). Hormones stimulating amino acid metabolism can indirectly affect cell volume by either dissipating the steady-state amino acid concentration gradient across the plasma membrane or causing the intracellular accumulation of a product derived from the added amino acid. Stimulation of glutaminase following addition of glutamine will e.g. result in marked intracellular accumulation of (almost impermeable) glutamate as soon as urea synthesis is blocked.

2.3 Cell Volume Control by Hormones

Among other actions, hormones are known to modify the activities of ion transporters and channels in the plasma membrane, to affect the cell membrane potential and to modulate Na^+-driven substrate transport (Fehlmann and Freychat 1981; Fehlmann et al. 1981; Lynch et al. 1986; Jakubowski and Jakob 1990); for review, see Moule and McGivan 1990). All these actions are expected to modify cell volume. Indeed, hormones were recently recognized as potent modulators of liver cell volume. Whereas anisoosmotic exposure and cumulative amino acid transport into hepatocytes lead primarily to cell swelling with secondary activation of volume-regulatory ion transporters, hormones primarily affect the activity of these volume-regulatory transport systems which secondarily produces cell volume changes. This important aspect on hormone action sheds a new light on metabolic hormone actions, because the cell volume change itself modifies cellular function. In liver, insulin stimulates Na^+/H^+ exchange (Jakubowski and Jakob 1990), Na-K-2Cl cotransport (Hallbrucker et al. 1991b,c) and the Na^+/K^+ ATPase (Fehlmann and Freychat 1981; Jakubowski and Jakob 1990), i.e. transport systems which are turned on for a cell volume-regulatory increase in a variety of tissues (see Chap. 1). The concerted action of these transporters leads to cellular accumulation of potassium, sodium and chloride and consequently cell swelling. Glucagon is known to activate Na^+/K^+ ATPase (Moule and McGivan 1990), but simultaneously decreases cellular K^+ in isolated perfused rat liver (vom Dahl et al. 1991a,c; Hallbrucker et al. 1991c), probably due to a simultaneous opening of Ba^{2+}- and quinidine-sensitive K^+ channels (Hallbrucker et al. 1991c). Thus, glucagon may lead to a depletion of cellular Na^+, K^+ and probably Cl^-, resulting in cell shrinkage. Indeed, in both perfused livers and isolated hepatocytes, insulin increases and glucagon decreases cellular volume and these changes are dependent on the activation of the respective K^+ transport systems

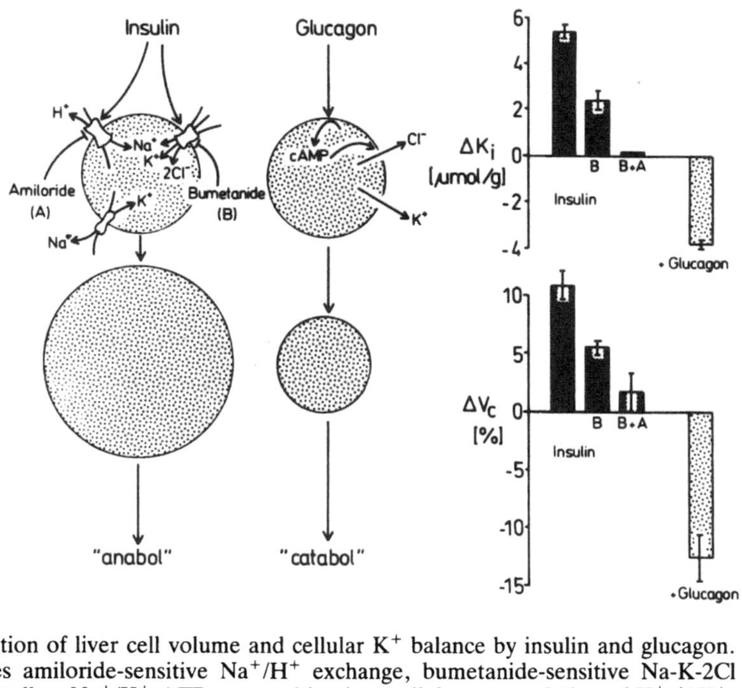

Fig. 2. Modulation of liver cell volume and cellular K^+ balance by insulin and glucagon. Insulin activates amiloride-sensitive Na^+/H^+ exchange, bumetanide-sensitive Na-K-2Cl cotransport as well as Na^+/K^+ ATPase, resulting in a cellular accumulation of K^+ (ΔK_i), Na^+ and Cl^- and cell swelling (ΔV_c). Both cellular K^+ accumulation and cell swelling by insulin are abolished in the presence of amiloride plus bumetanide. Conversely, glucagon and cAMP deplete cellular K^+ by activating Cl^- channels and quinidine/Ba^{2+}-sensitive K^+ channels, thereby inducing cell shrinkage. The hormone-induced cell volume alterations act like another signal which participates in mediating the hormone effects on hepatic metabolism: cell swelling triggers an anabolic, cell shrinkage a catabolic pattern of cellular function (see also Fig. 5)

(Häussinger et al. 1991; Hallbrucker et al. 1991b,c; vom Dahl et al. 1991a,c): insulin activates both loop diuretic-sensitive Na-K-2Cl cotransport and amiloride-sensitive Na^+/H^+ exchange, whereas glucagon stimulates cellular potassium release through barium- and quinidine-sensitive K^+ channels (Fig. 2). As shown in Fig. 2, insulin-induced cell swelling and cellular K^+ accumulation is abolished in the presence of bumetanide plus amiloride. Insulin-induced cell swelling is counteracted by further addition of glucagon. Similar to glucagon, also cyclic AMP shrinks liver cells in the intact organ (Hallbrucker et al. 1991b,c; vom Dahl et al. 1991a,c). Remarkably, half-maximal insulin-induced cell swelling and half-maximal glucagon-induced cell shrinkage in perfused rat liver are found at hormone concentrations normally present in portal venous blood in vivo (vom Dahl et al. 1991c; Fig. 3). Accordingly, one is tempted to speculate that physiological fluctuations of the portal insulin and glucagon concentrations may modify liver cell volume also in vivo. In perfused rat liver, vasopressin exhibits only minor effects on cell volume, but shrinks the cells when added on top of glucagon or cAMP (vom Dahl et al. 1991c). Adenosine leads to cell shrinkage, whereas phenylephrine slightly increases liver cell volume. The effects of a variety of hormones on liver cell volume are summarized in Table 1. Cell volume modulation by these hormones at least in part involves alterations of cellular K^+ and Na^+ balance and the cell volume effects exerted by different hormones are largely additive (vom Dahl et al. 1991c). It should be emphasized that cell volume changes markedly under the influence of hormones. For example, insulin plus phenylephrine induces a 20% cell swelling, whereas vasopressin plus glucagon leads to about 20% cell shrinkage. In order to achieve comparable changes in cell volume by anisoosmotic exposure extracellular osmolarity changes of about 100 mOsmol/l are required (vom Dahl et al. 1991c).

Hormone-induced cell volume changes were recently recognized to play a crucial role in mediating some metabolic responses to the hormones (Häussinger et al. 1991; Hallbrucker et al. 1991b; vom Dahl et al. 1991a): insulin and glucagon, at least in part, regulate hepatic proteolysis by stimulating carriers at the cell membrane, which modify cell volume (see Sect. 4) and the opposing effects of both hormones on proteolysis can be explained by their opposing effects on cellular volume.

Hormones can also affect cell volume by altering amino acid transport across the plasma membrane. These effects not only include alterations of the electrochemical Na^+ gradient as a driving force for Na^+-dependent amino acid transporters, but also induction of amino acid transport systems. For example, insulin-induced intracellular K^+ accumulation by activation of Na^+/K^+ ATPase will lead to hyperpolarization of the membrane potential. Hyperpolarization as well as the known cycloheximide-sensitive induction of the Na^+-dependent amino acid transport system A by insulin (Fehlmann et al. 1981; Le Cam and Freychat 1978) will stimulate sodium-coupled amino acid transport and accordingly tends to swell the cells. Glucagon similarly

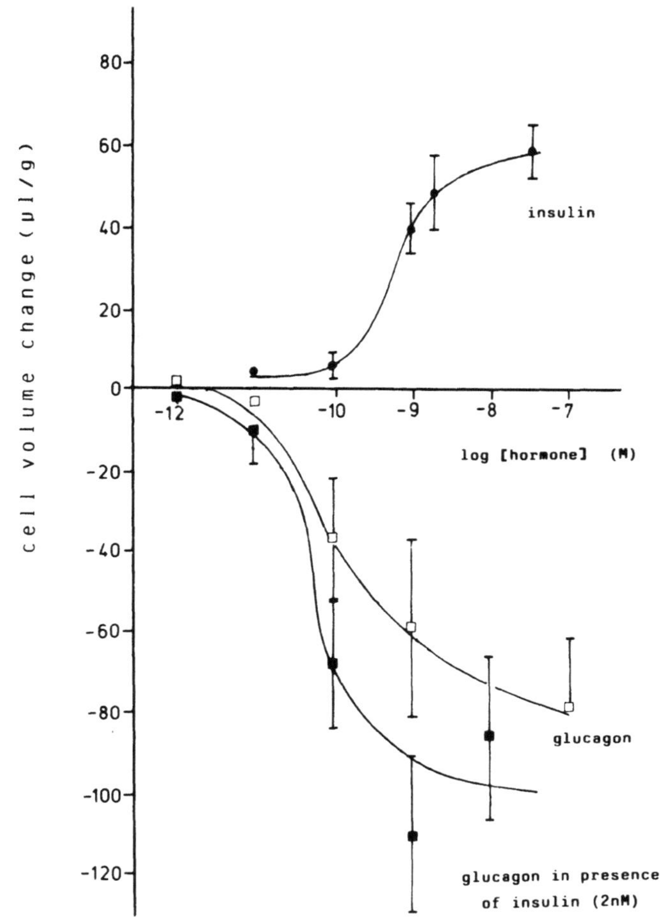

Fig. 3. Effect of glucagon and insulin on liver cell volume. In the absence of hormones cell volume was 559 ± 7 µl/g liver. (vom Dahl et al. 1991c)

stimulates amino acid uptake. Whereas this tends to swell the cell, glucagon in fact shrinks the cells due to a glucagon-induced net K^+ release and dissipation of the amino acid concentration gradient across the plasma membrane due to a glucagon-induced stimulation of amino acid breakdown.

3 Modulation of Hepatic Metabolism by Cell Volume

One consistent feature of volume regulation during cumulative substrate uptake or anisoosmotic exposure in liver cells is that volume-regulatory mechanisms never achieve a complete recovery of cell volume; instead, the

Table 1. Effects of hormones and amino acids on liver cell volume

	Change in intracellular water space (%)
Insulin (35 nM)	+12 ± 1 (15)
Glucagon (100 nM)	−14 ± 3 (6)
Dibutyryl-cAMP (50 μM)	−9 ± 1 (4)
Adenosine (50 μM)	−6 ± 3 (3)
Vasopressin (15 nM)	−2 ± 2 (4)
Phenylephrine (5 μM)	+8 ± 1 (4)
ATP (20 μM)	−6 ± 2 (6)
Bradykinin (100 nM)	+8 ± 1 (4)
Angiotensin (15 nM)	0 ± 3 (4)
Serotonin (5 μM)	−5 ± 1 (4)
Prostaglandin $F_{2\alpha}$ (2 μM)	0 ± 1 (3)
Vasopressin (15 nM) + glucagon (100 nM)	−23 ± 3 (4)
Vasopressin (35 nM) + cAMP (50 μM)	−16 ± 1 (6)
Adenosine (50 μM) + phenylephrine (5 μM)	−1 ± 3 (3)
Insulin (35 nM) + glucagon (100 nM)	+3 ± 3 (4)
Insulin (35 nM) + cAMP (50 μM)	−4 ± 2 (3)
Insulin (35 nM) in presence of bumetanide (5 μM)	+6 ± 1 (3)
Glutamine (1 mM)	+5 ± 1 (3)
Glutamine (3 mM)	+10 ± 1 (3)
Glycine (2 mM)	+5 ± 0 (3)
Glycine (2 mM) + glutamine (2 mM)	+13 ± 2 (3)
*Glycine (2 mM)	+11 ± 0 (3)
Alanine (2 mM)	+6 ± 0 (7)
*Alanine (2 mM)	+14 ± 2 (6)

The intracellular water space ("liver cell volume") was determined in the perfused rat liver using a [^{3}H]inulin/[^{14}C]urea washout technique as described in (vom Dahl et al. 1991d). Without effector addition, the intracellular water space was 559 ± 7 μl/g (n = 88), the data are given as % change of intracellular water space as they occur after a 30-min exposure to hormones and amino acids. Negative values indicate cell shrinkage, positive values cell swelling. Data are from liver from fed rats, except "which denotes 24-h starvation.

cells are left in either a slightly swollen or shrunken state. Current evidence suggests that the extent of this remaining volume deviation somehow influences cell function. Accordingly, it was suggested that cell volume alterations may represent a new principle of metabolic control (Häussinger and Lang 1991a,b), which becomes even more interesting in view of the recent demonstration that hormones such as insulin and glucagon markedly affect liver cell volume. Anisoosmotic cell volume changes are accompanied by a variety of alterations of metabolic liver cell function, which persist throughout anisotonic exposure and are maintained even after completion of volume-regulatory ion fluxes (Table 2). Several long-known, but mechanistically poorly understood effects of amino acids, which could not be related to their metabolism, such as stimulation of glycogen synthesis (Katz et al. 1976; Chen and Lardy 1985; Rognstad 1985; Lavoinne et al. 1987) or

Table 2. Liver cell swelling as an anabolic signal

Cell swelling	
Stimulates	Inhibits
Protein synthesis	Proteolysis
Glycogen synthesis	Glycogenolysis
Lactate uptake	Glycolysis
Aminoacid uptake	
Glutaminase	Glutamine synthesis
Glycine oxidation	
Ketoisocaproate oxidation	
Urea synthesis from amino acids	Urea synthesis from NH_4^+
Glutathione efflux	
Taurocholate excretion into bile	
Pentose phosphate shunt	

inhibition of proteolysis (Hopgood et al. 1977; Schworer and Mortimore 1979; Seglen et al. 1980; Mortimore and Pösö 1984, 1987; Pösö et al. 1982a,b; Seglen and Gordon 1984; Leverve et al. 1987) can simply be mimicked by hypoosmotic swelling of the cells to a similar extent as the amino acids (Baquet et al. 1990; Häussinger et al. 1990b, 1991; Hallbrucker et al. 1991a). Thus, the above-mentioned amino acid effects are probably mediated by amino acid-induced cell swelling. Because liver cell swelling was shown to inhibit glycogenolysis, glycolysis (Graf et al. 1988; Lang et al. 1989; vom Dahl et al. 1991b) and proteolysis (Häussinger et al. 1990b, 1991; Hallbrucker et al. 1991a) and to stimulate glycogen synthesis (Baquet et al. 1990), amino acid uptake (Häussinger and Lang 1990; Häussinger et al. 1990a,c; Bode and Kilberg 1991), ureogenesis from amino acids (Häussinger and Lang 1990; Häussinger et al. 1990c) and also protein synthesis (Stoll et al. 1992), it seems that cell swelling acts like an "anabolic signal". In line with this, the "anabolic hormone" insulin induces hepatocyte swelling, whereas the "catabolic hormone" glucagon induces cell shrinkage. This view can be extended further, if one takes into account the relationship between cell growth and differentiation and the activity of Na^+/H^+ exchange, i.e. a system which is affected not only by hormones and growth factors, but also serves cell volume-regulatory increase. Because the activity of Na^+-dependent amino acid transport systems in the plasma membrane is now recognized to exert marked effects on cellular function via cell volume changes, amino acid transporters may be seen not only as amino acid-translocating systems, but also as transmembrane signalling systems (Häussinger and Lang 1991b).

3.1 Cell Volume, Amino Acid and Ammonial Metabolism

Amino acids are not only potent modulators of liver cell volume, but conversely, there is evidence for a control of amino acid transport by the cell volume. Amino acid transport system N was shown to be activated by amino acids in a cycloheximide-insensitive way (Weissbach and Kilberg 1984). This stimulation was not explained by trans-stimulation and besides substrates of system N alanine, serine and the non-metabolizable analogue amino-isobutyrate, i.e. substrates of the Na^+-dependent system A, were also effective, whereas substrates of Na^+-independent system L, such as leucine, were ineffective (Weissbach and Kilberg 1984). It now appears that this amino acid-dependent short-term stimulation of amino acid transport is due to cell swelling (Bode and Kilberg 1991). Hypoosmotic cell swelling increases hepatic net alanine uptake in starved rats (Häussinger and Lang 1990), whose intracellular degradation rate was shown to be controlled by transport rather than by metabolism (Sips et al. 1980; Groen et al. 1982). Similar observations were made regarding hepatic glutamine metabolism (Häussinger et al. 1990a). A swelling-induced hyperpolarization of the membrane (Graf et al. 1988) may also explain why sinusoidal glutathione efflux from the liver is increased following hypoosmotic liver cell swelling (Häussinger et al. 1990c), because glutathione release was shown to be under the control of the membrane potential (Fernandez-Checa et al. 1988). In addition to this short-term control of transport by cell volume, there is some evidence that cell volume also influences the expression of amino acid transporters in the plasma membrane. In cultured human fibroblasts, amino acid deprivation leads to a cycloheximide-sensitive adaptive increase in the activity of transport system A. This adaptive increase is potentiated by hyperosmotic cell shrinkage and counteracted by hypoosmolar cell swelling (Gazzola et al. 1991). It remains to be established whether such a regulation also occurs in liver.

Hypoosmotic cell swelling also exerts marked effects on amino acid metabolism. In rat liver perfused with a physiological portal glutamine concentration (0.6 mM), exposure to hypoosmotic perfusion fluid switches hepatic glutamine balance from net release to net uptake. This is due to a stimulation of flux through glutaminase in periportal hepatocytes and a simultaneous inhibition of glutamine synthesis in perivenous hepatocytes (Häussinger et al. 1990a). Conversely, hyperosmotic cell shrinkage inhibits flux through glutaminase, whereas flux through glutamine synthetase is stimulated. Cell swelling-induced activation of glutaminase is most likely due to simultaneous mitochondrial swelling, which probably alters the attachment of the enzyme to the inner mitochondrial membrane (McGivan et al. 1984). Likewise, hypoosmotic swelling stimulates glycine oxidation in perfused rat liver (Häussinger et al. 1992a) and in isolated mitochondria (Brosnan et al. 1990). There is good evidence that stimulation of glycine and glutamine oxidation by glucagon, cAMP or Ca^{2+} mobilizing hormones is due to hormonally induced increases in mitochondrial matrix volume (see Halestrap, this Vol.). Here, it should be mentioned that glucagon and cAMP lead to mitochondrial swelling (Halestrap 1989), although these agents simultaneously induce marked liver cell shrinkage. In perfused rat liver stimulation of glutamine breakdown by glucagon and hypoosmotic cell swelling was not additive (unpubl. result), but both maneuvers were additive to the ammonia stimulation of glutaminase (Häussinger et al. 1983, 1990a). These findings suggest that alterations of amino acid metabolism following anisotonic cell volume changes can at least in part be explained by parallel alterations of mitochondrial matrix volume. Indeed, morphometric studies revealed mitochondrial swelling in isolated hepatocytes following hypoosmotic exposure (Pfaller et al. 1993).

When livers from fed, but not from 24-h starved, rats are perfused with an amino acid mixture roughly mimicking physiological portal concentrations in the post-absorptive state, hypoosmotic liver swelling stimulates urea synthesis and ammonia formation from the amino acids (Häussinger et al. 1990c; Häussinger and Lang 1990). When, however, ammonia is used as sole substrate for urea synthesis in livers from fed rats, hypoosmotic cell swelling markedly inhibited urea and glutamine synthesis, although swelling of isolated mitochondria stimulates citrulline synthesis (for review, see Halestrap 1989). Inhibition of urea synthesis from ammonia in the intact liver following cell swelling is probably due to a block of the urea cycle at the step of argininosuccinate synthesis, because it is accompanied by an increase in citrulline tissue levels, whereas the intracellular concentrations of aspartate, glutamate and malate decrease (Häussinger et al. 1990c). Simultaneously, under these conditions the lactate/pyruvate ratio increased, whereas the β-hydroxybutyrate/acetoacetate ratio decreased. These findings were interpreted to reflect a cell swelling-induced disturbance of the transfer of reducing equivalents via the malate/aspartate shuttle and an impaired aspartate regeneration during cell swelling. In line with this, the swelling-induced inhibition of urea synthesis from NH_4Cl was overcome by addition of lactate and pyruvate. These substrates allowed intramitochondrial regeneration of oxaloacetate via pyruvate carboxylase, a reaction which was shown to be stimulated by swelling of isolated mitochondria, too (for review, see Halestrap 1989).

3.2 Cell Volume and Protein Turnover

3.2.1 Proteolysis

Hepatic proteolysis is under the control of amino acids (Hopgood et al. 1977; Schworer and Mortimore 1979; Seglen et al. 1980; Pösö et al. 1982a,b; Mortimore and Pösö 1984, 1987; Seglen and Gordon 1984; Leverve et al. 1987) and hormones, such as insulin and glucagon (Mortimore and Mondon 1970; Neely et al. 1974; Hopgood et al. 1977, 1981; Draznin and Trowbridge 1982), but the underlying mechanisms remained obscure (Mortimore and Pösö 1987). It recently became clear that hypoosmotic cell swelling inhibits proteolysis in liver whereas, conversely, hyperosmotic cell shrinkage stimulates protein breakdown under conditions when the proteolytic pathway is not yet fully activated (Häussinger et al. 1990b, 1991; Hallbrucker et al. 1991a; vom Dahl et al. 1991a). The known antiproteolytic effect of several amino acids can in part be ascribed to an amino acid-induced cell swelling (Häussinger et al. 1990b, 1991; Hallbrucker et al. 1991a). A close linear relationship is observed between the amino acid-induced degree of cell swelling and the inhibition of proteolysis. The antiproteolytic effect of glutamine and glycine can be fully explained by the glycine- or glutamine-induced cell swelling, since inhibition of proteolysis by these amino acids is quantitatively mimicked by hypoosmotic cell swelling when induced to the same extent as the above-mentioned amino acids (Häussinger et al. 1990b, 1991; Hallbrucker et al. 1991a). Inhibition of proteolysis by glutamine and glycine is additive to the same extent as additivity is observed with respect to cell swelling induced by both amino acids. In livers from fed rats, the antiproteolytic effect of glycine is only about one-third compared to that found in livers from starved rats. This is explained by an approximately three-fold swelling potency of glycine during starvation, as a consequence of a higher activity of glycine-transporting amino acid transport system A (Hayes and McGivan 1982). On the other hand, glutamine-transporting system N is not adaptively up-regulated during starvation and consequently glutamine-induced cell swelling and proteolysis inhibition are only slightly different in the fed and starved state respectively (Hallbrucker et al. 1991a). In addition, there is no influence of the nutritional state on the degree of cell swelling during hypoosmotic exposure and on the accompanying inhibition of proteolysis. Also alanine, proline and serine lead to cell swelling, but the inhibition of proteolysis induced by these amino acids is about twice as strong as one would predict from anisotonic swelling experiments producing the same degree of cell swelling as these amino acids (Hallbrucker et al. 1991a). Accordingly, about 50% of the antiproteolytic effect of alanine, proline and serine can be explained on the basis of cell volume changes. On the other hand, inhibition of proteolysis by leucine, phenylalanine or a complete amino acid mixture can only in small part be explained by alterations of cell volume; here, apparently other mechanisms come into play.

3.2.2 Protein Synthesis

Cell volume not only affects proteolysis, but also protein synthesis. Hyperosmotic cell shrinkage (by increasing the extracellular osmolarity by 120 mOsmol/l) inhibits protein synthesis by 60–70% in isolated hepatocyte suspensions incubated with amino acids at twice the physiological concentration (Stoll et al. 1992). On the other hand, hypoosmotic exposure does not significantly stimulate protein synthesis. Apparently, protein synthesis is already maximally active and cannot be further enhanced under control conditions. Here, it must be kept in mind that already during normotonic incubation the hepatocytes are already preswollen by the amino acids present in the incubation mixture. However, a stimulation of protein synthesis by hypoosmotic or glutamine-induced cell swelling becomes apparent when the pathway was preinhibited. Regarding a role of cell volume in the hormonal control of protein synthesis, see Section 5.

In perfused rat liver, hypoosmotic exposure significantly increases the amount of silver-stainable nucleolar organizers (K.W. Schmid, A. Hittmair, C. Hallbrucker and D. Häussinger, unpubl. result), suggestive of an increased synthesis of rRNA (Schwarzacher et al. 1978).

Like proteolysis, mRNA breakdown has been suggested to involve the autophagic pathway and to underlie a control by amino acids and hormones similar to proteolysis (Lardeux and Mortimore 1987; Balavoine et al. 1990). The remarkable similarities between the factors controlling both autophagic proteolysis and mRNA degradation are suggestive of an inhibition of mRNA degradation due to cell swelling, induced by amino acids or hormones. Accordingly, cell swelling may prolong the half-life of certain mRNA species. This could explain why hypoosmotic exposure increases the tissue mRNA levels for β-actin in perfused rat liver, whereas mRNA levels for albumin and tyrosine aminotransferase remain unchanged (Schulz et al. 1991). These findings suggest that cell volume changes may not alter protein synthesis in general, but may affect the expression of specific proteins only.

Thus, currently available evidence suggests that cell volume changes may also play a role in the regulation of gene expression. It is well conceivable that alterations of the intracellular ionic composition are involved, which are long known to affect chromatine structure (Leake et al. 1972).

The finding that hypoosmotic, hormone- (see Sect. 5) and amino acid-induced cell swelling inhibits proteolysis in liver whereas, conversely, cell shrinkage stimulates proteolysis and inhibits protein synthesis, may shed new light on the understanding of catabolic states and the known anabolic effect of insulin and amino acids such as glutamine (Rennie et al. 1986; McLennan et al. 1987, 1988; Roth et al. 1990). Protein turnover in skeletal muscle is complexly regulated and glutamine seems to play an important role. In skeletal muscle glutamine is transported by a Na^+-dependent carrier, which builts up a remarkably high intra/extracellular glutamine concentration gradient. A relationship between the intramuscular glutamine concentration, on the one hand, and protein turnover, on the other, was described (McLennan et al. 1987, 1988). Protein catabolic states, such as sepsis, are characterized by a marked lowering of intracellular glutamine concentrations in skeletal muscle (for review, see Roth et al. 1990). It is well conceivable that these largely empirical observations find their explanation in cell volume changes resulting from alterations of the intra/extracellular glutamine gradient. Thus, like in liver (see Sects. 3.4 and 4), cell shrinkage in glutamine-depleted muscle could in part be responsible for the increased protein breakdown found under these conditions.

3.3 Cell Volume and Carbohydrate Metabolism

In perfused rat liver, cell swelling inhibits and cell shrinkage stimulates lactate, pyruvate and glucose release from the liver, pointing to an inhibition (stimulation) of glycogenolysis during cell swelling (shrinkage) (Graf et al. 1988; Lang et al. 1989). Indeed, glycogen phosphorylase a activity increases (decreases) in isolated hepatocytes from fed rats following hyperosmotic (hypoosmotic) exposure (Stoll and Häussinger, unpubl. observ.). The effects of cell volume alterations on carbohydrate metabolism seem to be modified by the presence of ammonia (Häussinger et al. 1990c). When livers were perfused with lactate and pyruvate, hypoosmotic cell swelling increased the net uptake of lactate and pyruvate (vom Dahl et al. 1991b). Subsequently, glycogen synthesis has indeed been shown to be stimulated during hypoosmotic cell swelling (Baquet et al. 1990). Several amino acids such as glutamine, proline and asparagine, but also the non-metabolizable amino acid analogue 2-aminoisobutyric acid (AIB) are known to stimulate glycogen synthesis in hepatocytes from starved rats (Katz et al. 1976; Chen and Lardy 1985; Rognstad 1985; Lavoinne et al. 1987). This stimulation could not be ascribed to a metabolite derived from the amino acids, but was suggested to be primarily due to an activation of glycogen synthase resulting largely from a stimulation of synthase phosphatase (Lavoinne et al. 1987). It now appears that the stimulation of glycogen synthesis by the above-mentioned amino acids is a consequence of amino acid-induced cell swelling (Baquet et al. 1990). The extent of stimulation of glycogen synthesis paralleled the degree of cell swelling elicited by the amino acids and the stimulation of glycogen synthesis was prevented by hyperosmolar incubation of the hepatocytes (Baquet et al. 1990). Only the proline stimulation of glycogen synthesis was about twice as high as expected when a similar degree of cell swelling was induced by hypoosmotic incubation. Interestingly, such a behaviour of proline was also found in studies on amino acid-induced inhibition of proteolysis, which can only in part be ascribed to cell volume alterations (Hallbrucker et al. 1991a). Here, apparently other mechanisms come into play. It also appears that the stimulation of glycogen synthesis from glucose by exposing liver cells to a high K^+ medium (Hue et al. 1975) can at least in part be ascribed to cell swelling (Baquet et al. 1990). Similarly, stimulation of glycogen synthesis by quinine, choline or mercaptopicolinate can be explained on the basis of an agonist-induced cell swelling (Baquet et al. 1990). The mechanism how cell swelling stimulates glycogen synthesis is unclear. Cell swelling by raising the extracellular K^+ concentration inactivates phosphorylase (Hue et al. 1975), whereas amino acid-induced cell swelling does not (Baquet et al. 1990).

3.4 Cell Volume and Ketogenesis

In perfused rat liver, anisoosmotic cell swelling slightly stimulates the oxidation of ketoisocaproate, whereas hyperosmotic cell shrinkage inhibits $^{14}CO_2$ production from [1-^{14}C]ketoisocaproate. A similar influence of anisotonic exposure was observed with respect to ketogenesis from ketoisocaproate (Häussinger et al. 1992a). Whether these observations can be explained by a swelling-induced stimulation of Na^+-dependent ketoisocaproate transport across the plasma membrane (Kilberg and Gwynn 1983) or are due to an activation of branched chain ketoacid dehydrogenase is not yet clear. Anisoosmotic exposure of perfused rat liver did not significantly affect ketogenesis from tyrosine (2 mM) and only a slight inhibition of ketone body formation from octanoate (0.1 mM) was found following hyperosmotic cell shrinkage. These data suggest that cell volume may exert some control on ketone body formation from ketoisocaproate, but probably not from tyrosine and octanoate. However, with all these substrates, hypoosmotic cell swelling increased the release of acetoacetate at the expense of β-hydroxybutyrate and, conversely, hyperosmotic cell shrinkage stimulated β-hydroxybutyrate release at the expense of acetoacetate. Accordingly, hypoosmotic (hyperosmotic) cell swelling (shrinkage) led to a decrease (increase) in the β-hydroxybutyrate/acetoacetate ratio. The cell swelling-induced decrease in the β-hydroxybutyrate/acetoacetate ratio may be explained by a stimulation of the respiratory chain, as found in isolated rat liver mitochondria following hypoosmotic incubation (see Halestrap, this Vol.). In the presence of glucagon the effects of anisoosmotic cell volume changes in the β-hydroxybutyrate/acetoacetate ratio are largely abolished (Häussinger et al. 1992a). This lack of additivity again suggests that mitochondrial swelling is involved in the action of both glucagon and hypoosmotic liver swelling.

The inhibition of hepatic ketogenesis by glutamine, proline, alanine and asparagine (Lavoinne et al. 1987; Baquet et al. 1991a) is probably not due to amino acid-induced cell swelling. Hypoosmotic cell swelling was without effect on ketogenesis from glucose (Baquet et al. 1991a), tyrosine and octanoate, however, ketogenesis from ketoisocaproate (unpubl. result) and endogenous substrates (Baquet et al. 1991a) was even stimulated. No effect on ketogenesis was observed with the non-metabolizable amino acid analogue 2-aminoisobutyrate, despite cell swelling (Baquet et al. 1991a).

3.5 Cell Volume and Lipogenesis

Hypoosmotic incubation of hepatocytes slightly stimulates lipogenesis from glucose (Baquet et al. 1991a). On the other hand, a more pronounced stimulation of lipogenesis occurs upon addition of glutamine, proline and alanine, but not aminoisobutyrate, histidine and asparagine (Lavoinne et al.

1987; Baquet et al. 1991a). No correlation was detectable between the potency of these amino acids to swell the hepatocytes, on the one hand, and their potency to stimulate lipogenesis, on the other (Baquet et al. 1991a). This suggested that volume changes do not play a major role in mediating the lipogenic effect of glutamine or other amino acids. However, recent studies showed that swelling of hepatocytes by either hypoosmotic exposure or addition of glutamine activates acetyl-CoA carboxylase (Baquet et al. 1991c).

4 Cell Volume and Bile Formation

Hepatic bile flow was shown to be linearly dependent on canalicular bile acid secretion, however, sizeable bile flow still occurs when bile acid secretion is extrapolated to zero. Accordingly, hepatic bile formation is thought to consist of two components: bile acid-independent and bile acid-dependent bile flow (for reviews, see Erlinger 1988; Sellinger and Boyer 1990). Bile acid-dependent bile flow is probably due to osmotic water flux into the canalicular lumen resulting from the osmotic activity of secreted bile acids, whereas bile acid-independent bile flow in part involves exocytosis in the canalicular domain as well as osmotic water flux due to active secretion of compounds distinct from bile acids such as glutathione or glutathione conjugates. Apart from this, bile flow is modified by solute and water fluxes via the paracellular pathway, which is controlled by tight junctional permeability, and the reabsorption or secretion of solutes and fluid by biliary epithelia. Despite this complexity, liver cell volume changes appear to be one major factor interfering with hepatic bile formation.

4.1 Bile Acid-Dependent Bile Flow

Hypoosmotic exposure of rat livers perfused with taurocholate (100 μM) markedly stimulates bile flow. This effect is not only due to an increased water shift into the biliary compartment due to an increased osmotic gradient between the canalicular and vascular/intracellular space, but also to a marked stimulation of taurocholate uptake and secretion into bile (Fig. 4). Conversely, hyperosmotic cell shrinkage lowers biliary excretion of taurocholate and bile flow (Fig. 4). Also, when isotonic cell swelling is induced by addition of amino acids, such as glutamine and glycine, bile flow as well as biliary taurocholate excretion is stimulated (Hallbrucker et al. 1992). Also insulin, which leads to liver cell swelling, stimulates taurocholate excretion and bile flow. There is a roughly linear relationship between biliary taurocholate excretion and the cell volume change under the influence of anisotonicity, glutamine, glycine and insulin (Fig. 5). On the other hand, cAMP which induces cell shrinkage (Hallbrucker et al. 1991c) was reported

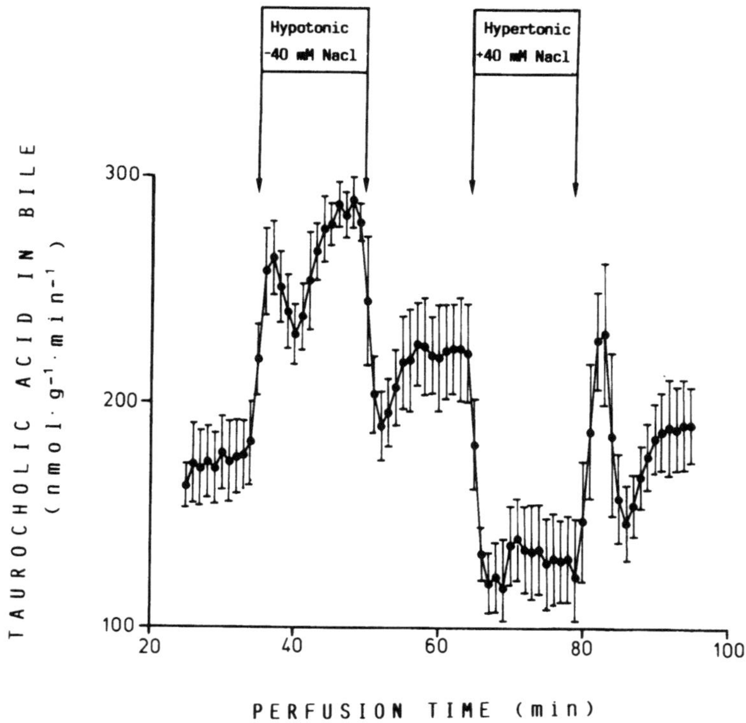

Fig. 4. Effect of anisoosmotic exposure on taurocholate excretion into bile in single-pass perfused rat liver. The influent perfusate contained taurocholate at a concentration of 100 µM and osmolarity changes were induced by addition or removal of NaCl from the influent perfusate. (Hallbrucker et al. 1992)

to enhance biliary taurocholate excretion (Hayakawa et al. 1990). Here, apparently mechanisms distinct from volume changes dominate.

In perfused rat liver the stimulation of taurocholate translocation from the sinusoidal space into the biliary compartment following hypoosmotic cell swelling is due to an increased V_{max} of the overall process: V_{max} doubles when a 15% cell swelling is induced by lowering the NaCl concentration by 40 mmol/l. Because canalicular bile acid secretion is much slower than Na^+-dependent bile acid uptake across the sinusoidal membrane, the canalicular secretion step is thought to be rate-controlling (for reviews, see Erlinger 1988; Sellinger and Boyer 1990). This suggests that cell swelling may recruit additional bile acid transporters in the canalicular membrane (Häussinger et al. 1992b, Boyer et al. 1992). This view is augmented by the observation that the swelling-induced stimulation of taurocholate secretion is blocked in presence of colchieine (unpublished).

A decrease in biliary taurocholate excretion and bile flow in perfused rat liver at high vascular taurocholate concentrations has been reported (Herz et al. 1975) and this cholestatic effect was ascribed to bile acid toxicity.

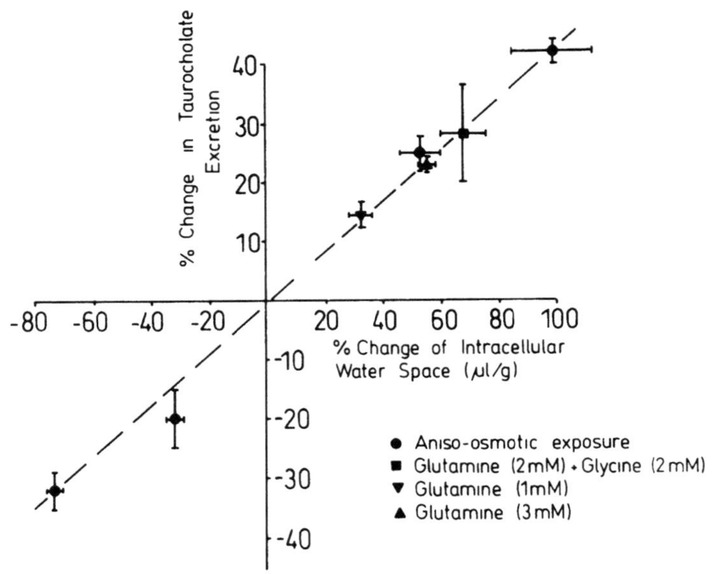

Fig. 5. Relationship between taurocholate excretion into bile and cell volume. Liver was single-pass perfused with taurocholate (100 µM); taurocholate excretion into bile under normotonic perfusion conditions with hormones and amino acids absent was 185 ± 9 nmol/min/g liver (n = 15); i.e. about 53% of the portal taurocholate load. Data on taurocholate excretion under the influence of various effectors on cell volume are given as a percentage hereof

Interestingly, in hypoosmotically swollen cells no signs of bile acid-induced cholostasis are observed at taurocholate concentrations up to 400 µM, whereas in hyperosmotically shrunken cells a cholestatic effect is already seen at concentrations of 100 µM (Häussinger et al. 1992b). Apparently, bile acid toxicity is related to cell volume. Further studies are required to establish the pathophysiological significance of these observations for the understanding of cholestatic disorders.

4.2 Bile Acid-Independent Bile Flow

Even in the absence of added bile acids, bile flow slightly decreases in perfused rat liver following hyperosmotic exposure and slightly increases during hypoosmotic exposure, addition of amino acids or insulin. When anisotonicity is induced by lowering or increasing the perfusate NaCl concentration by 40 mM, bile flow alterations were less than 20%, whereas hypertonicity induced by addition of mannitol or sucrose produced a much stronger inhibition of bile flow. With mannitol as the extracellular osmolyte the decrease in bile flow was transient, but it was sustained, if sucrose was used (Haddad et al. 1989). Mannitol, unlike sucrose, probably easily permeates liver cell membranes and thus does not lead to sustained osmotic stress. Reduction in extracellular osmolarity to isotonic levels by removal of either substance leads to a transient enhancement of bile flow.

The influence of anisoosmotic cell volume alterations (by altering the NaCl concentration in the perfusate) on the paracellular and transcytotic pathway was studied in liver perfusion experiments monitoring the biliary appearance of horseradish peroxidase injected as a bolus into vascular perfusate. Under these conditions, the appearance of horseradish peroxidase in bile is biphasic (Lowe et al. 1988; Hayakawa et al. 1990; Llopis et al. 1991): a first peak appearing about 6 min after bolus injection is thought to reflect the paracellular pathway, whereas a second broad peak appearing about 25 min after the bolus injection probably represents the transcytotic pathway. Neither hypoosmotic nor hyperosmotic exposure significantly affected the appearance of kinetics of horseradish peroxidase in bile. This suggests that cell volume may not interfere with tight junctional permeability, at least when studied with the bolus technique using horseradish peroxidase.

5 Role of Cell Volume in Hormone Action

5.1 Hormonal Control of Protein Turnover

As discussed in Section 2, hormones lead to persistent liver cell volume changes within minutes and current evidence suggests that hormone-induced cell volume changes participate in mediating some of the hormones' effects on metabolism. This has recently been studied in detail with respect to the long-known, but mechanistically unclear, hormonal control of hepatic proteolysis (Mortimore and Mondon 1970; Neely et al. 1974; Hopgood et al. 1977, 1981; Draznin and Trowbridge 1982; Mortimore and Pösö 1987). In amino acid-free liver perfusions, the net amount of K^+ accumulated or released by insulin or glucagon is linearly related to their effects on proteolysis (Hallbrucker et al. 1991b). This relationship is also maintained when the insulin-induced net K^+ accumulation inside the liver cell is modified by inhibitors of Na-K-2Cl cotransport, anisotonic cell swelling, the nutritional state or glucagon. The antagonism between insulin and glucagon regarding proteolysis can be quantitatively explained on the basis of opposing net K^+ shifts induced by these hormones (Hallbrucker et al. 1991b). A decisive role of the hormone-induced intracellular K^+ accumulation in the antiproteolytic action of insulin is also suggested by the finding that intracellular K^+ accumulation under the influence of the K^+ channel blocker Ba^{2+} inhibits hepatic proteolysis to the same extent observed when a similar amount of K^+ accumulates under the influence of insulin. Whereas the antiproteolytic effect of insulin and Ba^{2+} is parallelled by cellular K^+

accumulation, inhibition of proteolysis following hypoosmotic exposure of perfused liver is accompanied by a decrease in cellular K^+. This suggested that alterations of the intracellular K^+ concentration cannot explain the antiproteolytic effect of both insulin and hypoosmotic exposure. It recently became clear that the strict relationship between hormone- or Ba^{2+}-induced changes in cellular K^+ balance and the effects on proteolysis is due to cell volume changes (Häussinger et al. 1991; Hallbrucker et al. 1991b; vom Dahl et al. 1991a). As shown in Fig. 6, there is a close relationship between the proteolytic activity in perfused liver and the alterations of cell volume: proteolysis is inhibited during cell swelling and stimulated following cell shrinkage. This relationship is maintained regardless of whether cell volume is modified by insulin, glucagon, cAMP, inhibitors of Na-K-2Cl cotransport, glutamine, glycine, Ba^{2+} or hypoosmotic exposure (Fig. 6). Apparently, the extent of cell volume change produced by these effectors is the common denominator for their effects on proteolysis. These findings suggest that hormone-induced cell volume changes act like a "second messenger", which mediates at least some hormone effects on metabolism (Häussinger and Lang 1991b; Häussinger et al. 1991).

Glucagon, like cyclic AMP, not only stimulates proteolysis, but also inhibits protein synthesis in rat liver (Woodside et al. 1974). The cAMP-induced inhibition of protein synthesis can at least in part be ascribed to cell shrinkage, because equipotent hyperosmotic cell shrinkage decreases protein synthesis in isolated rat hepatocytes to a comparable extent (Stoll et al. 1992). Vasopressin, which affects neither protein synthesis nor cell volume, however, potentiates the cAMP effect on both cell shrinkage and inhibition of protein synthesis (Stoll et al. 1992). A roughly linear relationship between cell shrinkage and inhibition of protein synthesis under the influence of hyperosmotic exposure, cAMP and vasopressin exists (Fig. 7). On the other hand, insulin or phenylephrine had no significant effect on protein synthesis. However, when phenylephrine is added together with cAMP, cAMP-induced inhibition of protein synthesis is potentiated, although phenylephrine counteracts cAMP-induced cell shrinkage. This may indicate that cell volume changes are only one factor among others mediating the hormonal control of protein synthesis.

5.2 Other Hormone Effects

Hyperosmotic liver cell shrinkage stimulates glycogenolysis and glycolysis and inhibits glycogen synthesis (see Sect. 3). This is also observed upon addition of glucagon or cAMP and both agents lead to marked cell shrinkage (Hallbrucker et al. 1991c). Accordingly, one is tempted to speculate that the hormone-induced alterations of cell volume may contribute not only to the known effects of glucagon and insulin on hepatic protein turnover, but also on carbohydrate metabolism (Fig. 7). On the other hand, cell swelling

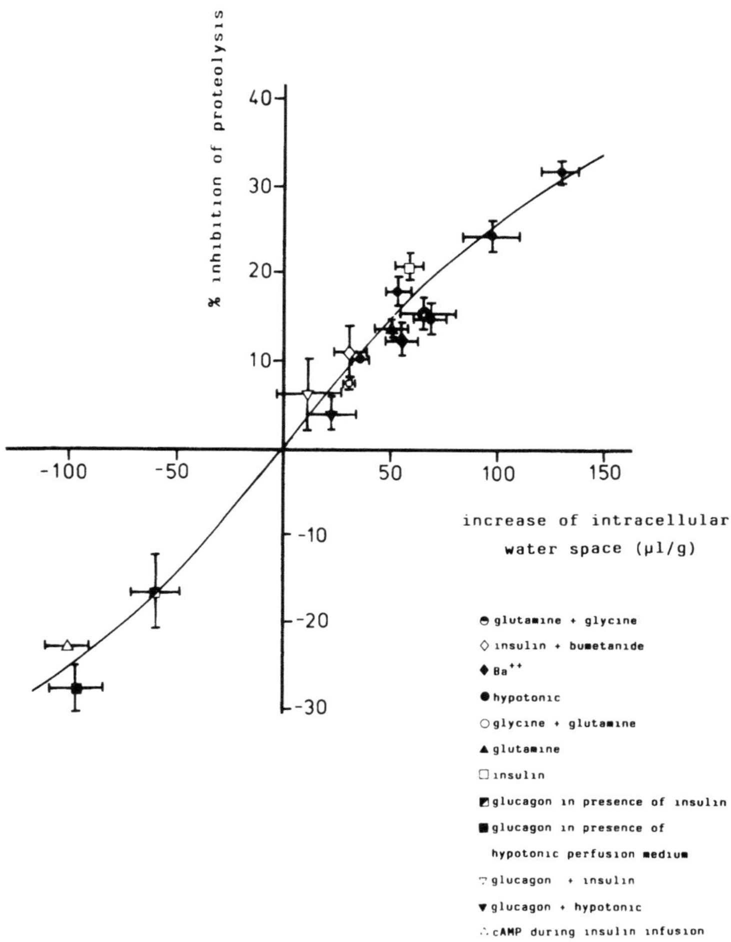

Fig. 6. Relationship between cell volume and proteolysis in liver. Cell volume in perfused liver was determined as intracellular water space (vom Dahl et al. 1991d) and proteolysis was assessed as [^{3}H]leucine release in effluent perfusate from perfused livers from rats, which were prelabelled in vivo by intraperitoneal injection of [^{3}H]leucine 16 h prior to the perfusion experiment. Cell shrinkage stimulates proteolysis, whereas cell swelling inhibits. It should be noted that proteolysis is already maximally activated in the absence of hormones and amino acids and cannot be further stimulated by hyperosmotic or glucagon-induced cell shrinkage. The proteolysis-stimulating effect of these cell-shrinking maneuvers, however, becomes apparent when proteolysis is preinhibited by either amino acids or insulin. Cell volume changes were induced by insulin, cAMP, glucagon, amino acids, Ba^{2+} or anisoosmotic exposure. (Häussinger et al. 1991)

under the influence of phenylephrine is accompanied by a stimulation of glycogenolysis, whereas hypoosmotic cell swelling actually inhibits it.

Similarly, the effects of insulin, but not of cAMP, on hepatic bile formation can in part be ascribed to cell volume changes. Like hypoosmotic exposure, cell swelling induced by insulin increases bile flow and taurocholate

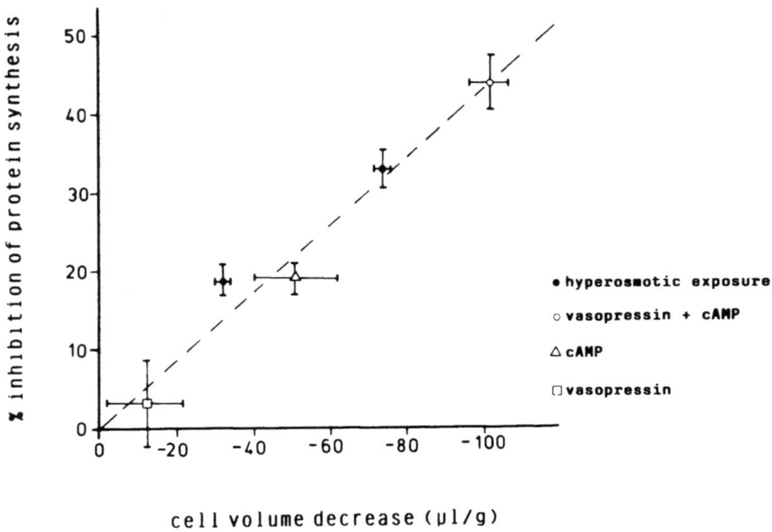

Fig. 7. Cell volume and protein synthesis. Protein synthesis was determined as [^{3}H]leucine (4 mM) incorporation into acid-precipitable material in isolated rat hepatocyte suspensions incubated with amino acids at twice their physiological concentration. Cell volume changes under the influence of effectors on protein synthesis were determined in isolated perfused rat liver in the absence of amino acids, because cell shrinkage of isolated cells cannot be quantitated reliably (see vom Dahl et al. 1991d)

excretion into bile (Hallbrucker et al. 1992). These insulin effects are reversed by further addition of glucagon, which is known to counteract insulin-induced cell swelling. However, cAMP alone stimulates biliary taurocholate excretion, although it shrinks the cells.

Hormones do not only alter cell volume, but also the volume of intracellular organelles (see Halestrap, this Vol.). It must be noted that the pathways discussed above, i.e. proteolysis, protein synthesis, glycogenolysis and glycogen synthesis, are predominantly cytosolic. These pathways are affected by cell volume changes regardless of whether cell volume is modulated by anisotonic exposure, hormones or amino acids. When, however, mitochondrial pathways such as glutamine or glycine breakdown are considered, the situation may become quite different. It should be noted that the currently available techniques for measuring cell volume alterations may predominantly reflect alterations of the cytosolic space, but allow no conclusions about mitochondrial volume changes. There is good evidence that hypoosmotic cell swelling leads to swelling of both the cytosol and mitochondria, however, hormones may affect the cytosolic and mitochondrial spaces in opposite directions. Although glucagon induces cell shrinkage, this hormone, like cAMP, actually swells the mitochondria (for review, see Halestrap 1989). Mitochondrial swelling also occurs under the influence of Ca^{2+} mobilizing hormones, such as phenylephrine or vasopressin (for review, see Halestrap 1989; Halestrap et al. 1990), although phenylephrine leads to cell swelling and vasopressin to slight cell shrinkage (Table 1). For the mechanisms leading to an increase in mitochondrial matrix volume by hormones and the functional consequences arising there from, the reader is referred to Halestrap (this Vol.). Thus, mitochondrial pathways, such as glutamine and glycine oxidation, are stimulated not only by hypoosmotic cell swelling, but also by cAMP, glucagon and Ca^{2+}-mobilizing hormones, despite opposing effects of these effectors on cell volume.

Cell differentiation and cell growth under the influence of growth factors and insulin are coupled to activation of amiloride-sensitive Na^+/H^+ exchange and bumetanide-sensitive NaCl-KCl cotransport, which leads to cell swelling (Hallbrucker et al. 1991b,c; Häussinger et al. 1991). In liver, both insulin-induced cell swelling and intracellular K^+ accumulation are inhibited by about 60% in the presence of bumetanide and insulin-induced cellular K^+ accumulation is completely blocked in the presence of bumetanide plus amiloride (Hallbrucker et al. 1991c). Similarly, in mouse 3T3 cells amiloride added together with bumetanide completely blocked the growth factor-induced exit from G_0/G_1-phase and entry into S-phase (Panet et al. 1986a,b). This may suggest a role of cell volume changes during the action of growth factors.

5.3 Relation to Intracellular Hormone Signalling

Hormone action involves a complex network of events including receptor regulation, liberation of second and third messenger molecules and the involvement of complex phosphorylation cascades (for reviews, see Gammeltoft and van Obberghen 1986; Cohen 1988; Nishizuka 1988; Yarden and Ullrich 1988; Berridge and Irvine 1989; Taylor 1990). The place of cell volume changes in the hierarchy or in the network of intracellular signalling events being initiated by hormone receptor activation is a matter of speculation. Cell swelling leads to membrane stretching, which itself could modify membrane-associated signalling systems. In rat liver, hypoosmotic cell swelling stimulates inositol-1,4,5-trisphosphate formation, increases $[Ca^{2+}]_i$, decreases pH_i, hyperpolarizes the cell membrane and lowers significantly the cAMP/cGMP ratio (see Sect. 7). It is not yet known whether such effects also occur during hormone-induced cell swelling, but the possibility must be considered that hormone-induced cell volume changes may participate in the liberation of second and third messengers observed under the influence of certain hormones. Na^+/H^+ exchange activity is, among other factors (Aronson 1985), regulated by phosphorylation/dephosphorylation and its activation by insulin and growth hormones is possibly due to phosphorylation (Sardet et al. 1990, 1991). Thus, activation of protein kinases following receptor stimulation probably precedes insulin-induced cell swelling. Also, adenylate cyclase activation seems to be required for glucagon-induced cell shrinkage, because cAMP mimicks the glucagon effect on cell volume. On the other hand, when receptor-gated ion channels are activated by hormones, cell volume changes could represent a rather early event in the intracellular signal transduction. Remarkably, the cell volume response to Ca^{2+}-mobilizing hormones is not uniform. Although phenylephrine, vasopressin and extracellular ATP stimulate phosphatidylinositol hydrolysis with a concomitant increase in $[Ca^{2+}]_i$, phenylephrine swells the cells, whereas ATP shrinks and vasopressin does not significantly alter cell volume when added in the absence of glucagon or cAMP (Table 1). This suggests that factors other than $[Ca^{2+}]_i$ or inositol-1,4,5-trisphosphate are involved in cell volume changes under the influence of Ca^{2+}-mobilizing hormones. In summary, currently available evidence suggests that cell volume changes under the influence of hormones act as "second or third" messengers mediating metabolic hormone effects, however, their interplay with other mechanisms of signal transduction remains to be established.

6 Signalling Systems

The mechanisms coupling cell volume changes to volume regulatory ion fluxes have not been fully elucidated as yet and may be different for different cell types. Extensive studies have been carried out regarding this question in Ehrlich ascites tumor cells (see Hoffmann, this Vol.), but little is known for hepatocytes. Several possible mechanisms can be envisaged. Cell swelling is accompanied by membrane stretching and in a variety of cells stretch-activated K^+ channels have been described (for review, see Lang et al., this Vol.) which could in part mediate volume regulatory K^+ efflux following cell swelling. In addition, participation of the cytoskeleton in cell volume regulation has been suggested in the liver (van Rossum et al. 1987). Alanine-induced cell swelling was shown to open Ca^+-activated K^+ channels in isolated hepatocytes (Bear and Petersen 1987) and more recently a transient rise in intracellular Ca^{2+} following hypotonic liver cell swelling was demonstrated in hepatocytes (Corasanti et al. 1990a; Baquet et al. 1991b). In liver, this Ca^{2+} transient is probably due to Ca^{2+} release from intracellular stores and the hypotonicity-induced stimulation of K^+ efflux from perfused livers is not blunted in the absence of extracellular Ca^{2+} (vom Dahl et al. 1991b). A rise in the cytosolic free Ca^{2+} concentration has been implicated for stimulation of volume-regulatory K^+ fluxes in a variety of other tissues, but various distinct mechanisms seem to be responsible for the increase in $[Ca^{2+}]_i$ in different tissues (see Hoffmann et al., this Vol.). In liver, hypoosmotic cell swelling leads to an increased formation of inositol 1,4,5-trisphosphate (Baquet et al. 1991b; vom Dahl et al. 1991b), which might trigger a rise in the intracellular Ca^{2+} concentration. In perfused rat liver, after in vivo prelabelling of phospholipids with $[^3H]$-myo-inositol, cell swelling, induced by lowering the perfusate osmolarity from 305 to 225 mOsmol l^{-1}, led to about three-fold stimulation of Li^+-sensitive [^{3}H]inositol release (vom Dahl et al. 1991b). The maximum of hypotonicity-induced [^{3}H]inositol release preceded maximal K^+ efflux by some 10–20 s and occurred after maximum cell swelling. The stimulation of [^{3}H]inositol release, however, was not observed when cell swelling was induced by starting from a preshrunken state (i.e. by decreasing the extracellular osmolarity from 385 to 305 mOsmol l^{-1}), although the volume-regulatory K^+ efflux observed under these conditions was virtually indistinguishable from that found when the osmolarity was lowered from 305 to 225 mOsmol l^{-1} (vom Dahl et al. 1991b). This might indicate that volume-regulatory K^+ efflux is not necessarily linked to swelling-induced alterations of phosphoinositide turnover.

Recent studies on volume regulation in Ehrlich ascites tumour cells have indicated an important role of leukotrienes and prostaglandins in triggering volume-regulatory ion fluxes (Lambert 1987; Lambert et al. 1987; see also Hoffmann et al., this Vol.). Although leukotrienes D_4 and C_4 and prostaglandins were shown to elicit complex K^+ fluxes in perfused rat liver (Häussinger et al. 1987, 1988; Wettstein et al. 1989), cell volume-regulatory K^+ efflux following anisotonic exposure was not affected in the presence of inhibitors of cyclooxygenase, phospholipase A_2 and lipoxygenase or leukotriene receptor antagonists (vom Dahl et al. 1991b). No effect of cell swelling on hepatic prostaglandin D_2 release was detectable, whereas a marked stimulation of prostaglandin D_2 release was observed upon addition of adenosine. These data suggest that eicosanoids probably do not play a crucial role in liver cell volume regulation and marked differences may exist between different cell types with respect to a role of eicosanoids in triggering volume-regulatory responses.

What are the mechanisms and signals which couple cell volume changes to persistent alterations of cellular function? This issue remains obscure, although the understanding of the mechanisms linking hormone-induced alterations of mitochondrial volume to alterations of mitochondrial function is growing (see Halestrap, this Vol.). Liver cell swelling increases $[Ca^{2+}]_i$ (Corasanti et al. 1990a), decreases pH_i (Gleeson et al. 1990), hyperpolarizes the cell membrane (Graf et al. 1988), stimulates inositol-1,4,5-trisphosphate formation (Baquet et al. 1991b; vom Dahl et al. 1991b) and lowers significantly the cAMP/cGMP ratio (unpubl. result). However, the significance of these alterations for the observed effects on hepatic metabolism is doubtful. For example, a swelling-induced rise in inositol-1,4,5-trisphosphate and intracellular Ca^{2+} is expected to stimulate glycogen phosphorylase, but cell swelling actually inhibits glycogenolysis. We speculate that alterations in the known cellular signalling systems may rather be an epiphenomenon of cell volume changes and suggest that the degree of hydratation and ionic composition in the cytosol as well as cytoskeletal effects may be a major event in triggering cellular function when cell volume changes.

7 Perspective

Current evidence suggests that cell volume alterations represent a new, not yet fully recognized principle of metabolic control; they act as a yet missing link (i.e. like a "second or third messenger") mediating hormone and amino acid effects. Accordingly, concentrative amino acid transport systems and hormone-modulated ion transporters in the plasma membrane can be viewed as a part of transmembrane signalling systems. The major open questions refer to the mechanisms linking cell volume changes to the alterations of metabolic cell function. It is hoped that this review will stimulate further work in this new and exciting area.

References

Aronson PS (1985) Kinetic properties of the plasma membrane Na^+/H^+ exchanger. Annu Rev Physiol 47: 454–560

Bakker-Grunwald T (1983) Potassium permeability and volume control in isolated rat hepatocytes. Biochim Biophys Acta 731: 239–242

Balavoine S, Feldmann G, Lardeux B (1990) Rates of RNA degradation in isolated rat hepatocytes. Effects of amino acids and inhibitors of lysosomal function. Eur J Biochem 189: 617–623

Baquet A, Hue L, Meijer AJ, van Woerkom GM, Plomp PJAM (1990) Swelling of rat hepatocytes stimulates glycogen synthesis. J Biol Chem 265: 955–959

Baquet A, Lavoinne A, Hue L (1991a) Comparison of the effects of various amino acids on glycogen synthesis, lipogenesis and ketogenesis in isolated rat hepatocytes. Biochem J 273: 57–62

Baquet A, Meijer AJ, Hue L (1991b) Hepatocyte swelling increases inositol 1,4,5-trisphosphate, calcium and cAMP concentration but antagonizes phosphorylase activation by calcium-dependent hormones. FEBS Lett 278: 103–106

Baquet A, Maisin L, Hue L (1991c) Swelling of rat hepatocytes activates acetyl-CoA carboxylase in parallel to glycogen synthase. Biochem J 278: 887–890

Bear CE (1990) A nonselective cation channel in rat liver cells is activated by membrane stretch. Am J Physiol 258: C421–C428

Bear CE, Petersen OH (1987) L-alanine evokes opening of single Ca^{2+}-activated K^+ channels in rat liver cells. Pflügers Arch Eur J Physiol 410: 342–344

Berridge MJ, Irvine RF (1989) Inositol phosphates and cell signalling. Nature 341: 197–205

Bode B, Kilberg M (1991) Amino acid dependent increase of hepatic system N activity is linked to cell swelling. J Biol Chem 266: 7376–7381

Boyer JL, Graf J, Meier PJ (1992) Hepatic transport systems regulating pH: cell volume and bile secretion. Ann Rev Physiol 54: 415–438

Brosnan JT, Jois M, Hall B (1990) Hormonal regulation of glycine metabolism. In: Lubec G, Rosenthal GA (eds) Amino acids: biochemistry, biology and medicine. ESCOM Science Publ BV, Amsterdam, pp 896–902

Chen KS, Lardy HA (1985) Multiple requirements for glycogen synthesis by hepatocytes isolated from fasted rats. J Biol Chem 260: 14683–14688

Cohen P (1988) Protein phosphorylation and hormone action. Proc R Soc Lond 234: 115–144

Corasanti JG, Gleeson D, Gautam A, Boyer JL (1990a) Effects of osmotic stresses on isolated rat hepatocytes: ionic mechanisms of cell volume regulation and modulation of Na^+/H^+ exchange. Renal Physiol Biochem 13: 164

Corasanti JG, Gleeson D, Boyer JL (1990b) Effects of osmotic stresses on isolated rat hepatocytes. I. Ionic mechanisms of cell volume regulation. Am J Physiol 258: G290–G298
Draznin B, Trowbridge M (1982) Inhibition of intracellular proteolysis by insulin in isolated rat hepatocytes. J Biol Chem 257: 11988–11993
Erlinger S (1988) Bile flow. In: Arias IM, Jacoby WB, Popper H, Schachter D, Shafritz DA (eds) The liver: biology and pathobiology. Raven Press, New York, pp 643–661
Fehlmann M, Freychat P (1981) Insulin and glucagon stimulation of Na^+/K^+ ATPase transport activity in isolated rat hepatocytes. J Biol Chem 256: 7449–7453
Fehlmann M, Morin O, Kitabgi P, Freychat P (1981) insulin and glucagon receptors of isolated rat hepatocytes: comparison between hormone binding and amino acid transport stimulation. Endocrinology 109: 253–261
Fernandez-Checa JC, Ren C, Awe TY, Ookhtens M, Kaplowitz N (1988) Effect of membrane potential and cellular ATP on glutathione efflux from isolated rat hepatocytes. Am J Physiol 255: G403–G408
Gammeltoft S, van Obberghen E (1986) Protein kinase activity of the insulin receptor. Biochem J 235: 1–11
Gazzola GC, Dall'Asta V, Nucci FA, Rossi PA, Bussolati O, Hoffmann EK, Guidotti GG (1991) Role of amino acid transport system A in the control of cell volume in cultured human fibroblasts. Cell Physiol Biochem 1: 131–142
Gleeson D, Corasanti JG, Boyer JG (1990) Effects of osmotic stresses on isolated rat hepatocytes. II. Modulation of intracellular pH. Am J Physiol 258: G299–G307
Graf J, Haddad P, Häussinger D, Lang F (1988) Cell volume regulation in liver. Renal Physiol Biochem 11: 202–220
Groen AK, Sips HJ, Vervoorn RC, Tager JM (1982) Intracellular compartmentation and control of alanine metabolism in rat liver parenchymal cells. Eur J Biochem 122: 87–93
Guidotti GG, Borghetti AF, Gazzola GC (1978) The regulation of amino acid transport in animal cells. Biochim Biophys Acta 515: 329–366
Haberich FJ, Aziz O, Nowacki PE (1966) Flüßigkeitsspeicherung in der Leber nach enteraler Wasserresorption. Pflügers Arch Eur J Physiol 285: 73–89
Haddad P, Graf J (1989) Volume-regulatory K^+ fluxes in the isolated perfused rat liver: characterization by ion transport inhibitors. Am J Physiol 257: G357–G363
Haddad P, Thalhammer T, Graf J (1989) Effect of hypertonic stress on liver cell volume, bile flow, and volume-regulatory K^+ fluxes. Am J Physiol 256: G563–G569
Häussinger D (1990) Nitrogen metabolism in liver: structural-functional organization and physiological implications. Biochem J 267: 281–290
Häussinger D, Lang F (1990) Exposure of perfused liver to hypotonic conditions modifies cellular nitrogen metabolism. J Cell Biochem 43: 355–361
Häussinger D, Lang F (1991a) The mutual interaction between cell volume and function: a new principle of metabolic control. Biochem Cell Biol 43: 1–4
Häussinger D, Lang F (1991b) Regulation of liver cell function and cell volume: a new principle for metabolic control. Biochim Biophys Acta 1071: 331–350
Häussinger D, Gerok W, Sies H (1983) Regulation of flux through glutaminase and glutamine synthetase in isolated perfused rat liver. Biochim Biophys Acta 75: 272–278
Häussinger D, Stehle T, Tranh-ThiT-A, Decker K, Gerok W (1987) Prostaglandin responses in isolated perfused rat liver: Ca^{2+} and K^+ fluxes, hemodynamic and metabolic effects. Biol Chem Hoppe-Seyler 368: 1509–1513
Häussinger D, Stehle T, Gerok W (1988) Effects of leukotrienes and the thromboxane A_2 analogue U-46619 in isolated perfused rat liver: metabolic, hemodynamic and ion flux responses. Biol Chem Hoppe-Seyler 369: 97–107
Häussinger D, Lang F, Bauers K, Gerok W (1990a) Interactions between glutamine metabolism and cell volume regulation in perfused rat liver. Eur J Biochem 188: 689–695
Häussinger D, Hallbrucker C, vom Dahl S, Lang F, Gerok W (1990b) Cell swelling inhibits proteolysis in perfused rat liver. Biochem J 272: 239–242

Häussinger D, Lang F, Bauers K, Gerok W (1990c) Control of hepatic nitrogen metabolism and glutathione release by cell volume regulatory mechanisms. Eur J Biochem 193: 891–898

Häussinger D, Stehle T, Lang F (1990d) Volume regulation in liver: further characterization by inhibitors and ionic substitution. Hepatology 11: 243–254

Häussinger D, Hallbrucker C, vom Dahl S, Decker S, Schweizer U, Lang F, Gerok W (1991) Cell volume is a major determinant of proteolysis control in liver. FEBS Lett 283: 70–72

Häussinger D, Stoll B, Morimoto Y, Lang F, Gerok W (1992) Anisoosmotic liver perfusion: redox shifts and modulation of α-Ketoisocaproate and glycine metabolism. Biol Chem Hoppe-Seyler 373: 723–734

Häussinger D, Hallbrucker C, Jaha N, Lang F, Gerok W. (1992b) Cell volume and bile acid excretion. Biochem J 288: 681–689

Halestrap AP (1989) The regulation of the matrix volume of mammalian mitochondria in vivo and in vitro and its role in the control of mitochondrial metabolism. Biochim Biophys Acta 973: 355–382

Halestrap AP, Davidson AM, Potter WD (1990) Mechanisms involved in the hormonal regulation of mitochondrial function through changes in the matrix volume. Biochim Biophys Acta 1018: 278–281

Hallbrucker C, vom Dahl S, Lang F, Häussinger D (1991a) Control of hepatic proteolysis by amino acids: the role of cell volume. Eur J Biochem 197: 717–724

Hallbrucker C, vom Dahl S, Lang F, Gerok W, Häussinger D (1991b) Inhibition of hepatic proteolysis by insulin: role of hormone induced alterations of cellular K^+balance. Eur J Biochem 199: 467–474

Hallbrucker C, vom Dahl S, Lang F, Gerok W, Häussinger D (1991c) Modification of liver cell volume by insulin and glucagon. Pflügers Arch Eur J Physiol 418: 519–521

Hallbrucker C, Lang F, Gerok W, Häussinger D (1992) Cell swelling increases bile flow and taurocholate excretion into bile in isolated rat liver. Biochem J 281: 593–595

Hayakawa T, Cheng O, Ma A, Boyer JL (1990) Taurocholate stimulates transcytotic vesicular pathways labeled by horseradish peroxidase in the isolated perfused rat liver. Gastroenterol 99: 216–223

Hayes MR, McGivan JD (1982) Differential effects of starvation on alanine and glutamine transport in isolated rat hepatocytes. Biochem J 204: 365–368

Henderson RM, Graf J, Boyer JL (1989) Inward-rectifying potassium channels in rat hepatocytes. Am J Physiol 256: G1028–G1035

Herz R, Paumgartner G, Preisig R (1975) Inhibition of bile formation by high doses of taurocholate in the perfused rat liver. Scand J Gastroenterologia 11: 741–746

Hopgood MF, Clark MG, Ballard FJ (1977) Inhibition of protein degradation in isolated rat hepatocytes. Biochem J 164: 399–407

Hopgood MF, Clark MG, Ballard FJ (1981) Protein degradation in hepatocyte monolayers: effects of glucagon, adenosine 3′:5′-cyclic monophosphate and insulin. Biochem J 196: 33–40

Hue L, Bontemps F, Hers HG (1975) The effect of glucose and of potassium ions on the interconversion of the two forms of glycogen phosphorylase and of glycogen synthetase in isolated rat liver preparations. Biochem J 152: 105–114

Jakubowski J, Jakob A (1990) Vasopressin, insulin and peroxide(s) of vanadate (pervanadate) influence Na^+ transport mediated by (Na^+/K^+) ATPase or Na^+/H^+ exchanger of rat liver plasma membrane vesicles. Eur J Biochem 193: 541–549

Katz J, Golden S, Wals PA (1976) Stimulation of hepatic glycogen synthesis by amino acids. Proc Natl Acad Sci USA 73: 3433–3437

Kilberg MS, Gwynn MB (1983) Plasma membrane transport of 2-ketoisocaproate by rat hepatocytes in primary culture. J Biol Chem 258: 11524–11527

Kilberg MS, Barber EF, Handlogten ME (1985) Characteristics and hormonal regulation of amino acid transport system A in isolated rat hepatocytes. Curr Top Cell Regul 25: 113–163

Kristensen LO (1980) Energization of alanine transport in isolated rat hepatocytes. J Biol Chem 255: 5236–5243

Kristensen LO, Folke M (1984) Volume-regulatory K^+ efflux during concentrative uptake of alanine in isolated rat hepatocytes. Biochem J 221: 265–268
Kristensen LO, Folke M (1986) Effects of perturbation of the Na^+-electrochemical gradient on influx and efflux of alanine in isolated rat hepatocytes. Biochim Biophys Acta 855: 49–57
Lambert IH (1987) Effect of arachidonic acid, fatty acids, prostaglandins and leukotrienes on volume regulation in Ehrlich ascites tumor cells. J Membr Biol 98: 207–221
Lambert IH, Hoffmann EK, Christensen P (1987) Role of prostaglandins and leukotrienes in volume regulation by Ehrlich ascites tumor cells. J Membr Biol 98: 247–256
Lang F, Stehle T, Häussinger D (1989) Water, K^+, H^+, lactate and glucose fluxes during cell volume regulation in perfused rat liver. Pflügers Arch Eur J Physiol 413: 209–216
Lardeux BR, Mortimore GE (1987) Amino acid and hormonal control of macromolecular turnover in perfused rat liver. J Biol Chem 262: 14514–14519
Lavoinne A, Baquet A, Hue L (1987) Stimulation of glycogen synthesis and lipogenesis by glutamine in isolated rat hepatocytes. Biochem J 248: 429–437
Leake RE, Trench ME, Barry JM (1972) Effect of cations on the condensation of hen erythrocyte nuclei and its relation to gene activation. Exp Cell Res 71: 17–26
Le Cam A, Freychet P (1978) Effect of insulin on amino acid transport in isolated rat hepatocytes. Diabetologia 15: 117–123
Leverve XM, Caro LHP, Plomp PJAM, Meijer AJ (1987) Control of proteolysis in perifused rat hepatocytes. FEBS Lett 219: 455–458
Llopis J, Kass GEN, Duddy SK, Farell GC, Gahm A, Orrenius S (1991) Mobilization of the hormone-sensitive Ca^{2+} pool increases hepatocyte tight junctional permeability in the perfused rat liver. FEBS Lett 280: 84–86
Lowe PJ, Miyai K, Steinbach JH, Hardison WGM (1988) Hormonal regulation of hepatocyte tight junctional permeability. Am J Physiol 255: G454–G461
Lynch C, Wilson PB, Blackmore PF, Exton JH (1986) The hormone-sensitive hepatic Na^+ pump. Evidence for regulation by diacylglycerol and tumor promotors. J Biol Chem 261: 14551–14556
Madshus LH (1988) Regulation of intracellular pH in eukaryotic cells. Biochem J 230: 1–11
McGivan JD, Bradford NM, Verhoeven AJ, Meijer AJ (1984) Liverglutaminase. In: Häussinger D, Sies H (eds) Glutamine metabolism in mammalian tissues. Springer, Berlin Heidelberg New York, pp 122–137
McLennan PA, Brown RA, Rennie MJ (1987) A positive relationship between protein synthetic rate and intracellular glutamine concentration in perfused rat skeletal muscle. FEBS Lett 215: 187–191
McLennan PA, Smith K, Weryk B, Watt PW, Rennie MJ (1988) Inhibition of protein breakdown by glutamine in perfused rat skeletal muscle. FEBS Lett 237: 133–136
Mortimore GE, Mondon CE (1970) Inhibition by insulin of valine turnover in liver. J Biol Chem 245:2375–2383
Mortimore GE, Pösö AR (1984) Mechanisms and control of deprivation-induced protein degradation in liver: role of glucogenic amino acids. In: Häussinger D, Sies H (eds) Glutamine metabolism in mammalian tissues. Springer, Berlin Heidelberg New York, pp 138–157
Mortimore GE, Pösö AR (1987) Intracellular protein catabolism and its control during nutrient deprivation and supply. Annu Rev Nutr 7: 539–564
Moule SK, McGivan JD (1990) Regulation of the plasma membrane potentialin hepatocytes – mechanism and physiological significance. Biochim Biophys Acta 1031: 383–397
Neely AN, Nelson PB, Mortimore GE (1974) Osmotic alterations of the lysosomal system during rat liver perfusion: reversible suppression by insulin and amino acids. Biochim Biophys Acta 338: 458–472
Nishizuka Y (1988) Studies and prospectives of the protein kinase C family for cellular regulation. Cancer 63: 1892–1903

Panet R, Amir I, Atlan H (1986a) Fibroblast growth factor induces a transient net K^+ influx carried by the bumetanide-sensitive transporter in quiescent BALB/c 3T3 fibroblasts. Biochim Biophys Acta 859: 117–121
Panet R, Snyder D, Atlan H (1986b) Amiloride added together with bumetanide completely blocks mouse 3T3-cellexit from G_0/G_1 phase and entry into S-phase. Biochem J 239: 745–750
Pfaller W, Willinger C, Stoll B, Hallbrucker C, Lang F, Häussinger D (1993) Structural reaction pattern of hepatocytes following exposure to hypotonicity. Cell Physiol, in press
Pösö AR, Wert JJ, Mortimore GE (1982a) Multifunctional control by amino acids of deprivation-induced proteolysis in liver. J Biol Chem 257: 12114–12120
Pösö AR, Schworer CM, Mortimore GE (1982b) Acceleration of proteolysis in perfused rat liver by deletion of glucogenic amino acids. Biochem Biophys Res Commun 107: 1433–1439
Rennie MJ, Hundal HS, Babij P, MacLennan P, Taylor PM, Watt PW, Jepson MM, Millward DJ (1986) Characteristics of a glutamine carrier in skeletal muscle have important consequences for nitrogen loss in injury, infection and chronic disease. Lancet 2: 1008–1012
Rognstad R (1985) Possible role for carbamoylphosphate in the control of glycogen synthesis. Biochem Biophys Res Commun 130: 229–233
Roth E, Karner J, Ollenschläger G (1990) Glutamine: an anabolic effector? J Parenter Ent Nutr 14: 130S–136S
Sackin H (1987) Stretch-activated potassium channels in renal proximal tubule. Am J Physiol 253: F1253–1262
Sardet C, Counillon L, Franchi A, Pouyssegur J (1990) Growth factors induce phosphorylation of the Na^+/H^+ antiporter, a glycoprotein of 110kD. Science 247: 723–726
Sardet C, Wakayabashi S, Fafournoux P, Counillon L, Pagès G, Pouyssegur J (1991) Molecular properties of Na^+/H^+ exchange. In: De Santo GP, Capasso G (eds) Acid-base balance: molecular, cellular and clinical aspects. Editoriale Bios, Cosenza, pp 13–20
Schulz W, Eickelmann C, Hallbrucker C, Sies H, Häussinger D (1991) Increase of β-actin mRNA upon hypotonic perfusion of perfused rat liver. FEBS Lett 292: 264–270
Schwarzacher HG, Mikelsaar AV, Schnedl W (1978) The nature of the Ag-staining of nuleolus organizer regions. Cytogenet Cell Genet 20: 24–39
Schworer CM, Mortimore GE (1979) Glucagon-induced autophagy and proteolysis in rat liver. Mediation by selective deprivation of intracellular amino acids. Proc Natl Acad Sci USA 76: 3169–3173
Seglen PO, Gordon PB (1984) Amino acid control of autophagic sequestration and protein degradation in isolated rat hepatocytes. J Cell Biol 99: 435–444
Seglen PO, Gordon PB, Poli A (1980) Amino acid inhibition of the autophagic/lysosomal pathway of protein degradation in isolated rat hepatocytes. Biochim Biophys Acta 630: 103–118
Sellinger M, Boyer JL (1990) Physiology of bile secretion and cholestasis. In: Popper H, Schaffner F (eds) Progress in liver disease, vol IX. WB Saunders, Philadelphia, pp 237–259
Shotwell MA, Kilberg MS, Oxender D (1983) The regulation of neutral amino acid transport in mammalian cells. Biochim Biophys Acta 737: 267–284
Sips HJ, Groen AK, Tager JM (1980) Plasma-membrane transport of alanine is rate-limiting for its metabolism in rat liver parenchymal cells. FEBS Lett 119: 271–274
Stoll B, Hallbrucker C, Lang F, Häussinger D (1992) Effect of anisoosmotic exposure and hormones on protein synthesis in isolated rat hepatocytes: regulation by cell volume? Biochem J 287: 217–222
Taylor CW (1990) The role of G proteins in transmembrane signalling. Biochem J 272: 1–13

van Rossum GDV, Russo MA, Schisselbauer JC (1987) Role of cytoplasmic vesicles in volume maintenance. Curr Top Membr Transp 30: 45–74

vom Dahl S, Hallbrucker C, Lang F, Gerok W, Häussinger D (1991a) Regulation of liver cell volume and proteolysis by glucagon and insulin. Biochem J 278: 771–777

vom Dahl S, Hallbrucker C, Lang F, Häussinger D (1991b) Role of eicosanoids, inositol phosphates and extracellular Ca^{++} in cell volume regulation of rat liver. Eur J Biochem 198: 73–83

vom Dahl S, Hallbrucker C, Lang F, Gerok W, Häussinger D (1991c) Regulation of cell volume in the perfused rat liver by hormones. Biochem J 280: 105–109

vom Dahl S, Hallbrucker C, Lang F, Häussinger D (1991d) A non-invasive technique for cell volume determination in isolated perfused rat liver. Biol Chem Hoppe-Seyler 372: 411–418

Weissbach L, Kilberg MS (1984) Amino acid activation of amino acid transport system N early in primary cultures of rat hepatocytes. J Cell Physiol 121: 133–138

Wettstein M, Gerok W, Häussinger D (1989) Metabolism and biliary excretion of cysteinyl leukotrienes in the perfused rat liver. Eur J Biochem 181: 115–124

Wettstein M, vom Dahl S, Lang F, Gerok W, Häussinger D (1990) Cell volume regulatory responses of isolated perfused rat liver: the effect of amino acids. Biol Chem Hoppe-Seyler 371: 493–501

Woodside KH, Ward WF, Mortimore GE (1974) Effects of glucagon on general protein degradation and synthesis in perfused rat liver. J Biol Chem 249: 5458–5463

Yarden Y, Ullrich A (1988) Growth factor receptor tyrosine kinases. Annu Rev Biochem 57: 443–478

Chapter 3

Volume Regulation in Transporting Epithelia

J. Graf[1], *W.B. Guggino*[2], and *K. Turnheim*[3]

Contents

[1]Department of General and Experimental Pathology, University Hospital, Währinger Gürtel 18–20, 1090 Vienna, Austria
[2]Department of Physiology, The Johns Hopkins University, School of Medicine, 725 N. Wolfe Street, Baltimore, Maryland 21205, USA
[3]Pharmakologisches Institut, Universität Wien, Währingerstraße 13a, 1090 Vienna, Austria

Advances in Comparative and Environmental Physiology, Vol. 14

1 Introduction

The cell volume is regulated by solute transport across the surface membranes and by metabolic generation or consumption of intracellular, osmotically active solutes. Many membrane transport mechanisms and metabolic pathways that are involved in volume regulation serve other vital cell functions also, including the control of the intracellular milieu (electrolyte concentrations, cell pH), energy, carbohydrate and protein metabolism, nutritive and secretory functions, excitability, cell growth, proliferation and differentiation. Cell volume regulation is thus integrally related to these functions. Furthermore, the mechanisms of homeostatic control and regulation through hormonal, neural and intracellular signals may modify membrane transport mechanisms and enzyme activities that are involved in volume regulation. Thus, both the mechanisms that determine the cell volume and their regulatory signals may be common to other cell functions.

Absorbing or secreting epithelial cells are polarized and are therefore faced with one more task: vectorial transport of solutes and water across these cells may be high (approx. the complete cell volume in 20 s in the kidney proximal tubule) and may be subject to brisk stimulation by hormones or neurotransmitters (salivary and lacrimal glands) so that uptake of solutes and water across one membrane must be balanced to their delivery across the opposing surface membrane, precisely and timely, in order to keep the cell volume constant. Balanced operation and coordinate upregulation or downregulation of apical and basolateral cell membrane transport mechanisms is thus required.

In the following an overview of the principal volume regulatory mechanisms present in transporting epithelia and a more detailed account of the specific mechanisms operating in absorbing (kidney, intestine) and secreting (liver) epithelial cells will be given. Intestinal transport will be discussed with reference to the balance of apical and basolateral membrane transport processes, Section 4 will include regulation of intracellular organic osmolytes, and the role of cytoplasmic membrane vesicles will be addressed in Section 5.

2 Mechanisms of Epithelial Cell Volume Regulation

2.1 The Steady State Cell Volume

Cell membranes are, in general, highly permeable to water (the hydraulic permeability, P_f, being in the order of 10^{-4} to 10^{-1} cm s^{-1}) and tolerate little tension (approx. 5 dyn cm^{-1}). (Notable exceptions to this generalization are the outer membrane of frog skin, the thick ascending limb of Henle and later parts of the nephron in the absence of vasopressin.) Therefore, cell membranes do not allow significant osmotic or hydrostatic pressure gradients to develop and cells must maintain their volume by the homeostatic control of the intracellular amount of osmotically active solutes (osmolytes) while internal and external osmolarities remain near equal.

The cell interior contains membrane-impermeant solutes (e.g. protein polyanions) that would induce cell swelling (Gibbs-Donnan effect) unless membrane-permeant osmolytes (mainly electrolytes) are continuously expelled from the cell (double Donnan system) (Tosteson and Hoffmann 1960). The continuous, active expulsion of osmolytes that leak back into the cell requires a constant supply of energy and any disturbance of this pump/leak equilibrium may result in cell swelling or shrinkage. Alternatively, shrunken or swollen cells may readjust their volume by modulation of the transport rates through the pump and leak components, resulting in regulatory cell volume increase (RVI) or decrease (RVD). In addition, volume regulation may involve intracellular generation or consumption of osmolytes by metabolic processes (cf. Sect. 3; Häussinger et al., this Vol.) or by pH-dependent dissociation of weak acids or bases (Andersen 1989).

2.2 General Principles and Energetic Aspects of Epithelial Electrolyte Transport

Na^+, K^+, Cl^- and HCO_3^- are the most abundant extracellular and intracellular electrolytes. In most absorbing or secreting epithelial cells, basolateral and apical-luminal membrane transport of these ions is the predominant mechanism for transporting fluid across the epithelium and for maintaining the intracellular amount and composition of electrolytes. In general, uptake of K^+ and extrusion of Na^+ by the basolateral membrane *Na^+/K^+-ATPase* is the *primary active*, energy-consuming transport process that establishes the high intracellular K^+ (approx. 80 to 130 mM) and low Na^+ concentration (approx. 10 to 40 mM). Both the turnover rate of the enzyme and the number of enzyme molecules within the membrane, together with the supply of ATP and of Na^+ and K^+ from inside and outside the cell membrane, determine the magnitude of these ion fluxes.

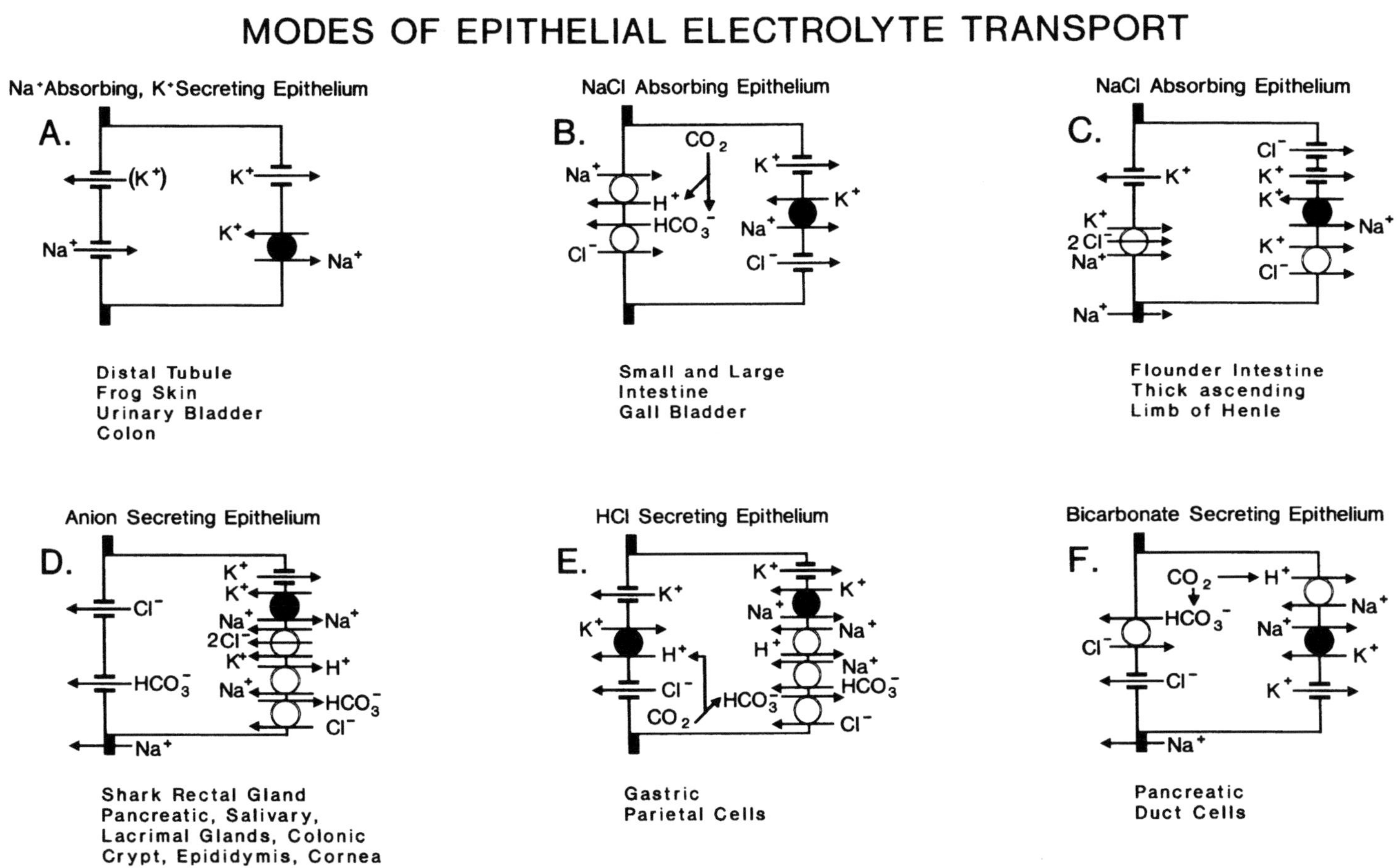
MODES OF EPITHELIAL ELECTROLYTE TRANSPORT
Na⁺Absorbing, K⁺Secreting Epithelium
A.
(K⁺)
Na⁺
K⁺
K⁺
Na⁺
Distal Tubule
Frog Skin
Urinary Bladder
Colon
NaCl Absorbing Epithelium
B.
CO2
Na⁺
H⁺
HCO3⁻
Cl⁻
K⁺
K⁺
Na⁺
Cl⁻
Small and Large
Intestine
Gall Bladder
NaCl Absorbing Epithelium
C.
K⁺
K⁺
2 Cl⁻
Na⁺
Na⁺
Cl⁻
K⁺
K⁺
Na⁺
K⁺
Cl⁻
Flounder Intestine
Thick ascending
Limb of Henle
Anion Secreting Epithelium
D.
Cl⁻
HCO3⁻
Na⁺
K⁺
K⁺
Na⁺
2Cl⁻
K⁺
Na⁺
Na⁺
H⁺
HCO3⁻
Cl⁻
Shark Rectal Gland
Pancreatic, Salivary,
Lacrimal Glands, Colonic
Crypt, Epididymis, Cornea
HCl Secreting Epithelium
E.
K⁺
K⁺
H⁺
Cl⁻
CO2
HCO3⁻
K⁺
K⁺
Na⁺
Na⁺
H⁺
Na⁺
HCO3⁻
Cl⁻
Gastric
Parietal Cells
Bicarbonate Secreting Epithelium
F.
CO2
H⁺
HCO3⁻
Cl⁻
Cl⁻
Na⁺
Na⁺
Na⁺
K⁺
K⁺
Pancreatic
Duct Cells

Fig. 1. Examples of the polar distribution of membrane transport mechanisms in transporting epithelia. Apical cell membranes are shown to the *left* and are separated from the basolateral membrane by tight junctions (*black bars*). Ion conductances (channels) are represented by interruptions of the membrane by *double bars*; *open circles* represent carrier molecules; and *filled circles* indicate transport ATPases. *Arrows* indicate the prevailing direction of transport. The references may be consulted for details or controversial aspects.

A The classical model of *epithelial Na^+ absorption* (Koefoed-Johnsen and Ussing 1958) includes apical conductive Na^+ entry and a basolateral Na^+/K^+ pump for Na^+ extrusion with K^+ being recycled at this membrane. Electroneutrality of transepithelial transport is provided by Cl^- transport through other structures. In other epithelia that employ this mode of transport, in particular the principal distal tubule cells, apical Na^+ entry is electrically coupled to K^+ secretion (see Dawson 1987; Field and Giebisch 1989; Donaldson et al. 1989).

B *Absorption of NaCl* is provided by apical Na^+/H^+ and Cl^-/HCO_3^- double exchange and is driven by the Na^+ gradient that is established by basolateral Na^+/K^+ ATPase. The double exchange is fuelled with H^+ and HCO_3^- by dissociation of carbonic acid through carboanhydrase. Basolateral Cl^- exit is by diffusion down its electrochemical gradient but, as shown in *C*, may involve KCl symport (see Powell 1979; Liedtke and Hopfer 1982; Weinman and Reuss 1982; Knickelbein et al. 1983, 1985; Reuss 1985, 1989). In sweat duct cells, NaCl uptake is associated with acid secretion through apical combination of Na^+ and Cl^- channels with Cl^-/HCO_3^- exchange and a H^+ pump (Quinton and Reddy 1989).

C *Apical entry of NaCl* is provided by furosemide-sensitive Na^+-K^+-$2Cl^-$ cotransport. Recycling of K^+ through apical and basolateral K^+ channels, respectively (see Greger 1985; Kinne et al. 1985; O'Grady et al. 1986, 1987; Field and Giebisch 1989).

D In *anion-secreting epithelia*, intracellular $[Cl^-]$ is accumulated above electrochemical equilibrium by basolateral carrier systems that are driven by the Na^+ gradient established through Na^+/K^+ ATPase. These carriers include Na^+/H^+ and Cl^-/HCO_3^- double exchange, Na^+-K^+-$2Cl^-$ cotransport or, in corneal epithelium (Candia 1989) Na-Cl symport. Apical outward diffusion of Cl^- is down its electrochemical gradient, the resulting luminal negative transepithelial potential providing for paracellular secretion of Na^+. Apical outward diffusion of HCO_3^- is included (epididymis: Wong 1988, 1990; salivary gland: Lau et al. 1990). For further references, see Greger et al. (1984), Epstein and Silva (1985), Silva et al. (1986), Petersen (1986), Nauntofte and Poulsen (1986), Turner et al. (1986), Pirani et al. (1987), Rick et al. (1987), Ozawa et al. (1988), Saito et al. (1988, 1990), Singh (1984), Case and Argent (1989), Kanno (1990).

E In *gastric HCl secretion* the apical H^+/K^+ pump provides for H^+ extrusion. This pump, together with basolateral Na^+/H^+ exchange and Na^+-HCO_3^- symport, allows intracellular alkalinization that energizes basolateral Cl^- uptake through Cl^-/HCO_3^- exchange. As in anion-secreting epithelia, intracellular accumulation of Cl^- allows apical outward diffusion down the electrochemical gradient (see Sachs et al. 1978; Paradiso et al. 1986; Forte et al. 1989; Machen et al. 1989; Demarest and Loo 1990; Frömter et al. 1990).

F In *bicarbonate-secreting epithelia*, Cl^-/HCO_3^- exchange is localized to the apical cell membrane and driven through intracellular alkalinization by basolateral Na^+/H^+ exchange, whereas Cl^- recycles through apical Cl^- channels (see Novak and Greger 1988a,b; Stuenkel et al. 1988; Novak 1990; or for turtle urinary bladder: Stetson et al. 1985)

Passive ion flux across *membrane ion channels* occurs by *electrodiffusion*. It is driven (or opposed) by the transmembrane electrical potential difference and the chemical concentration difference of the particular ionic species. Besides these driving forces, the number of ion channels within the membrane and their single-channel conductance determine the magnitude of diffusive ion flux across the cell membrane. Both these latter components may be subject to regulation by extracellular signals and their transduction pathways.

The *outward-directed K^+-concentration gradient* and the high relative membrane K^+ permeability, provided by membrane K^+ channels, establish the *intracellular negative membrane potential* (approx. −40 to −70 mV) that serves as an energy source for diffusion of other ions through membrane channels (outward diffusion of Cl^-, inward diffusion of Na^+) or for electrogenic (rheogenic), carrier-mediated transport processes. The predominant location of these ion channels at the apical cell membrane thus provides for electrogenic transcellular transport of K^+ or Cl^- into the lumen and/or for Na^+ absorption. In "leaky" epithelia, the resulting transcellular ionic current is shunted by paracellular diffusion of the respective counterion across the tight junctions, thus providing overall electroneutral transepithelial transport. In "tight" epithelia, inward and outward ionic currents across either cell membrane must be of equal amplitude. Electrical coupling between diffusive ion fluxes is thus obtained, e.g. between apical membrane Na^+ influx and K^+ efflux (cf. Fig. 1A). Mainly basolateral location of K^+ channels allows K^+ to leak out of the cell across the same membrane in which it is pumped. No transepithelial transport of this ion results therefore if pump and leak fluxes have the same rate, a situation equivalent to a non-polarized cell. In epithelial cells, basolateral recycling of K^+ through this pump-leak system must be increased, however, when transcellular transport of Na^+ through Na^+ extrusion by Na^+/K^+ ATPase is stimulated (cf. Schultz 1989a,b).

The *inward-directed Na^+-concentration gradient*, provided by the operation of the Na^+/K^+ pump, is the second major energy source for membrane transport of solutes against their chemical concentration difference and/or uphill the electrical membrane potential difference. Transduction of the potential energy of the transmembrane electrochemical Na^+ gradient to driving the transport of other solutes uphill their electrochemical potential difference is provided by *membrane carrier molecules* that couple the influx of Na^+ to the membrane transport of other solutes. By cotransport (symport) with Na^+, these *secondary active* transport systems allow the intracellular accumulation of the other solutes (e.g. intracellular accumulation of Cl^- by Na^+-Cl^- symport or by Na^+-K^+-$2Cl^-$ symport; Na^+-dependent uptake of amino acids or glucose), whereas countertransport (antiport, exchange) allows the expulsion of solutes (e.g. intracellular alkalinization by Na^+/H^+ exchange). In principle, carrier-mediated transport exhibits saturation kinetics, while the direction and rate of transport is determined by the magnitude and direction of the electrochemical gradients of the individual, participating solutes, their affinity to the carrier molecule, their stoichiometry of binding (charge transfer by electrogenic carriers), the rate of membrane translocation of the substrate-loaded carrier and the rate of return of the unloaded carrier, and the number of carrier molecules available. Transport rates may thus be regulated by modifying these components.

Secondary active transport systems driven by the Na^+ gradient thus establish transmembrane electrochemical gradients of other ions, whose energy may be again used to drive, by carrier-mediated coupling of fluxes, the uphill transport of further solutes, these being conventionally termed *tertiary active* transport systems (e.g. Cl^-/HCO_3^- exchange). Furthermore, transmembrane pH gradients, established by H^+ pumps or by secondary or tertiary active transport of H^+ or HCO_3^-, may lead to asymmetric membrane distribution of weak acids or bases by *non-ionic diffusion*. In this, the undissociated acid or base equilibrates across the membrane by diffusion and, by pH-dependent dissociation, bases accumulate in the more acid compartment, whereas acids accumulate on the more alkaline side.

Epithelial transport of electrolytes is thus provided by the combined operation of ion pumps (transport ATPases), carriers and ion channels. Their asymmetric distribution to the apical and basolateral cell membrane serves to build up an electrochemical gradient for a given ionic species across one cell membrane. This gradient dissipates across the opposite cell membrane resulting in transcellular ion flux. In order to maintain cell volume in transporting epithelia steady state ion fluxes through these individual transport systems in the apical and basolateral cell membrane must be precisely balanced and coordinately regulated when secretory or absorptive transport rates change. In order to illustrate this concept, the following gives some examples of epithelial electrolyte transport.

2.3 The Polar Distribution of Membrane Ion Transport Systems in Absorbing and Secreting Epithelial Cells

From a large repertory of membrane ion channels, carriers and pumps, available in the genome, absorbing and secreting epithelial cells selectively express and activate those transport mechanisms that are needed for their specific functions. Channels, carriers and pumps are inserted into the apical and basolateral cell membranes to support vectorial transcellular transport for absorption or secretion of electrolytes. Figure 1 gives examples of epithelial electrolyte transport and is intended to illustrate how specific transport functions are supported by appropriate combinations of apical and basolateral transport systems.

Uptake of K^+ and extrusion of Na^+ is mainly provided by the primary active Na^+/K^+ pump that generates the respective ion gradients and is thus responsible for establishing the membrane potential that regulates conductive ion fluxes (e.g. Cl^- efflux) and for providing energy for secondary active transport systems (see above).

Uptake of NaCl into the cell can be obtained by several mechanisms. These include (1) double exchange of Na^+/H^+ and Cl^-/HCO_3^-supported by the dissociation of carbonic acid through carbonic anhydrase (e.g. apical membrane in the small and large intestine and gallbladder, Fig. 1B; basolateral membrane in epididymis, Fig. 1D, or gastric mucosa, Fig. 1F); (2) Na^+-Cl^- symport (e.g. basolateral membrane in choroid plexus, Christense et al. 1989) or (3) a combination of Na^+-$2Cl^-$-K^+ symport with a K^+ channel (e.g. apical membrane in the thick ascending limb of Henle or flounder intestine, Fig. 1C; or basolateral membrane in epididymis, shark rectal gland and lacrimal, salivary and pancreatic acinar cells, Fig. 1D). Furthermore, Na^+ uptake is provided by Na^+ channels (e.g. apical membrane of principal, distal, tubule cells, Fig. 1A), by a multitude of coupled transport systems for organic solutes (e.g. amino acids, glucose), by Na^+/Ca^{2+} exchange (both not shown), or by Na^+-$(HCO_3^-)_n$ symport (e.g. basolateral membrane in frog oxyntic cells, Fig. 1E). This latter electrogenic transport system ($n > 1$) may also operate in the outward direction, depending on the concentration gradients of the participating ions across the cell membrane, the stoichiometry of the transporter and the magnitude of the membrane potential.

Cl^- efflux occurs mainly through Cl^- channels, provided the intracellular Cl^- concentration is elevated above electrochemical equilibrium through coupled uptake systems at the contralateral cell membrane (e.g. Fig. 1B–D,F). Cl^- extrusion may be aided by coupling to efflux of K^+ through K^+-Cl^- symport (e.g. flounder intestine, Fig. 1C). Passive fluxes of Cl^- or K^+ through channels may be required for recycling of the respective ions in order to sustain the activities of Cl^-/HCO_3^- exchange (e.g. pancreatic duct cell, Fig. 1F) or of the K^+/H^+-pump respectively (Fig. 1E).

Several carrier-mediated transport systems regulate the intracellular pH (pH_i) and, analogous to the preservation of cell volume, transcellular transport of acid or base equivalents across the apical and basolateral cell membranes must be appropriately balanced in order to maintain pH_i. These carriers are driven by the transmembrane electrochemical potential differences of the participating solutes and may operate in either direction, depending on the deviation from electrochemical equilibrium (e.g. defined for electroneutral Cl^-/HCO_3^- exchange by the concentration ratios $[Cl^-]_i/[Cl^-]_o = [HCO_3^-]_i/[HCO_3^-]_o$). As shown in Fig. 1, translocation of acid or base equivalents by membrane carriers may involve Na^+/H^+ antiport, Cl^-/HCO_3^- antiport or Na^+-$(HCO_3^-)_n$ symport. Furthermore, HCO_3^--permeable anion channels (e.g. epididymis, Fig. 1D) or ATP-driven pumps for K^+/H^+ exchange (Fig. 1E) and non-coupled H^+ extrusion (e.g. apical membrane of intercalated, distal tubule cells) may be involved.

As discussed below, many of these apical and basolateral transport systems that are involved in transcellular transport may change their transport rate, either by changes in substrate load or by modulation of their activity. Thus, modulation of transcellular transport may transiently change the cell volume until it is restored by volume regulatory mechanisms.

2.4 Factors that Cause Epithelial Cell Swelling or Cell Shrinkage

Perturbations of the cell volume may occur either under isotonic conditions by changes in transport rates of individual transporters or by exposure of cells to anisotonic media, provided the membrane water permeability is high. In general, cells respond with volume regulation under both conditions, indicating that the cell volume per se and not the tonicity of the intracellular milieu activates these responses.

2.4.1 Isotonic Cell Swelling

Under isoosmotic conditions cell swelling or shrinkage is obtained by disturbing the equilibrium between uptake and extrusion of osmolytes. Cell swelling may thus result from increased substrate load for *organic solute* transporters that provide for sufficient transport capacity. In epithelial cells, examples for cell swelling and subsequent RVD include Na^+-driven uptake of amino acids (Grasset et al. 1983; Giraldez and Sepúlveda 1987; Beck and Potts 1990; MacLeod and Hamilton 1991a), galactose (Lau et al. 1984) or glucose (Tune and Burg 1971; Furlong and Spring 1990), or Cl^--dependent uptake of fumarate in the rabbit proximal tubule (Schild et al. 1990). In toad bladder, acetic acid uptake by non-ionic diffusion, intracellular dissociation and sequential activation of Na^+/H^+ exchange and Na^+/K^+ ATPase results in the accumulation of K^+-acetate and cell swelling (Lewis et al. 1985). In liver cells, amino acid uptake was reduced by preventing the subsequent RVD (blocking K^+ channels and hyperpolarization; Wondergem and Castillo 1988). Analogously, proximal tubular cells shrink by apical removal of glucose and alanine (Lapointe et al. 1990) and exhibit a regulatory decrease in basolateral K^+ conductance. These studies support the general concept that uptake of organic solutes by apical Na^+ cotransport is linked to basolateral activation of the Na^+/K^+ pump and K^+ conductance and that cell swelling may mediate this functional coupling of apical and basolateral transport processes (see Schultz 1989a; Lang and Rehwald 1992).

Isoosmotic cell swelling is also obtained when transmembrane ion gradients are disturbed so as to favour solute entry. Thus, raising the *external K^+ concentration* at the expense of Na^+ will increase K^+ uptake and reduce the internal negative cell membrane potential (provided the membrane K^+ conductance of the respective cell membrane is high) and this depolarization would favour anion uptake and cell swelling. Cell swelling may further lead to opening of Cl^--channels (see below) which would facilitate the uptake of KCl. Examples for K^+-induced cell swelling include: frog skin (Ussing et al. 1965), *Necturus* gallbladder (Davis and Finn 1985b, 1988; Hermansson and Spring 1986), shark rectal gland (Kleinzeller and Mills 1989) and mouse proximal tubule (Völkl and Lang 1990). In addition, reducing the outward-directed K^+ gradient could favour entry of NaCl by Na^+-K^+-Cl^- symport (e.g. dark vestibular cells; Wangemann and Marcus 1990). Salt uptake and cell swelling are also obtained by increasing the membrane Na^+ permeability by the pore-forming polyene amphotericin B (e.g. Germann et al. 1986).

When apical and basolateral membrane conductance is predominantly provided by anions (Cl^-) and cations (K^+), respectively, cell swelling may result from passing transepithelial current. Outward current flow through amphibian skin thus results in swelling of mitochondria-rich cells due to a gain in KCl (Foskett and Ussing 1986; Larsen et al. 1987).

Numerous studies have shown that inhibition of active sodium extrusion across the basolateral cell membrane results in cell swelling. Thus, application of *ouabain* leads to cell swelling in mitochondria-rich cells of toad skin (Larsen et al. 1987), shark rectal gland (Kleinzeller et al. 1990) or *Necturus* gallbladder (Ericson and Spring 1982a,b; Persson and Spring 1982; Jensen et al. 1984; Davis and Finn 1985b; Spring and Siebens 1988). Prostaglandin E2, a putative endogenous inhibitor of active Na^+ transport, also elicits cell swelling (Strange 1989). Swelling may be prevented by simultaneous inhibition of apical solute entry (e.g. application of bumetanide in papillary renal epithelium; Sands et al. 1986). Cell swelling by ouabain may also be self-limited by compensatory reduction of ion permeabilities (e.g. frog skin; MacRobbie and Ussing 1961; Ussing 1987, 1990) or inhibition of coupled NaCl uptake (gallbladder; Jensen et al. 1984). Feedback downregulation of apical Na^+ entry after inhibition of the basolateral Na^+/K^+ pump (Lewis and Diamond 1976; Helman et al. 1979) could thus be mediated through variation of the cell volume (see Lewis and Donaldson 1989; Palmer et al. 1989).

2.4.2 Isotonic Cell Shrinkage

Isotonic cell shrinking is induced by a *reduction in the solute entry* into the cell or by stimulation of extrusion. In NaCl-absorbing epithelia cell shrinking has been observed by apical replacement of Na^+ or Cl^- by non-absorbed ionic species or by application of transport inhibitors (Persson and Spring 1982; Jensen et al. 1984; Davis and Finn 1985b; Spring and Ussing 1986; Larsen et al. 1987). In gallbladder, reduced uptake of NaCl and cell shrinking are also obtained by depriving the Na^+/H^+ and Cl^-/HCO_3^- double-exchange transport (Fig. 1) of its supply of H^+ and HCO_3^- by inhibition of carboanhydrase (Persson and Larson 1986). To compensate the reduced apical solute entry, basolateral Cl^- conductance was found to decrease (Stoddard and Reuss 1989). In toad urinary bladder (Lewis et al. 1985) and frog skin (Costa et al. 1987) inhibition of basolateral Cl^- entry results in cell shrinkage and a decrease in ion permeabilities. Chloride-secreting frog cornea epithelial cells shrink by inhibiting salt uptake with loop diuretics (Rick et al. 1987).

Stimulation of glandular secretion is generally associated with a decrease in cell volume, indicating that the primary effect of hormonal or neuronal stimulation is upon solute release. Salivary gland acinar cells shrink upon stimulation and, during continuous production of saliva, maintain a volume below control (Foskett and Melvin 1989; He et al. 1989; Soltoff et al. 1989, 1990; Foskett 1990; Nakahari et al. 1990). It appears that with muscarinic, α-adrenergic, purinergic and substance P stimulation, fluid secretion and volume reduction are elicited by a rise in intracellular $[Ca^{2+}]$ (Foskett and Melvin 1989; Soltoff et al. 1989, 1990; Foskett 1990) but, upon β-adrenergic stimulation, a rise in cAMP and decrease in cell Cl^- content have also been observed (He et al. 1989). Similarly, in pancreatic acinar cells, muscarinic stimulation or application of cholecystokinin-pancreozymin opens Ca^{2+}-activated K^+, non-selective cation and Cl^- channels (e.g. Petersen and Maruyama 1984; Petersen 1986) and cell volume may then be maintained by Ca^{2+} stimulation of basolateral uptake of NaCl through Na^+/H^+ and Cl^-/HCO_3^- double exchange (Muallem and Loessberg 1990). Furthermore, human gastric parietal cells, whose apical cell membrane Cl^- channel is activated through cAMP by histamine (Sandle et al. 1990) also shrink upon stimulation (Helander et al. 1986). Numerous studies indicate that stimulation of epithelial transport is associated with the exocytic insertion of membrane transporters (e.g. Lienhard 1983; Clausen and Dixon 1986; Wade 1986; Maruyama 1989), a mechanism that may also be involved in volume regulatory responses.

Disturbances of the *cytoskeleton* in epithelial cells may result in cell swelling (e.g. Kleinzeller et al. 1990). The mechanism of this phenomenon is unclear however, as hypotonic cell swelling does not appear to be limited by mechanical constraints, which are possibly provided by anchoring the cell membrane to the cytoskeleton. In MDCK cells, cytochalasin B stimulated cell shrinkage, similar to the effect of dibutyryl cAMP (Mills and Skiest 1985; Mills and Lubin 1986).

Changes in the osmolarity of the external medium generally result in cell swelling or shrinkage. In most instances, these perturbances in cell volume are less physiologically significant, and in most isotonically transporting epithelia these volume changes may be limited by the small variations of blood plasma osmolarity. A notable exception are cells that reside in the renal medulla or blood cells that pass through its vasculature. Cell volume changes in this tissue and volume regulatory responses will therefore be discussed separately. Nonetheless, as discussed below, volume regulatory mechanisms have been mainly studied by osmotic perturbances.

Several reviews on cell volume regulation have appeared that also address features peculiar to epithelial cells: Larson and Spring (1987),

Hoffmann and Simonsen (1989), Lang et al. (1988); evolutionary aspects: Schultz (1989a); apical and basolateral mechanisms: Reuss (1988); possible transduction pathways: Eveloff and Warnock (1987); synthesis and release of organic solutes: Spring and Siebens (1988); ion channels: Lewis and Donaldson (1990), Palmer and Sackin (1988); Kidney: Montrose-Rafizadeh and Guggino (1990).

2.5 Mechanisms of Electrolyte Transport in Regulatory Cell Volume Decrease (RVD)

As shown in Fig. 2A swollen epithelial cells may down-regulate their volume by activation (and inhibition) of a number of transport mechanisms that are also involved in transcellular transport.

A rather ubiquitous finding is that hypotonically swollen epithelial cells reduce their volume by releasing KCl through an increase in membrane *K^+ and Cl^- permeabilities*. In some instances, the specific ion channels involved have been identified. Release of K^+ and/or Cl^- occurs frequently across the basolateral cell membrane and transporting epithelia that invoke these mechanisms include: frog skin (Ussing 1986; Costa et al. 1987), frog urinary

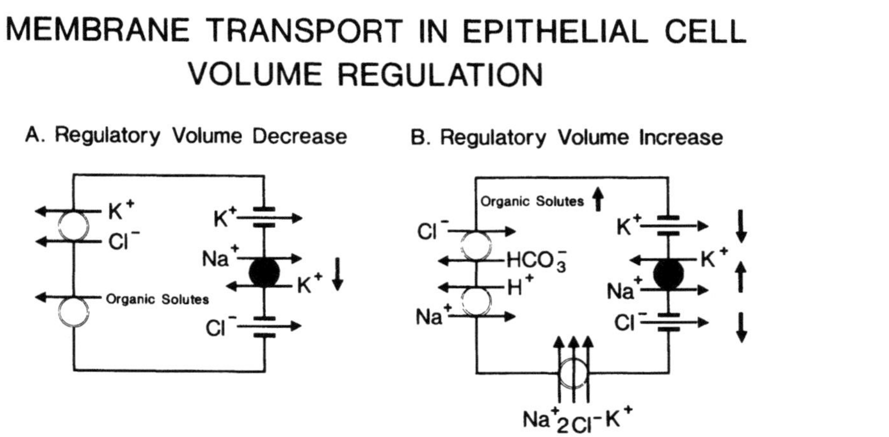

Fig. 2. Volume regulatory mechanisms in epithelial cells involve principally the same membrane transport systems that provide for transcellular electrolyte transport (cf. Fig. 1). In regulatory volume decrease (RVD) those mechanisms prevail that allow the release of KCl. In regulatory volume increase (RVI) mechanisms that provide NaCl entry are activated, intracellular Na^+ is exchanged by K^+ through the Na^+/K^+ pump and conductive efflux of K^+ and Cl^- is reduced, resulting in the accumulation of KCl. Individual epithelia regulate either apical or basolateral transport or both, cell polarity is therefore omitted from the scheme. See text for apical and basolateral membrane transport mechanisms involved in RVD and RVI in individual epithelia, for transport mechanisms that are specifically activated during volume regulation, for interference of volume regulation with transcellular transport, and for release and synthesis of organic solutes during RVD and RVI, respectively.

bladder (Davis and Finn 1985a), small and large intestine enterocytes (e.g. Lau et al. 1984; Germann et al. 1986; Dawson 1987; Hazama and Okada 1988; Montero and Ilundáin 1989; Wiener and Turnheim 1990; MacLeod and Hamilton 1991b), renal proximal tubule (Welling et al. 1985; Kirk et al. 1987; Welling and O'Neil 1989; Sackin 1990; Tauc et al. 1990; Völkl and Lang 1990; Schild et al. 1991), MDCK cells (Simmons 1984; Völkl et al. 1988), airway epithelium (Butt et al. 1990), ciliary epithelium (Farabakhsh and Fain 1987; Yantorno et al. 1989) and lens fibers (Beebe et al. 1990). In turtle colon, the swelling-induced basolateral K^+ conductance differed from control K^+ conductance with respect to ion selectivity and sensitivity to inhibitors, indicating that a particular type of K^+ channel is activated (Germann et al. 1986; Richards and Dawson 1986). In tight epithelia (frog skin and toad urinary bladder), cell swelling and the increase in K^+ conductance are associated with an increase in apical Na^+ entry and stimulation of transepithelial Na^+ transport (Finn and Reuss 1975; Lewis et al. 1985; Costa et al. 1987). Analogous observations have been made for *Necturus* kidney proximal tubules (Lopes and Guggino 1987). Conversely, inhibition of apical Na^+ entry inhibits basolateral K^+ conductance (Davis and Finn 1982, 1985b; the significance of parallel changes in apical Na^+ and basolateral K^+ conductances is discussed e.g. in Lang et al. 1986; Schultz and Hudson 1986). Activation of apical Cl^- conductance has been observed in *Necturus* enterocytes (Giraldez and Sepúlveda 1987; Giraldez et al. 1988) and choroid plexus (Christensen et al. 1989) and cAMP-activated Cl^- channels appear involved in RVD in airway epithelium (McCann et al. 1989). Activation of apical K^+ channels has been studied in renal cells (e.g. Ubl et al. 1988a,b; Taniguchi and Guggino 1989; Dubé et al. 1990; Filipovic and Sackin 1991). In *Necturus* gallbladder KCl release has been attributed to a rise in K^+ conductance and basolateral *KCl cotransport* (Larson and Spring 1984), but may involve a bumetanide-sensitive anion channel (Furlong and Spring 1990). Increase in the release of K^+ to reduce cell volume may be supported by a simultaneous decrease in K^+ uptake through inhibition of *Na^+/K^+ ATPase* (Montero and Ilundáin 1989). In contrast, gallbladder cells swollen through pump inhibition by ouabain do not exhibit RVD, but reduce their volume after additional swelling in hypotonic medium. Ouabain may thus alter ionic gradients so as to shift the "set point" for the volume regulatory response (Spring and Siebens 1988), but inhibition of RVD by ouabain in hypotonically swollen cells has also been observed (Farahbakhsh and Fain 1987). *K^+/H^+ exchange*, i.e. loss of K^+-intracellular acidification-titration of organic anions, may play a role in RVD in frog skin (Ussing 1990) and corneal epithelium (Bonanno 1991). Conversely, hypotonic cell swelling was shown to activate the apical *K^+/H^+ pump* in gastric parietal cells (Sernka 1990).

Other mechanisms of epithelial cell RVD involve the *release of organic solutes* (see Beck et al. 1990) such as amino acids (e.g. Sánchez Olea et al. 1991) or sorbitol (e.g. Bevan et al. 1990) and are reviewed in Section 4.

The role of cytoplasmic vesicles in RVD (see van Rossum et al. 1987) has been mainly studied in liver and is discussed in Section 5.

2.6 Mechanisms of Electrolyte Transport in Regulatory Volume Increase (RVI)

As seen in Fig. 2B, shrunken cells employ several membrane transport mechanisms in order to gain intracellular osmolytes and return their volume towards control. Cells exposed to a medium that was made hypertonic by the addition of membrane-impermeant solutes thus need to accumulate osmolytes in excess of the intracellular content present under isotonic conditions. Some epithelial cells lack this ability and remain shrunken in hypertonic medium (e.g. principal frog skin cells; Ussing 1985, 1986). RVI may be elicited, however, when the cellular content of osmolytes is first reduced by swelling the cells in hypotonic medium and allowing the release of osmolytes to occur during subsequent RVD. These cells shrink osmotically below the control volume when returned to normal isotonic medium and exhibit RVI and return to the control volume as the normal intracellular content of the osmolytes is restored (pseudoregulatory volume increase). Osmolytes gained through RVI are mainly K^+ and Cl^-. In most instances, uptake of NaCl is primarily initiated and is followed by replacement of intracellular Na^+ by K^+ through Na^+/K^+ ATPase activity. Simultaneously, exit of KCl may be reduced through down-regulation of the respective ion conductances.

Uptake of NaCl during RVI may involve activation of apical membrane *Na^+/H^+ and Cl^-/HCO_3^-* double exchange (*Necturus* gallbladder; Ericson and Spring 1982a,b; Marsh and Spring 1985; Reuss 1985) and depends on the supply of H^+ and HCO_3^- by carboanhydrase (Persson and Larson 1986). The same mechanism operates in RVI at the basolateral membrane in the mouse ascending limb (Hebert 1986a,b; Sun et al. 1990). Apical membrane *Na^+-K^+-Cl^- cotransport* was found to provide for RVI in rabbit ascending limb cells (Eveloff and Calamia 1986) and retinal pigment epithelium (Adorante and Miller 1990) and, in a basolateral localization, this transporter supports pseudoregulatory volume increase in frog skin (Ussing 1985, 1986). Stimulation of basolateral *Na^+/K^+ ATPase* during RVI has been demonstrated in chicken enterocytes (Montero and Ilundáin 1989) and thick limb cells (Hebert 1986). In rabbit urinary bladder basolateral *K^+ and Cl^- conductances* decrease during RVI (Donaldson et al. 1989), a decrease in Cl^- conductance has been shown for gallbladder (Stoddard and Reuss 1989) and a decrease in K^+ conductance in hypertonic medium in chicken enterocytes (Montero and Ilundáin 1989), proximal tubular cells (Lapointe et al. 1990) and in toad urinary bladder (Lewis et al. 1985) and frog skin (Costa et al. 1987; Leibowich et al. 1988), but RVI after hypertonic shrinkage was absent in the latter epithelium. Furthermore, in tight epithelia, a reduction in cell volume by reducing basolateral entry of Cl^- results in the closure of basolateral K^+ channels in association with inhibition of apical Na^+ entry (toad urinary bladder: Lewis et al. 1985; Leibowich et al. 1988; frog skin: Costa et al. 1987; Finn and Reuss 1975) indicating that transepithelial transport of Na^+ is closely linked to cell volume regulation (Lewis and Donaldson 1989; see also below). Intracellular alkalinization of osmotically shrunken frog skin epithelial cells (Civan et al. 1988) may contribute to RVI by allowing for an increased dissociation of weak organic acids.

In glandular cells that are shrunken by stimulation of fluid secretion, cell volume is maintained by a simultaneous increase in solute uptake. Uptake mechanisms include stimulation of basolateral Na^+-K^+-$2Cl^-$ cotransport, Na^+/H^+ exchange and Na^+/K^+ ATPase in parotid acinar cells (Soltoff et al. 1989) and of Na^+/H^+ and Cl^-/HCO_3^- double exchange in pancreatic acinar cells (Muallem and Loessberg 1990), respectively. The stimulus-associated rise in intracellular [Ca^{2+}] may mediate both the increase in apical ion secretion and in basolateral uptake but, independent of the stimulus, changes in the cell volume could mediate coupling between these transport processes.

The role of synthesis of intracellular *organic solutes* in RVI is discussed below in connection with volume regulation in the renal medulla.

2.7 Volume Sensing and Signal Transduction Mechanisms in Transporting Epithelia

In order to regulate their volume, cells must perceive deviations from normal size and, by appropriate signals, modify membrane transport processes. It appears that, analogous to recognition of cell deformation in mechanoreceptors (e.g. Sachs 1987), stretching of the cell membrane directly activates cell membrane ion channels. In swollen cells, opening of ion channels (activation of dormant channels, modulation of the close/open kinetics) could directly provide for ion efflux (Sackin 1989) but, apparently more general, *stretch-activated ion channels* are non-selective and/or Ca^{2+} conducting and thus allow for Ca^{2+} entry. The rise in [Ca^{2+}] inside the cell membrane may activate/deactivate or modulate ion transport mechanisms, in particular Ca^{2+}-activated K^+ channels. This sequence of events appears to apply for the choroid plexus (Christensen 1987; Christensen and Zeuthen 1987) and *Necturus* proximal tubule basolateral (Sackin and Palmer 1987; Palmer and Sackin 1988) and apical cell membrane (Filipovic and Sackin 1991). In opossum kidney cells stretch-activated, non-selective and Ca^{2+}-activated channels have been observed (Ubl et al. 1988a,b). In addition, directly stretch-activated non-selective channels have been identified in frog kidney cells (Hunter 1990), indirect stretch activation of K^+ channels appears responsible for RVD in thick ascending limb (Taniguchi and Guggino 1989) and cultured proximal tubule cells (Dubé et al. 1990).

Disturbances of the *cytoskeleton* have been shown to change cell volume (e.g. Kleinzeller et al. 1990), its integrity appears required for normal volume regulatory responses to take place (e.g. van Rossum and Russo 1981; Foskett and Spring 1985; Mills 1987; van Rossum et al. 1987; Cornet et al. 1988), and rearrangement of microfilaments upon hypotonic stress has been observed (Cornet et al. 1987; Kleinzeller and Mills 1989). In contrast, cytochalasin B supports cell shrinkage in MDCK cells. These cells also shrink by applying cAMP and form microfilament bundles (Mills and Lubin 1986; Mills and Skiest 1985). As the cell membrane and its transporters are anchored to the cytoskeleton and microtubules and microfilaments serve many intracellular functions (e.g. mechanical stability, contraction, assembly-disassembly, regulation of intracellular vesicular traffic, regulation of membrane insertion and retrieval or exocytosis and endocytosis, respectively), the specific roles of cytoskeletal elements in epithelial cell volume regulation remain largely undetermined. Interestingly, interaction of G-proteins and microfilaments in hypotonically induced uncoupling of pancreatic acinar cells has been observed (Ngezahayo and Kolb 1990).

Ion channels, carriers, membrane pumps and cell metabolism in absorbing and secreting epithelia are subject to manifold hormonal regulations. Accordingly, the same *intracellular signals* that are generated by hormone-membrane receptor interaction have been considered to modify membrane transport in cell volume regulation. Clear evidence for their specific role in volume regulation in transporting epithelia is still scarce (cf. Hoffmann et al., this Vol.). In principal, membrane transporters may be subject to short-term regulation by factors such as pH_i, $[Ca^{2+}]_i$, voltage, G-proteins, eikosanoids or covalent modification (cAMP-, cGMP- or Ca-calmodulin-activated protein kinases, protein kinase C, phosphatases, methylation) or by long-term regulation (synthesis, degradation, membrane insertion and retrieval). As indicated above for stretch-activated channels, $[Ca^{2+}]_i$ appears to play a major role in RVD as reviewed by Pierce and Politis (1990). In transporting epithelia, RVD was shown to depend on Ca^{2+} and inhibition by trifluoperazine indicates a role of calmodulin (Foskett and Spring 1985). The rise in intracellular $[Ca^{2+}]$ during RVD was measured in some epithelial cells (e.g. frog urinary bladder: Wong and Chase 1986; human enterocytes: Hazama and Okada 1990; rabbit proximal tubule: Suzuki et al. 1990). Transduction of this signal to its effector may involve direct modulation, Ca-calmodulin, protein phosphorylation, and microfilaments, and results in an increase in K^+ and Cl^- efflux (e.g. frog urinary bladder: Davis and Finn 1987; human enterocytes: Hazama and Okada 1988, 1990) or in a release of organic solutes (e.g. sorbitol: Bevan et al. 1990). In MDCK cells, *cAMP* may support RVD in addition to $[Ca^{2+}]_i$ (Mills and Skiest 1985; Mills 1987; Allen and Mills 1988) and indirect evidence suggests that cAMP is involved in RVD in mouse medullary thick ascending limb (Hebert 1986a) and airway epithelium (McCann et al. 1989). By associating mechanisms that are involved in volume regulation in non-polarized cells and signals that regulate transepithelial transport, rather than by experimental evidence, other transduction pathways in epithelial cell volume regulation may be taken into consideration. These transduction signals include eicosanoids (see Lambert et al. 1987; Lambert 1989; Lote and Haylor 1989), cGMP (O'Grady et al. 1987), protein phosphorylation by protein kinase C (Civan et al. 1985; Grinstein et al. 1986) or insertion of membrane transporters (Hoffmann et al. 1986; Wills et al. 1991).

3 Intestinal Transport and Volume Regulation

Homeostatic mechanisms maintain intracellular electrolyte composition and volume during changes in transepithelial Na^+ transport.

3.1 Introduction

Compared to the cellular content of Na^+, the variations of Na^+ transport across intestinal epithelia can be considerable. Assuming an intracellular space of 20 μl/cm^2 of tissue (Turnheim et al. 1986; Roden and Turnheim unpubl.) and an intracellular Na^+ concentration of 10 mmol/l, the Na^+ content of the epithelium may be estimated to be 0.2 μEq/cm^2. Maximal rates of transcellular Na^+ transport are 3–5 μEq/cm^2/h in rabbit ileum (Schultz and Zaluski 1964), 5–6 μEq/cm^2/h in rabbit descending colon (Turnheim et al. 1986), and 14 μEq/cm^2/h in rabbit gallbladder (Moran et al. 1986). Hence, the intracellular Na^+ would be exchanged 15–70 times every hour. This turnover most likely represents an underestimate, as not all epithelial cells participate in Na^+ absorption.

Transcellular Na^+ transport is a two-step process, first Na^+ enters "downhill" across the apical (or luminal) cell membrane and is then extruded against an electrochemical gradient across the basolateral membrane. Four main mechanisms of apical Na^+ entry can be distinguished, three carrier-mediated or coupled transport systems (Na^+/H^+ exchange, Na^+-Cl^- or Na^+-K^+-$2Cl^-$ cotransport, and Na^+ entry coupled to that of organic solutes such as sugars or amino acids) and one channel-mediated entry mechanism (Schultz 1986; Powell 1987; see also Fig. 1). The biochemical equivalent of the uphill basolateral Na^+ extrusion mechanism, the "Na^+ pump", is the ouabain-sensitive Na^+/K^+ ATPase.[1] This pump appears to

[1] However, it cannot be totally excluded that there is a second basolateral Na^+ pump which is not dependent on K^+ and is not inhibited by cardiac glycosides (Proverbio et al. 1989).

exchange 3 Na^+ for 2 K^+ under most conditions (Nielsen 1979; Jorgensen 1980; Kirk et al. 1980; Eaton et al. 1982; Turnheim et al. 1983). The K^+ taken up during Na^+ extrusion leaves the cell via K^+-selective channels in the basolateral membrane.

This series arrangement of downhill apical Na^+ entry and the basolateral Na^+ pump – K^+ leak combination can account for transcellular Na^+ absorption as well as the characteristic intracellular ion composition (low Na^+ and high K^+) and the electrical cell membrane potential difference (Koefoed-Johnsen and Ussing 1958). However, problems may arise with respect to maintenance of the intracellular composition and cell volume when the rate of transcellular transport is varied. Increased apical entry of solutes would threaten the electrolyte and volume homeostasis of the cells, if the basolateral extrusion mechanism were not stimulated in parallel. Indeed, mechanisms seem to exist that closely and rapidly coordinate the Na^+ transport functions of the apical and basolateral membranes, so that excessive changes in intracellular Na^+ activity, $(Na^+)_i$, are prevented when the rate of transcellular Na^+ transport is altered (Schultz 1981; Palmer et al. 1989; Turnheim 1991). Further, inasmuch as increased transcellular Na^+ transport is accompanied by increased basolateral K^+ uptake via the Na^+/K^+ ATPase, cell K^+ may rise when transcellular Na^+ transport is stimulated. This rise in cell K^+ is prevented by a parallel increase in basolateral membrane K^+ conductance when transcellular Na^+ transport and the activity of the Na^+/K^+ ATPase are stimulated (Schultz 1981; Schultz and Hudson 1986). The mechanisms that prevent inordinate changes in intracellular ion composition and hence cell volume during changes in transcellular Na^+ transport have been termed "homocellular regulation" (Schultz 1981) or "membrane cross-talk" (Diamond 1982).

3.2 Regulation of Apical Na^+ Entry by Intracellular Na^+

Procedures that increase the intracellular Na^+ concentration in epithelia reduce the rate of apical Na^+ entry. This is to be expected, as the driving force for entry is reduced. However, in epithelia with channel-mediated apical Na^+ entry, an elevation in intracellular Na^+ was also repeatedly shown to reduce the conductance of the apical membrane (Lewis et al. 1976; Turnheim et al. 1978; Palmer et al. 1980; Chase and Al-Awqati 1981), whereas an increase in apical membrane conductance would be expected with increased $(Na^+)_i$ according to the Goldman-Hodgkin-Katz constant field equation. A decrease in $(Na^+)_i$, on the other hand, is associated with an increase in apical Na^+ entry and apical membrane Na^+ conductance (Leblanc and Morel 1975; Turnheim et al. 1978). This phenomenon, which was termed "feedback inhibition", is a result of a decrease in the density of conducting or open Na^+ channels in the apical membrane when $(Na^+)_i$ increases and vice versa (van Driessche and Erlij 1983; Fisher et al. 1986).

Elevations in $(Na^+)_i$ also decrease the rate of apical Na^+ entry via the Na^+-sugar or Na^+-amino acid cotransport mechanisms (Aronson and Sacktor 1975; Kessler and Semenza 1983; Kimmich and Randles 1988).

The physiological utility of the inverse relation between $(Na^+)_i$ and apical membrane Na^+ permeability is obvious: as intracellular Na^+ increases, the rate of apical Na^+ entry decreases and consequently the cell is protected from excessive changes in $(Na^+)_i$ and volume. Indeed, steady-state $(Na^+)_i$ remains relatively constant despite wide variations in the rate of transcellular Na^+ transport (Lee and Armstrong 1972; Wills and Lewis 1980; White and Imon 1983; Hudson and Schultz 1984; Moran et al. 1986; Turnheim et al. 1986, 1987; Clauss et al. 1987).

The mechanism of the down-regulation of apical Na^+ permeability by increases in $(Na^+)_i$ is not resolved. It is unlikely that Na^+ itself is the mediator of feedback inhibition. Other factors that may participate in the autoregulation of apical Na^+ entry include intracellular Ca^{2+}, pH, and G-proteins (see Turnheim 1991).

3.3 Na^+/K^+ ATPase Activity and Na^+ Transport

Relatively invariant intracellular Na^+ concentrations appear advantageous for cell homeostasis, however, the question arises how the basolateral pump is activated under conditions of increased transcellular Na^+ transport, when $(Na^+)_i$ remains more or less constant. Studies in which $(Na^+)_i$ was continuously measured using Na^+-selective microelectrodes have shown that following stimulation of transcellular Na^+ transport by luminal addition of galactose to the small intestine (Hudson and Schultz 1984) or bicarbonate to the gallbladder (Moran et al. 1986), there is an initial increase in $(Na^+)_i$ to a maximum at 2–4 min, but then $(Na^+)_i$ declines and is not different from the control value after 10–15 min despite continued, increased Na^+ transport (Fig. 3). The initial increase in $(Na^+)_i$ most likely reflects the large rate of apical Na^+ entry, whereas the subsequent decrease in $(Na^+)_i$ may be the result of increased basolateral Na^+ exit.

Stimulation of Na^+-sugar or Na^+-amino acid cotransport is known to cause an increase in cell volume (Csaky and Esposito 1969; Armstrong et al. 1970; Hudson and Schultz 1988; MacLeod and Hamilton 1991a). Hence, one of the mechanisms by which the activity of the basolateral Na^+ pump may be increased in the face of unchanged levels of intracellular Na^+ is cell swelling. There is evidence in some epithelial cells (Ussing 1965; Lipton 1972; Ericson and Spring 1982b; Lewis et al. 1985) and non-epithelial cells (Cala 1977; Pollack et al. 1981; Grinstein et al. 1984; Siebens and Kregenow 1985; Venosa 1991) that pump activity is increased by osmotic cell enlargement, but that it is decreased by osmotic shrinkage.

Stimulation of the Na^+ pump will not only decrease $(Na^+)_i$ but should also cause cell shrinkage as more Na^+ is extruded from the cell than K^+

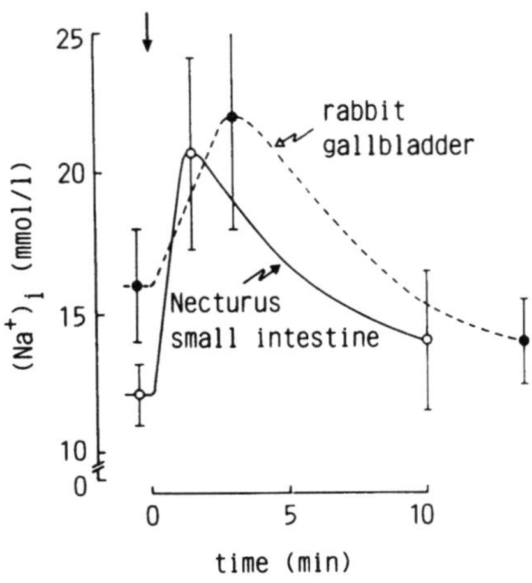

Fig. 3. Effect of stimulation of transcellular Na^+ transport by addition of 15 mmol/l galactose in *Necturus* small intestine (○) or 20 mmol/l HCO_3^- in rabbit gallbladder (●), both added at the *arrows*, on intracellular Na^+ activity, $(Na^+)_i$. Transcellular Na^+ transport was stimulated by galactose 3–4 fold (from 0.5 to 1.9 μEq/cm²/h) and by HCO_3^- 2-fold (from 7.7 to 14.2 μEq/cm²/h). (After Hudson and Schultz 1984; Moran et al. 1986)

taken up per cycle. In other words, the pump could have a dehydrating effect.

Aldosterone stimulates channel-mediated epithelial Na^+ transport (Garty 1986), and this stimulation is also associated with a significant increase in cell volume during the early phase of the aldosterone effect (O'Neil 1989). There is no or only a small increase in $(Na^+)_i$ when Na^+ transport across colonic epithelium is stimulated by aldosterone (Turnheim et al. 1986, 1987), but the number of Na^+ pump sites in the basolateral membrane and their Na^+ turnover rate are increased (Roden and Turnheim 1988). The Na^+/K^+ ATPase consists of α- and β-subunits, and three isoforms of the α-subunit have been identified (Rossier et al. 1987). Hence, it is possible (and the topic of intensive research) that different isoforms with different pump activities exist during various states of hormone-regulated transport.

Cell swelling may not only influence basolateral Na^+ extrusion but also apical Na^+ entry. The number of conducting Na^+ channels in the apical membrane of cultured renal A6 cells was markedly higher when the cells were bathed in hypotonic solutions (Wills et al. 1991). Further, there is evidence that stretch, caused by hydrostatic pressure or osmotic swelling, leads to expansion of the apical membrane area by fusion of cytoplasmatic vesicles that contain Na^+ channels (Lewis and DeMoura 1982, 1984).

3.4 Relation Between Basolateral K^+ Conductance and Na^+ Transport

It was shown for many epithelia that apical Na^+ permeability and basolateral K^+ conductance change in parallel. Subsequent to stimulation of apical Na^+ entry, basolateral K^+ conductance increases (Helman and Fisher 1977; Higgins et al. 1977; Davis and Finn 1982; Gunther-Smith et al. 1982; Albus et al 1983; Brown et al. 1983; Grasset et al. 1983; Hudson and Schultz 1984; Messner et al. 1985; Moran et al. 1986; Turnheim et al. 1987). In fact, a linear relation between basolateral membrane conductance and the rate of Na^+ transport (Frömter and Gebler 1977; Thomas et al. 1983) appears to exist. As eloquently discussed earlier (Schultz 1981, 1989a; Schultz and Hudson 1986), the increase in basolateral K^+ conductance may serve to enhance the recycling of K^+ to the serosal side of the tissue and hence prevent an increase in intracellular K^+ when transcellular Na^+ transport and cellular uptake of K^+ via the Na^+,K^+ exchange pump are increased ("pump-leak parallelism"). In addition, increasing basolateral K^+ conductance will hyperpolarize the cell interior and therefore increase or restore the electrical driving force for apical Na^+ entry.

An example of the time-dependent changes in basolateral conductance in response to a marked rise in apical Na^+ entry is given in Fi. 4. Addition of alanine to the luminal solution, which enhances transcellular Na^+ transport almost three-fold, rapidly depolarizes the apical membrane and increases the conductance of the apical membrane of *Necturus* small intestine, as reflected by a decrease in the ratio of apical and basolateral membrane resistances (Rm/Rs). This initial effect is followed by a gradual increase in basolateral K^+ conductance, as apparent from the slow increase in Rm/Rs. As a consequence of the increase in basolateral K^+ conductance, the apical membrane is repolarized[2], hence, the electrical driving force for apical Na^+ entry is restored. An explanation for this relation between Na^+ transport and the basolateral K^+ conductance may be provided by the recent observation that single K^+-channels from basolateral membranes of *Necturus* small intestine reconstituted into planar lipid bilayers have the properties of an "inward rectifier", i.e. the probability that the channel is in the open state (Po) increases with increasing hyperpolarization of the membrane (Costantin et al. 1989). Assuming that these channels have the same properties in the intact epithelium, they would be predominantly closed when the cell is depolarized by the high rate of alanine-stimulated

---

[2] An increase in basolateral membrane K^+ conductance will hyperpolarize the basolateral membrane. In "leaky" epithelia, i.e. in those with a high paracellular or shunt conductance, the apical and basolateral membranes are electrically coupled, hence, the apical membrane is also hyperpolarized when the basolateral K^+ conductance increases (Schultz 1972).

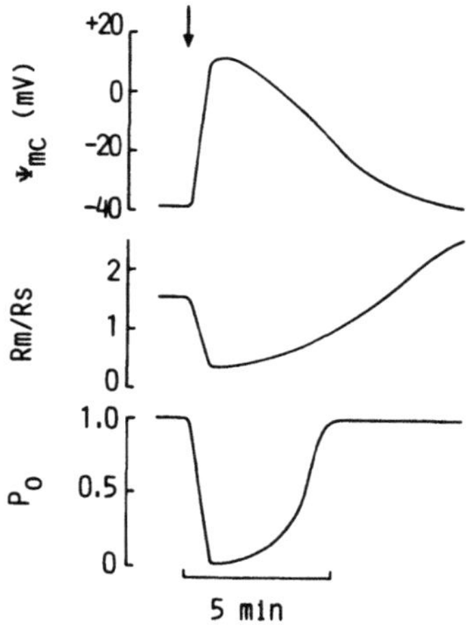

Fig. 4. Changes in the electrical potential difference across the apical membrane (Ψ_{mc}) and the ratio of the electrical resistances of the apical and basolateral cell membranes (*Rm*/*Rs*) after addition of 10 mmol/l alanine to the solution bathing the luminal surface of isolated epithelia of *Necturus* small intestine. Also given is the open-time probability of single basolateral K^+ channels (P_o) corresponding to the values of Ψ_{mc}. (After Gunther-Smith et al. 1982; Costantin et al. 1989)

electrogenic Na^+ entry. This would prevent excessive K^+ loss during depolarization. The increase in apical Na^+ entry will increase $(Na^+)_i$ (see Fig. 3), which activates the basolateral Na^+/K^+ exchange pump. As the pump is taken to be electrogenic because of the Na^+-K^+ stoichiometry of 3:2, its activation will hyperpolarize the basolateral membrane, which in turn will reactivate the K^+ channels (Costantin et al. 1989). In short, this system would function as a positive feedback between pump rate and basolateral K^+ leak.

The messengers and signals that coordinate Na^+ transport and basolateral K^+ conductance are not entirely clear at present. Several possible candidates were recently discussed by Dawson and Richards (1990). It is interesting to note that some of the same factors may be involved that are also considered to control apical Na^+ entry (Ca^{2+}, pH, cell swelling).

3.5 Volume Regulation and Transport

As mentioned above, activation of K^+ channels participates in the regulatory volume decrease following exposure of cells to hypotonic solutions. Since basolateral K^+ channels are also activated when transcellular Na^+ transport is increased, it is quite possible that the mechanisms of membrane cross-talk during variations in Na^+ transport and the mechanisms of the volume

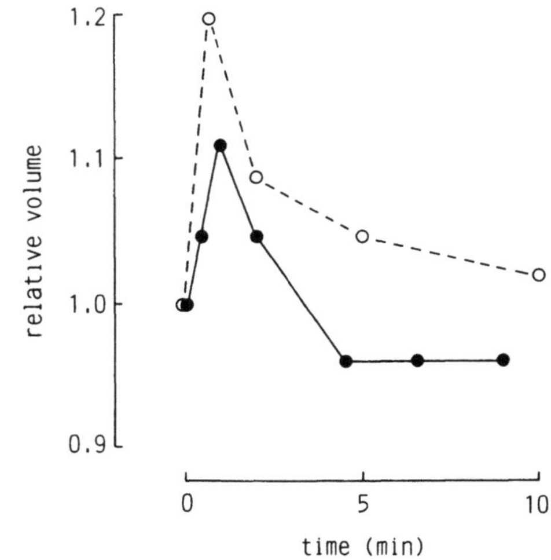

Fig. 5. Volume changes in isolated jejunal enterocytes from guinea pigs after addition of 20 mmol/l alanine, a Na^+-cotransported solute (•), or after exposure to a hypotonic solution (0.7 × isotonic, ○). Note that the time course of the transport-induced volume changes is similar to the changes in the electrical membrane potential difference and membrane conductances illustrated in Fig. 4. (After MacLeod and Hamilton 1991a,b)

regulatory responses to changes in external osmolarity are identical or related.

The time courses of cell volume decrease in guinea pig enterocytes following cell swelling, due either to Na^+-alanine cotransport or exposure to a hypotonic solution, are very similar (Fig. 5). The mechanisms of preserving or restoring cell volume following swelling as a result of transport stimulation or decreased extracellular osmolarity are also similar with respect to the inhibitory effect of K^+ and Cl^- channel blockers. In both cases cell shrinkage after the initial swelling was prevented by the K^+-channel inhibitors Ba^{2+} and quinine[3] and the Cl^- channel blocker 9-phenyl-anthracene (MacLeod and Hamilton 1991a,b). Also, in *Necturus* small intestine the changes induced by transport or hypotonicity are similar. Addition of galactose to the luminal solution causes a slow increase in

[3] A Ca^{2+}-activated, Ba^{2+}- and quinine-sensitive K^+ channel from basolateral membranes of rabbit colon was recently reconstituted into planar lipid bilayers (Turnheim et al. 1989), and a K^+ channel with similar properties was also detected in this tissue using the patch-clamp technique (Loo and Kaunitz 1989). A rise in intracellular Ca^{2+} is suggested to be responsible for the activation of volume-regulatory K^+ channels in epithelial cells (Hazama and Okada 1990).

basolateral K^+ conductance, and exposure of the epithelium to a hypotonic solution has the same effect. In both instances the increase in basolateral K^+ conductance is inhibited by Ba^{2+} (Lau et al. 1984). The notion that the transport-induced increase in basolateral K^+ conductance is indeed brought about by cell swelling is supported by the finding that this transport-induced effect is abolished when the galactose-containing solution is 20% hypertonic (Lau et al. 1986). The dependence of the transport-induced increase in cell membrane K^+ conductance on cell swelling was also demonstrated in isolated hepatocytes (Bakker-Grunwald 1983; Kristensen and Folke 1984). In short, there is increasing evidence that the autoregulatory changes taking place in an epithelial cell confronted with wide variations of transport load are in fact volume-regulatory responses (Schultz 1989a,b).

Inspection of Fig. 5 clearly shows that transport-induced cell swelling is transient in isolated epithelial cells. If this is also the case in intact epithelia[4], then the volume-regulatory mechanisms may be responsible only for the initial adaptation of the cell to the increase in transport, whereas in the steady-state of elevated transport, when the cell has more or less returned to control values of volume, other mechanisms must be responsible for the increases in Na^+ pump rate and in the basolateral membrane K^+ conductance.

4 Control of Renal Cell Volume: The Renal Medulla Role of Organic Solutes in Cell Volume Regulation

4.1 Introduction

Most cells in the kidney have mechanisms for controlling cell volume in response to changes in extracellular osmolality (see Montrose-Rafizadeh and Guggino 1990, for a review). Because the extracellular osmolality in the renal cortex does not change significantly during normal variations in the state of hydration of the body, mechanisms for controlling the volume of cells in the renal cortex (Sackin 1987; Ubl et al. 1988a,b; Welling and Linshaw 1988; Lohr et al. 1989; Marin et al. 1989; Rome et al. 1989; Strange 1989; Dube et al. 1990; McCarty and O'Neil 1990; Natke 1990; Völkl and Lang 1990; Beck et al. 1991) function primarily to maintain a steady – state volume following changes in transepithelial transport (see Sect. 3 for a treatment of isosmotic volume regulation). In contrast, cells in the renal medulla are normally exposed to large changes in extracellular osmolality.

[4] In rabbit cortical collecting duct aldosterone caused an increase in cell volume of 44% at 8 h, but after 24 h of continuous aldosterone exposure the cell volume had returned to the initial control value (O'Neil 1989).

These cells must control cell volume both in response to changes in net transepithelial transport and to variations in extracellular osmolality. The following section will focus on the regulatory mechanisms with control cell volume in the unique environment of the renal medulla.

4.2 Control of Cell Volume in the Renal Medulla

Urine osmolality in mammals can vary enormously depending upon the state of hydration. In rats, for example, urine osmolality can vary between 170, following a large water intake, to about 3817 mOsmol/kg water following severe dehydration (data from Saikia 1991 and Valtin 1966; also see Burger 1987, for a review). The ability to excrete large amounts of dilute urine during periods of extreme hydration and small amounts of highly concentrated urine during periods of water deprivation allows mammals to adapt to such diverse environments as freshwater streams and deserts. Mammals vary the amount and osmolality of urine functions to balance water loss with intake in order to maintain a constant osmolality of the extracellular body fluids. Remarkably, despite large differences in the state of hydration, the osmolality of extracellular fluids between animals living in different terrestrial environments does not differ.

Extracellular osmolality is tightly controlled by antidiuretic hormone (ADH). The major action of this hormone is to modulate the water permeability of the apical cell membrane of the collecting ducts. In the absence of ADH, the water permeability of the apical cell membrane of the collecting ducts is extremely low (see Robertson and Berl 1991). When the apical water permeability is low, collecting ducts continue to reabsorb NaCl, reducing urine osmolality much below that of extracellular fluids. In the presence of ADH the water permeability of collecting ducts increases and water is removed from the urine into the interstitium of the kidney. The osmolality of luminal fluid in distal nephron segments located in the renal cortex is in equilibrium with that of the cortical interstitium (see Robertson and Berl 1991). In order for water reabsorption to occur in distal nephron segments located in the medulla and papilla, the osmolality of the interstitium of the medulla and papilla must be higher than that of the renal cortex. For this reason, a cortical to medullary increase in renal interstitial osmolality is characteristic of mammals (see Schmidt-Nielson 1975). For example in a dehydrated rat the osmolality of the renal papilla can rise to greater than 3500 mOsmol compared to 300 mOsmol in the cortex (see Beck et al. 1985; and Schmolke et al. 1989), a greater than ten-fold gradient.

Clearly, the renal medulla is a unique place in the body, an environment with a very high osmolality. It is also a place which is subjected to large changes in volume flow and osmolality when living environments switch from diuresis to antidiuresis conditions (see Beck et al. 1985; and Schmolke et al. 1989). These unique characteristics require that the cells in the renal medulla have mechanisms for coping with the high osmolality in steady state and with large changes in osmolality depending upon the degree of hydration.

4.3 Coping with a High Extracellular Osmolality

In the medullary interstitium of an antidiuretic rat the total osmolality is about 2500–3800 mOsmol/kg. About 60% of the total mOsmol/kg can be accounted for by urea. The remaining milliosmoles are contributed by electrolytes, primarily NaCl (see Beck et al. 1984; Garcia-Perez and Burg 1990, 1991). Thus, in the medullary interstitium, urea and NaCl are the primary osmolytes. All the cells in the renal medulla are in osmotic equilibrium with the extracellular fluids. Despite the high osmolality of the renal inner medulla, intracellular electrolyte concentrations are similar to cells in other parts of the body (see Table 1; Beck et al. 1980, 1985; Boron and Sackin 1983; for a review of intracellular composition of renal cells).

The total intracellular electrolyte concentration in the inner medulla accounts for only about 300 mOsmol. Since urea is permeable across cell membranes, it can be assumed that the intracellular urea concentration is the same as that in the interstitium. This means that in cells of the inner medulla, urea and electrolytes account for only a portion of the total intracellular osmolality. This leaves about 700–1700 mOsmol of additional solutes which are needed to match interstitial osmolality in the antidiuretic state (see Garcia-Perez and Burg 1990, 1991). The difference is made up by organic osmolytes (see Table 2) which fit into three general classes, polyhydric alcohols, amino acids and polyamines (for reviews, see Balaban and Burg 1987; Garcia-Perez and Burg 1990, 1991).

The total concentrations of osmolytes presented in Table 2 were determined for whole renal medulla which includes intra- and extracellular fluids. However, medullary organic osmolytes are primarily intracellular (see Garcia-Perez and Burg 1990, 1991). Assuming a ratio of intracellular to extracellular water of 3/2, the total osmolality of intracellular osmolytes

Table 1. Electrolyte concentrations (mmol/kg wet weight) of cells in the renal outer and inner medulla

Segment	Concentration			Reference
	Na^+	K^+	Cl^-	
Thick ascending limb	9.8	106	20	Sudo and Morel (1984)
Medullary collecting duct (light cells)	11.1	106	26	Beck et al. (1989)
Medullary collecting duct (dark cells)	8.9	116	45	Beck et al. (1989)
Papillary collecting duct	28	135	76	Beck et al. (1980)

Elemental concentrations measured with electron probe analysis in antidiuretic rats

Table 2. Tissue concentrations of organic osmolytes (mmol/kg wet weight) in the inner medulla or rat during antidiuresis

Osmolyte	Concentration	Reference
Sorbitol	4–80	Gullans et al. (1988, 1989); Sudo and Morel (1984)
Inositol	10–15	Gullans et al. (1988, 1989); Sudo and Morel (1984)
Betaine	40	Sudo and Morel (1984)
Glycerol phosphorylcholine	40	Gullans et al. (1988, 1989); Sudo and Morel (1984)
Total amino acids	25–40	Gullans et al. (1988)

approaches the concentration needed to balance extracellular osmolality. Polyols, methylamines and amino acids are considered non-perturbing solutes that do not interfere with protein function. This is why accumulation of these organic solutes during dehydration occurs in such varied organisms such as bacteria, plants and yeast. In addition to being non-toxic to cells, methylamines in renal medullarly cells may also protect intracellular proteins from the denaturing effects of urea (Yancey and Burg 1989).

In order to be important in controlling cell volume in anisotonic solutions, the content of an intracellular solute must vary directly with changes in extracellular osmolality. Many studies have shown that sorbitol, inositol, betaine, and glycerol phosphorylcholine concentrations do indeed vary considerably between diuretic and antidiuretic states (see Garcia-Perez and Burg 1990, 1991 for reviews), suggesting that these solutes are important in regulating cell volume. Although abundant in renal medullary cells, the role of intracellular amino acids is not so clear. Intracellular amino acid concentrations measured in rat inner medulla do not vary appreciably following changes in interstitial osmolality (Gullans et al. 1988). This may suggest that amino acids act as intracellular osmolytes in the steady state but do not have a role in controlling cell volume following changes in osmolality. In contrast, in some medullary cell culture preparations, intracellular amino acid concentrations are sensitive to changes in osmolality (Roy and Sauve 1987; Olea et al. 1991), suggesting that they may be important in modulating cell volume under some conditions. Certainly, more work is needed to define the precise functional role of amino acids in the control of cell volume in the renal medulla.

4.4 The Response to Diuresis

In diuresis, interstitial Na^+ decreases to about one-half and urea to about 5% of the concentrations found in the antidiuretic animal (see Beck et al. 1980, 1985; Burger 1987; Garcia-Perez and Burg 1990, 1991). The resulting decrease in interstitial osmolality causes significant cell swelling. Renal cells

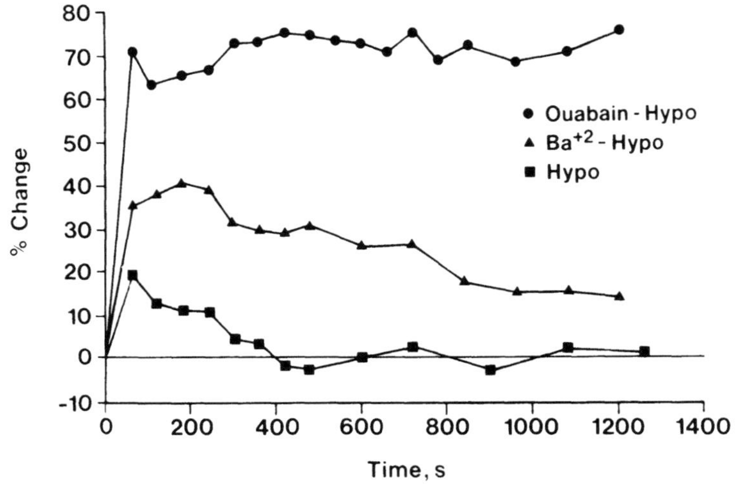

Fig. 6. Cell volume regulation following a reduction in lumen and bath osmolality by 85 mOsmol in the perfused thin descending limb of Henle's loop. Either Ba^{2+} (9 mM) or ouabain (10^{-4} M) was applied in the bath before the solutions were made hyposmotic at time = 0. Note that thin descending limb cells swell initially in hyposmotic solutions and gradually return to their original volume. Partial inhibition by Ba^{2+} and complete inhibition by ouabain of volume regulation demonstrate the importance of K^+ efflux pathways and the basolateral Na/K pump in volume-regulatory decrease. (After Lopes et al. 1988)

are surrounded on their basal side by a basement membrane which restricts changes in cell length and outer tubular diameter during changes in cell volume. Thus, when cell swelling occurs, the cells increase in height primarily by bulging into the lumen, reducing the inner tubular diameter (Amzel and Guggino 1985). Since tubular fluid flow is critically dependent upon a patent lumen, mechanisms to control cell volume in hyposmotic media are very important for normal renal function. Thus, cells in the renal medulla possess powerful volume regulatory mechanisms which restore cell volume very rapidly (see Fig. 6). These regulatory mechanisms function to rapidly lower intracellular inorganic and organic solutes.

Following a reduction in extracellular osmolality, medullary cells swell. Swelling increases the K^+ conductance and either the HCO_3^-, Cl^- or both conductances (see Fig. 7 and Montrose-Rafizadeh and Guggino 1990, for a review). The increased permeability facilitates a rapid loss of K^+ and anions. Loss of K^+ and anions in response to a reduction in osmolality is common in renal medullary cells occurring in the thin descending (Lohr et al. 1989) and ascending limbs (A. Lopes, pers. comm.) of Henle's loop and in medullary thick ascending limb cells either in the intact tubule or in cells in culture (Hebert and Sun 1988; Montrose-Rafizadeh and Guggino 1991). Details of the signal which increases the K^+ and anion permeabilities are not well understood. One possibility is a direct stretch activation of ion channels

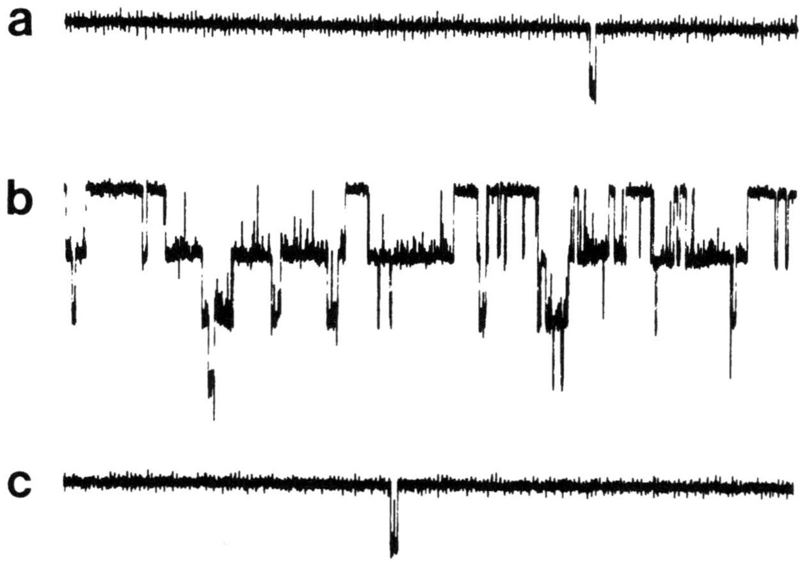

Fig. 7. Activation of Ca^{2+}-activated K^+ channels by low osmolality in cultured rabbit medullary thick ascending limb cells. *a* Cell-attached patch recording at control osmolality (293 mOsmol). *b* Cell-attached patch recording in the same cells in low osmotic solution in which the NaCl concentration was reduced by 50 mM. Note the dramatic increase in the Ca^{2+}-activated K^+ channel open probability in hyposmotic solutions. *c* is a recording from the same cells following a return to control osmolality. (After Taniguchi and Guggino 1989)

either in the apical or basolateral cell membranes. For example, swelling induces activation of Ca^{2+} channels in medullary thick ascending limb cells causing an increase in intracellular [Ca^{2+}]. The increased [Ca^{2+}] stimulates Ca^{2+}-activated K^+ channels in the apical cell membrane of the medullary thick ascending limb (Fig. 7; Taniguchi and Guggino 1989; Montrose-Rafizadeh and Guggino 1991). Direct stretch of the membrane may also induce the activation of "G" proteins or cause the release of inositol trisphosphate which could activate additional ionic pathways (for a review, see Montrose-Rafizadeh and Guggino 1990).

The release of electrolytes from medullary cells during regulatory volume decrease is a dissipating process. The Na/K pump operates to pump K^+ into the cell and establish the anion gradients across the cell membrane. Although it is well established that active transcellular transport of ions occurs only in the thick ascending and collecting ducts, all cells in the medulla including the thin descending (Lopes et al. 1988) and ascending limbs (A. Lopes pers. comm.) of Henle's loop possess Na/K pumps which are essential in volume regulation (for a review see Montrose-Rafizadeh and Guggino 1990).

The permeability of medullary cells to organic osmolytes also increases dramatically following a reduction in extracellular osmolality (Siebens and

Spring 1989; Garcia-Perez and Burg 1990, 1991; Moriyama et al. 1990; Garty et al. 1991; Yamauchi et al. 1991). For example, when cultured papillary cells (PAP-HT25) or MDCK cells are adapted to hyperosmotic solutions and subsequently exposed to hyposmotic solutions resembling that of the cortex (approx. 300 mOsmol), intracellular sorbitol, inositol, betaine, and glycerol phosphorylcholine rapidly leave causing intracellular osmolyte concentrations to drop within minutes (Moriyama et al. 1989; Siebens and Spring 1989; Garty et al. 1991; Yamauchi et al. 1991).

The mechanism of sorbitol release has been studied in detail. When cultured papillary epithelial cells are grown in 500 mOsmol and rapidly exposed to 300 mOsmol, a large transient increase in [^{14}C]sorbitol movement across the cell membrane can be detected; a clear demonstration of increased permeability (Siebens and Spring 1989). The large, swelling-induced sorbitol permeability decreases rapidly upon return to a hyperosmolar solution. Related sugars and polyols do not compete with sorbitol transport, suggesting that the increased sorbitol permeability is via a specific transporter which has been called a sorbitol permease (Siebens and Spring 1989; Garty et al. 1991). Quinidine blocks both the transporter and the ability of papillary cells to regulate volume in hyposmotic solutions, demonstrating the importance of the permease in volume regulation. Similar permeases are induced by cell swelling which are specific for the other organic osmolytes. The signal which modulates the magnitude of the organic permeases is unknown. However, it is likely that stimulation is controlled by some of the same modulators which regulate electrolyte permeability.

Osmolyte loss is also a dissipating process. Increases in membrane permeability lead to the rapid loss of osmolytes down favorable concentration gradients. Primary or secondarily active processes lead to the accumulation of organic solutes in the antidiuretic state (see below).

4.5 Response to Antidiuresis

As interstitial osmolality increases during antidiuresis, cells shrink and must accumulate solutes to return the volume back to normal. Most cells in the renal medulla regulate volume very slowly in hyperosmotic solutions. For example, thin descending limb cells exposed to hyperosmotic solutions shrink and cannot regulate their volume rapidly (see Lopes et al. 1988, for an example). Two exceptions are the mouse medullary thick ascending limb and rat medullary collecting duct (Sun and Hebert 1989). Neither segments regulate cell volume in the absence of antidiuretic hormone (Hebert 1986a,b). However, when exposed to ADH, volume regulation occurs very rapidly. Both segments regulate cell volume in hyperosmotic solutions by activating parallel Na/H and Cl/HCO_3 transporters in the basolateral cell membrane. These transporters operate together to cause the net influx of

NaCl. Interestingly, mouse medullary thick ascending limb cells do not increase the influx of NaCl via apical Na/K/Cl cotransport which is responsible for the net absorption of NaCl, the primary function of the thick ascending limb (Hebert 1986a,b). The use of different transporters allows both the mouse medullary thick ascending limb and the rat medullary collecting duct to keep net transepithelial transport and volume regulatory functions separate. The use of completely separate cotransporters to transport solutes and regulate volume may only be specific to the mouse and rat. For example, in the rabbit medullary thick ascending limb the rate of NaCl transport via the apical cotransporter increases in hyperosmotic solutions (Eveloff and Calamia 1986). Perhaps, in the rabbit, both apical and basolateral cell membranes participate in the volume-regulatory process.

The primary response of most cells in the renal papilla to an increase in extracellular osmolality is to increase intracellular organic osmolyte concentrations (Bagnasco et al. 1986; Balaban and Burg 1987; Gullans et al. 1988, 1989; Blumenfeld et al. 1989; Wirthensohn et al. 1989; Yancey and Burg 1989; Garcia-Perez and Burg 1990, 1991; Yancey et al. 1990). The accumulation of organic osmolytes is a slow process taking hours and days, involving both uptake from the bath and de novo synthesis, depending on the osmolyte. For example, when papillary epithelial cells (PAP-HT25) are exposed to hyperosmotic media, the synthesis of both sorbitol from glucose and glycerol phosphorylcholine from choline is increased, leading to increased intracellular concentrations of both solutes (Bagnasco et al. 1987; Nakanishi and Burg 1989; Uchida et al. 1989; Wirthensohn et al. 1989; Crowley et al. 1990). On the other hand, inositol and betaine are accumulated from the media (Uchida et al. 1989; Garcia-Perez and Burg 1990, 1991).

The synthesis of sorbitol is catalyzed by aldose reductase whose cDNA sequence is known (Garcia-Perez et al. 1989). Transcription of aldose reductase mRNA occurs rapidly following exposure to hyperosmotic solutions, suggesting that transcriptional promoters in the gene turn on rapidly. Aldose reductase gene promoters are thought to detect changes in intracellular electrolyte concentration rather than osmolality (Moriyama et al. 1989). Transcription is followed by an increase in translation which ultimately leads to increases in protein levels. The rate of synthesis of aldose reductase protein reaches a maximum at about 24 h followed by a gradual reduction. The reduction in protein levels after 24-h continuous exposure to hyperosmotic solutions may be controlled by a negative feedback mechanism which senses an increase in intracellular sorbitol. This mechanism would assure that inositol levels do not overshoot those required to match the increase in osmolality. Although the enzymatic processes involved in glycerol phosphorylcholine synthesis are well known, the mechanism of the regulation of synthesis in response to changes in osmolality have not been determined.

Both inositol and betaine are transported into renal medullary cells via Na-dependent cotransporters (see Yamauchi et al. 1991). In cultured MDCK cells placed in hyperosmotic solutions the influx of both osmolytes increases dramatically, reaching a maximum within 24 h. Concentrations of inositol and betaine can increase 25- and 16-fold, respectively, in hyperosmotic solutions. Subsequently, transport rates decrease slowly with continued exposure to hyperosmolality. Replaced in isotonic media, Na-dependent transport rates diminish more rapidly, reverting to baseline values in about 3 days.

4.6 Integrated Response

Clearly, the renal medulla is a special environment which normally has an osmolality well above that of other tissues in the body. It is likely that most of the time the body is chronically in an antidiuretic state. Given that the response to hyperosmotic solutions is slow, requiring both uptake and synthesis of solutes, the resting state probably is one in which osmolyte transporters and synthesis pathways along with osmotically sensitive electrolyte transporters are adapted for accumulation of organic and inorganic solutes. This helps to both maintain cell volume and protect medullary cells from the protein denaturing effects of high concentrations of urea. The accumulation of both organic and inorganic solutes requires energy to synthesize inositol and glycerol phosphorylcholine and to transport solutes into the cells. Transport of solutes depends ultimately upon the action of Na/K pumps.

In diuresis when osmolality can fall rapidly, large solute efflux pathways for organic as well as inorganic solutes are activated, triggered most likely by an increase in intracellular Ca^{2+}. If the diuresis is short-lived, then the uptake and synthesis pathways normally operative in the antidiuretic state may remain intact and respond readily to a return to an increased osmolality. This would involve a reduction in the efflux pathways and a re-accumulation of intracellular solutes. If the diuresis is sustained, uptake and synthesis pathways are down-regulated requiring days to recover.

5 Regulation of Liver Cell Volume

The following sections summarize and update previous reviews on hepatocyte volume regulation (Macknight et al. 1974; Graf et al. 1988; Macknight 1988) putting emphasis on regulation of membrane transport processes, whereas metabolic alterations during cell volume homeostasis are reviewed by Häussinger et al. (this Vol.).

5.1 Epithelial Transport in Liver

Within the liver acini, bile canaliculi form an extensively branched meshwork of narrow tubes (approx. 1 μm diameter) that are drained into the intrahepatic bile ductules located in the periportal fields. In cross sections, their circumference is formed by the apical membranes of two or three hepatocytes, the lumen being sealed from the intercellular spaces by tight junctions. Eighty-seven percent of the hepatocyte plasma membrane forms the basolateral domain that allows for extensive substrate exchange with the bloodstream. The polar distribution of membrane electrolyte transport systems correponds to that shown in Fig. 1F and includes a basolateral Cl^- conductance and Na^+-HCO_3^- symport, the latter serving as an alkali loading system. It is unclear, however, whether HCO_3^- secretion contributes to canalicular bile formation because of the high electrolyte permeability of the tight junctional seals. Bile flow is provided by the canalicular secretion of bile salts which provide the osmotic gradient to drive transepithelial water flow. In the absence of bile acids bile secretion appears to be maintained by the canalicular secretion of other organic anions, mostly glutathione. At the basolateral cell membrane, uptake of many organic solutes is driven by the inward sodium gradient through cotransport systems (e.g. bile acids, amino acids, fatty acids). Extrusion of organic anions across the apical membrane is supported by the membrane potential and/or provided by specific transport ATPases. Basolateral receptor-mediated and fluid-phase endocytosis, intracellular vesicular transport and apical exocytosis provide for the transcytotic secretion of high molecular weight compounds (for references, see Graf 1990; Boyer et al. 1992).

5.2 Changes in Liver Cell Volume Due to Altered Membrane Transport

As outlined above, steady-state cell volume is maintained through the operation of Na^+/K^+ ATPase and through the magnitude of leak fluxes that dissipate the membrane potential or ion gradients established by the pump. Consequently, *maneuvers that inhibit ATP synthesis* (cold exposure, oxygen deprivation) result in cell swelling with loss of K^+, cell membrane depolarization and intracellular accumulatuion of NaCl. Interestingly, recovery from the swollen state is slowed by ouabain, but not abolished. This ouabain-insensitive volume decrease is independent of external K^+ but requires Cl^-, is sensitive to furosemide and inhibited by cytochalasins. Morphologically, it is associated with the formation of pericanalicular vacuoles in the apical cytoplasm that appear to empty their contents into the canalicular lumen. It is thus conceivable that osmolytes and water are secreted into these vacuoles and that the vacuoles are moved to the apical membrane with the aid of microfilaments and deliver their contents by exocytosis (van Rossum and

Russo 1981; van Rossum et al. 1987). It remains unclear, however, which ATP-dependent mechanism is responsible for solute transport across the vacuolar membrane and, after their insertion into the apical membrane, for transport into the lumen. Massive biliary secretion of ouabain itself (Graf et al. 1972), ATP-dependent secretion of endogenous metabolites (see Boyer et al. 1992) or a ouabain-insensitive and K^+-independent Na^+ pump (Proverbio et al. 1988; for possible roles of microfilaments other than in exocytosis, see Kleinzeller and Ziyadeh 1990) may be involved.

In isotonic media, liver cells shrink when extracellular Cl^- is replaced by a poorly permeant anion and swell when extracellular Na^+ is substituted by K^+. In both instances membrane depolarization is seen, favouring K^+ release in the former case and Cl^- uptake in the latter (see Graf et al. 1988, 1990). Exposure of liver cells to media that contain a high concentration of *amino acids* that are taken up by Na^+-mediated cotransport leads to cell swelling. The sequence of events is envisaged as follows (Kristensen 1980, 1986; van Dyke and Scharschmidt 1983; Kristensen and Folke 1984): the high capacity of the electrogenic cotransport with Na^+ leads to membrane depolarization that supports conductive uptake of Cl^- and the gain of NaCl, together with the amino acid, results in cell swelling. Due to the rise in intracellular $[Na^+]$, activation of Na^+/K^+ ATPase ensues and, as a volume-regulatory response, K^+ conductance increases (Bear and Petersen 1987), leading to membrane hyperpolarization and conductive release of K^+ and Cl^-. The hyperpolarization also supports further amino acid uptake (Wondergem and Castillo 1988) and continuation of gluconeogenesis and glycogen synthesis ensues (Wartford 1990).

Furthermore, cellular K^+ accumulation or K^+ release by *insulin* and *glucagon*, respectively, is associated with corresponding cell volume changes (vom Dahl et al. 1991; see also Häussinger et al. this Vol., for metabolic correlates).

5.3 Regulatory Volume Decrease After Hypotonic Stress

Due to high osmotic membrane water permeability, liver cells rapily shrink or swell when exposed to anisotonic media (Graf et al. 1990). Numerous recent studies, mostly performed in the rat, have been concerned with the alterations of membrane transport and of cell metabolism during RVD after hypotonic stress: Graf et al. (1988), Haddad and Graf (1989), Lang et al. (1989), Häussinger et al. (1990a,b), Häussinger and Lang (1990), Wartford (1990), Corasanti et al. (1990), Gleeson et al. (1990), Bear (1990), Haddad et al. (1991). The following gives a brief summary of these data: membrane distension by cell swelling leads to opening of non-selected stretch-activated cation channels that may allow for influx of Ca^{2+}, resulting in a rise in intracellular $[Ca^{2+}]$. Membrane K^+ conductance rises, resulting in K^+ release and membrane hyperpolarization. Membrane Cl^- conductance also rises, facilitating efflux of Cl^- that is supported by transient membrane hyperpolarization. Intracellular pH decreases, perhaps due to decreased activity of Na^+/H^+ exchange, release and dilution of intracellular HCO_3^-. Metabolic changes associated with cell swelling include: activation of glycogen synthesis, decrease in glucose release, decrease in glycolysis and in release of lactate and pyruvate, stimulation of lipogenesis, stimulation of protein synthesis/decrease in protein catabolism, increase in amino acid catabolism and stimulation of glutaminase/inhibition of glutamine synthetase and decrease in urea production from ammonia (see Häussinger et al., this Vol.). Thus far, it is unclear which intracellular signals mediate these volume-regulatory or volume-induced responses. For rat liver it is uncertain whether internal $[Ca^{2+}]$ accounts for the changes in membrane transport (see Boyer et al. 1992) and whether the pattern of metabolic changes largely deviates from effects induced by hormones that involve Ca^{2+} in their transduction pathway (see Exton 1988). Some metabolic changes could be explained by internal acidification (see Häussinger et al., this Vol.), but a fall in pH_i decreases rather than increases K^+ conductance.

5.4 Regulatory Volume Decrease, Bile Production and Vesicular Transport

In the isolated perfused rat liver, hypotonic cell swelling and subsequent RVD are associated with characteristic changes in bile flow and vesicular transport. Shifting the perfusion to hypotonic medium results in instantaneous, but transient, stimulation of bile flow followed by a second phase of more prolonged choleresis. The first stimulation is apparently due to rapid transepithelial osmotic equilibration by water flow into the bile canaliculi. Maximal bile flow in the second phase is 3 to 5 min after initiation of hypotonic stress and thus corresponds in time with the maximal release of K^+ during RVD. The increase in bile flow correlates with the exocytotic canalicular secretion of horse-radish peroxidase which was used as a fluid-phase marker for the transcytotic vesicular pathway. Furthermore, secretion of taurocholate into bile is stimulated during RVD. These effects are blunted by colchicine and also by blocking K^+ conductance and RVD with barium. RVD thus appears to involve microtubule-dependent exocytosis at the canalicular membrane that may also serve insertion into the apical cell membrane of bile acid carriers and K^+ channels (Bruck et al. 1990). RVD is also associated with the exocytotic release of FITC-dextran into the perfusate when the liver has been loaded with this fluid-phase marker prior to initiation of hypotonic stress (Fig. 8; H. Klapper, R. Fuchs, J. Graf unpubl.). These observations indicate that, during RVD, membrane recycling from an endosomal compartment is increased at the basolateral membrane domain, a mechanism that could serve the insertion of K^+ and Cl^- channels and thus be responsible for the increase in K^+ and Cl^- conductances. Thus,

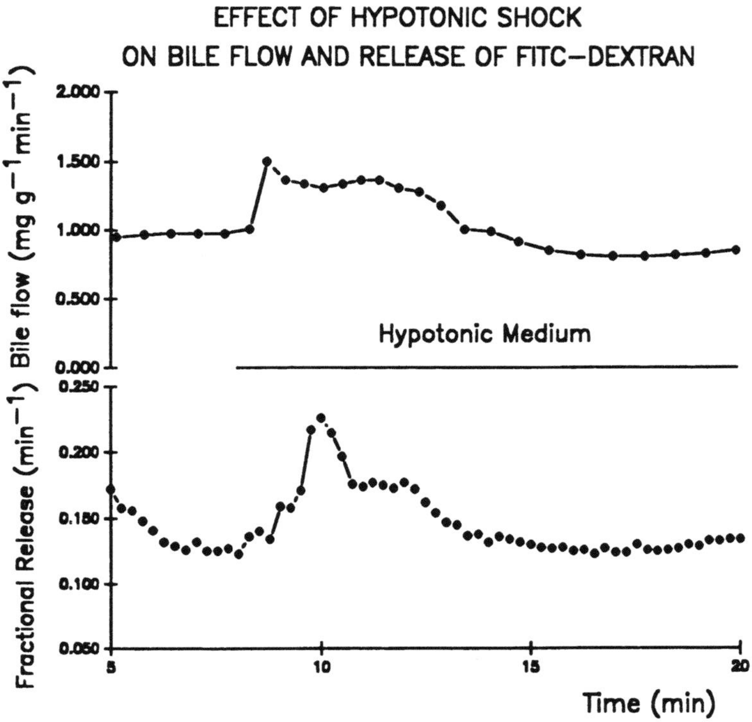

Fig. 8. Bile flow and release of the fluid-phase marker FITC-dextran were determined in the isolated perfused rat liver during control perfusion and after shifting the perfusion to a hypotonic medium (omission of 40 mM NaCl). Note the biphasic increase in bile flow (*upper trace*). Efflux of dextran into the perfusion medium was determined after loading the organ with the marker for 5 min and then shifting to a marker-free medium (time 0). The beginning of the *lower trace* shows the tail of the rapid washout of the extracellular space followed by cellular release at a rate of approx. 12% of cellular content per min. Note transient doubling of the efflux rate after shifting to hypotonic perfusion

the concept of regulation of transport by membrane insertion and retrieval (e.g. Lienhardt 1983) may not only apply for modulation of transepithelial transport (e.g. Mircheff et al. 1989), but also for volume-regulatory responses or for the correlated activity of these transport processes (cf. O'Neil 1989; Wills et al. 1991).

5.5 Regulatory Volume Increase

RVI in liver cells is elicited by returning the cells to isotonic medium following their swelling and subsequent RVD in hypotonic medium (pseudo-regulatory volume increase). By combining data obtained from measurements of cell volume, cell pH, ion fluxes and effects of transport inhibitors and of altered medium ion composition, liver cells appear to gain KCl (and water) by initial activation of basolateral Na^+/H^+ exchange and, apparently of minor importance, by Na^+-K^+-Cl^- symport. The rise in intracellular [Na^+] appears to be the stimulus for increased activity of Na^+/K^+ ATPase, leading to exchange of intracellular Na^+ by K^+. Simultaneously, Ba^{2+}-sensitive K^+ conductance is reduced, preventing loss of K^+ and causing membrane depolarization. Intracellular alkalinization by Na^+/H^+ exchange appears to activate the apical Cl^-/HCO_3^- exchange, resulting in uptake of Cl^-, its release being reduced through membrane depolarization (for references, see Graf et al. 1988; Lang et al. 1989; Häussinger et al. 1990b; Corasanti et al. 1990; Gleeson et al. 1990). As discussed by Häussinger (this Vol.), liver cell shrinkage is associated with metabolic changes that include stimulation of glycogenolysis, glucose release, glycolysis and release of lactate and pyruvate.

6 Conclusion

Cell volume in transporting epithelia is regulated by gain or loss of intracellular osmolytes. The regulatory mechanisms involve modification of apical and basolateral membrane transport and metabolic changes of the intracellular amount of organic solutes. In secretory and absorptive epithelia the rate of transcellular transport may vary over a large range depending on hormonal or neural stimuli or on substrate load. In these epithelia, up-regulation or down-regulation of transport across apical and basolateral membranes is appropriately matched in order to maintain cell volume and intracellular ion composition within a narrow range. Epithelial cells in the renal medulla are exposed to large variations of the extracellular osmolality and maintain their volume through appropriate metabolic generation, uptake or release of organic solutes which are not substrates for transcellular transport.

Regulation of cell volume requires mechanisms that sense changes in cell size and that generate signals to alter membrane transport and cell metabolism.

We have reviewed the mechanisms that provide for transcellular transport and that are altered during epithelial cell volume regulation and have summarized the adaptive regulation of organic solutes in renal medullary cells. Depending on the mechanisms of their activation, these volume-regulatory responses may be initiated instantaneously (ion channels activated by membrane stretch, alteration of transport through changes in substrate load or by allosteric modification) or may require minutes to hours for full activation (modulation of membrane transport proteins through intracellular signals, membrane insertion of transporters, changes in translation or transcription). Although phenomenological descriptions of a multitude of cellular responses to deviations from normal cell volume are available for various transporting epithelia, information on volume-sensing mechanisms and on signals that activate these volume-regulatory responses remains largely fragmentary.

References

Adorante JS, Miller SS (1990) Potassium-dependent volume regulation in retinal pigment epithelium is mediated by Na,K,Cl cotransport. J Gen Physiol 96: 1153–1176

Albus M, Bakker R, Van Heukelom R (1983) Circuit analysis of membrane potential changes due to electrogenic sodium-dependent sugar transport in goldfish intestinal epithelium. Pflügers Arch Eur J Physiol 398: 1–9

Allen JC, Mills JW (1988) Two lines of MDCK epithelial cells with different volume and ion responses to calcium ionophore A23187. In Vitro Cell Dev Biol 24: 588–592

Amzel M, Guggino WB (1985) Relationship between cell volume and ion transport in the early distal tubule of *Amphiuma* kidney. J Membr Biol 86: 31–58

Andersen OS (1989) Elementary aspects of acid-base permeation and pH regulation. Ann NY Acad Sci 574: 333–353

Armstrong WMcD, Musselman DL, Reitzung HC (1970) Potassium and water content of isolated bullfrog small intestinal epithelia. Am J Physiol 219: 1023–1026

Aronson PS, Sacktor B (1975) The Na^+ gradient-dependent transport of D-glucose in renal brush-border membranes. J Biol Chem 250: 5032–6039

Bagnasco S, Balaban RS, Fales HM, Yang Y-M, Burg M (1986) Predominant osmotically active organic solutes in rat and rabbit renal medullas. J Biol Chem 261: 5872–5877

Bagnasco S, Uchida S, Balaban RS, Kador PF, Burg M (1987) Induction of aldose reductase and sorbitol in renal inner medullary cells by elevated extracellular NaCl. Proc Natl Acad Sci 84: 1718–1720

Bakker-Grunwald T (1983) Potassium permeability and volume control in isolated hepatocytes. Biochim Biophys Acta 731: 239–242

Balaban RS, Burg M (1987) Osmotically active organic solutes in the renal inner medulla. Kidney Int 31: 562–564

Bear CE (1990) A nonselective cation channel in rat liver cells is activated by membrane stretch. Am J Physiol 258: C421–C428

Bear CE, Petersen OH (1987) L-alanine evokes opening of single Ca^{2+}-activated K^+ channels in rat liver cells. Pflügers Arch Eur J Physiol 410: 342–344

Beck JS, Potts DJ (1990) Cell swelling cotransport activation and potassium conductance in isolated perfused rabbit kidney proximal tubules. J Physiol (Lond) 425: 363–378

Beck F, Bauer R, Bauer U, Mason J, Dörge A, Rick R, Thurau K (1980) Electron microprobe analysis of intracellular elements in the rat kidney. Kidney Int 17: 756–763

Beck F, Dörge A, Rick R, Thurau K (1984) Intra- and extracellular element concentrations of rat renal papilla in antidiuresis. Kidney Int 25: 397–403

Beck F, Dörge A, Thurau K (1985) Osmoregulation of renal papillary cells. Pflügers Arch Eur J Physiol 405 (Suppl 1): S28–S32

Beck F, Dörge A, Ring T, Sauer M (1989) Element composition of tubule cells in the inner stripe of the renal outer medulla. Miner Electrolyte Metab 15: 144–149

Beck FX, Dörge A, Thurau K, Guder WG (1990) Cell osmoregulation in the countercurrent system of the renal medulla: the role of organic osmolytes. In: Beyenbach KW (ed) Comparative physiology, vol 4: Cell volume regulation. Karger, Basel, pp 132–158

Beck JS, Breton S, Laprade R, Giebisch G (1991) Volume regulation and intracellular calcium in the rabbit proximal convoluted tubule. Am J Physiol 260: F861–F867

Beebe DC, Parmelee JT, Belcher KS (1990) Volume regulation in lens epithelial cells and differentiating lens fiber cells. J Cell Physiol 143: 455–459

Bevan C, Theiss C, Kinne RK (1990) The role of Ca^{2+} in sorbitol release from rat inner medullary collecting duct cells under hypoosmotic stress. Biochem Biophys Res Commun 170: 563–568
Blumenfeld JD, Hebert SC, Heilig CW, Balschi JA, Stromski ME, Gullans S (1989) Organic osmolytes in inner medulla of Brattleboro rat: effects of ADH and dehydration. Am J Physiol 256: F916–F922
Bonanno JA (1991) K^+-H^+, a fundamental cell acidifier in corneal epithelium. Am J Physiol 260: C618–625
Boron WF, Sackin H (1983) Measurement of intracellular ionic composition and activities in renal tubules. Annu Rev Physiol 45: 483–496
Boyer JL, Graf J, Meier PJ (1992) Hepatic transport systems regulating pH_i, cell volume and bile secretion. Annu Rev Physiol (in press)
Brown PD, Burton KA, Sepúlveda FV (1983) The transport of sugars or amino acids increases potassium efflux from isolated enterocytes. FEBS Lett 163: 203–206
Bruck R, Haddad P, Graf J, Boyer JL (1990) Regulatory volume decrease following hypotonic stress stimulates bile flow, bile acid excretion and exocytosis in the isolated perfused rat liver. Hepatology 12: 888
Burger RE (1987) Composition of renal medullary tissue. Kidney Int 31: 556–561
Butt AG, Clapp WL, Frizzell RA (1990) Potassium conductance in tracheal epithelium activated by secretion and cell swelling. Am J Physiol 25: C630–C638
Cala PM (1977) Volume regulation by flounder red blood cells in anisotonic media. J Gen Physiol 67: 537–552
Candia OA (1989) Regulatory mechanisms of chloride transport in corneal epithelium. Ann NY Acad Sci 574: 416–427
Case RM, Argent BE (1989) Pancreatic secretion of electrolytes and water. In: Schultz SG (ed) Handbook of physiology, Section 6, vol III. American Physiological Society, Bethesda, pp 383–417
Chase HS Jr, Al-Awqati Q (1981) Regulation of the sodium permeability of luminal border of toad bladder by intracellular sodium and calcium. J Gen Physiol 77: 693–712
Christensen O (1987) Mediation of cell volume regulation by Ca influx through stretch activated channels. Nature 330: 66–68
Christensen O, Zeuthen T (1987) Maxi K^+ channels in leaky epithelia are regulated by intracellular Ca^{2+}, pH and membrane potential. Pflügers Arch Eur J Physiol 408: 249–256
Christensen O, Simon M, Randlev T (1989) Anion channels in a leaky epithelium. A patch clamp study of choroid plexus. Pflügers Arch Eur J Physiol 415: 37–46
Civan MM, Rubenstein D, Mauro T, O'Brien TG (1985) Effects of tumor promoters on sodium ion transport across frog skin. Am J Physiol 248: C457–C465
Civan MM, Cragoe EJ Jr, Peterson-Yantorno K (1988) Intracellular pH in frog skin: effects of Na^+, volume and cAMP. Am J Physiol 255: F126–F134
Clausen C, Dixon TE (1986) Membrane electrical parameters in turtle bladder measured using impedance-analysis techniques. J Membr Biol 92: 9–19
Clauss W, Dürr D, Guth S, Skadhauge E (1987) Effects of adrenal steroids on Na transport in lower intestine (coprodeum) of the hen. J Membr Biol 96: 141–152
Corasanti JG, Gleeson D, Boyer JL (1990) Effects of osmotic stresses on isolated rat hepatocytes. I. Ionic mechanisms of cell volume regulation. Am J Physiol 258: G290–G298
Cornet M, Delpire E, Gilles R (1987) Study of microfilaments network during volume regulation process of cultured PC 12 cells. Pflügers Arch Eur J Physiol 410: 223–225
Cornet M, Delpire E, Gilles R (1988) Relation between cell volume control, microfilaments and microtubules networks in T2 and PC12 cultured cells. J Physiol (Paris) 83: 43–49
Costa PM, Fernandes PL, Ferreira HG, Ferreira KT, Giraldez F (1987) Effect of cell volume changes on membrane ionic permeabilities and sodium transport in frog skin (*Rana ridibunda*). J Physiol (Lond) 393: 1–17

Costantin J, Alcalen S, Otero ADS, Dubinski WP, Schultz SG (1989) Reconstitution of an inwardly rectifying potassium channel from the basolateral membranes of *Necturus* enterocytes into planar lipid bilayers. Proc Natl Acad Sci USA 86: 5212–5216
Crowley BD, Ferraris JD, Carper D, Burg M (1990) In vivo osmoregulation of aldose reductase mRNA protein, and sorbitol in renal medulla. Am J Physiol 258: F154–F161
Csaky TZ, Esposito G (1969) Osmotic swelling of intestinal epithelial cells during active sugar transport. Am J Physiol 217: 753–755
Davis CW, Finn AL (1982) Sodium transport inhibition by amiloride reduces basolateral membrane potassium conductance in tight epithelia. Science 216: 525–527
Davis CW, Finn AL (1985a) Cell volume regulation in frog urinary bladder. Fed Proc 44: 2520–2525
Davis CW, Finn AL (1985b) Effects of mucosal sodium removal on cell volume in *Necturus* gallbladder epithelium. Am J Physiol 249: C304–C312
Davis CW, Finn AL (1987) Interaction of sodium transport, cell volume, and calcium in frog urinary bladder. J Gen Physiol 89: 687–702
Davis CW, Finn AL (1988) Potassium induced cell swelling in *Necturus* gallbladder epithelium. Am J Physiol 254: C643–C650
Dawson DC (1987) Properties of epithelial potassium channels. Curr Top Membr Transp 28: 41–72
Dawson DC, Richards NW (1990) Basolateral K conductance: role in regulation of NaCl absorption and secretion. Am J Physiol 259: C181–C195
Demarest JR, Loo DDF (1990) Electrophysiology of the parietal cell. Annu Rev Physiol 52: 307–319
Diamond JM (1982) Transcellular cross-talk between epithelial cell membranes. Nature 300: 683–685
Donaldson PJ, Chen LK, Lewis SA (1989) Effects of serosal anion composition on the permeability properties of rabbit urinary bladder. Am J Physiol 256: F1125–F1134
Dubé L, Parent L, Sauvé R (1990) Hypotonic shock activates a maxi K^+ channel in primary cultured proximal tubule cells. Am J Physiol 259: F348–F356
Eaton DC, Frace AM, Silverthorn SU (1982) Active and passive Na^+ fluxes across the basolateral membrane of rabbit urinary bladder. J Membr Biol 67: 219–229
Epstein FH, Silva P (1985) Na-K-Cl cotransport in chloride transporting epithelia. Ann NY Acad Sci 456: 187–197
Ericson A-C, Spring KR (1982a) Coupled NaCl entry into *Necturus* gallbladder epithelial cells. Am J Physiol 243: C140–C145
Ericson AC, Spring KR (1982b) Volume regulation by *Necturus* gallbladder: apical Na^+-H^+ and Cl^--HCO_3^- exchange. Am J Physiol 243: C146–C150
Eveloff J, Calamia J (1986) Effect of osmolarity on cation fluxes in medullary thick ascending limb cells. Am J Physiol 250: F176–F180
Eveloff JL, Warnock DG (1987) Activation of ion transport systems during cell volume regulation. Am J Physiol 252: F1–F10
Exton JH (1988) Role of inositol triphosphate and diacylglycerol in the regulation of liver function. In: Arias IM, Jakoby WB, Popper H, Schachter D, Shafritz DA (eds) The liver: biology and pathology. Raven, New York, pp 785–792
Farahbakhsh NA, Fain GL (1987) Volume regulation of non-pigmented cells from ciliary epithelium. Invest Ophthalmol Visual Sci 28: 934–944
Field MJ, Giebisch G (1989) Mechanisms of segmental potassium reabsorption and secretion. In: Seldin DW, Giebisch G (eds) The regulation of potassium balance. Raven, New York, pp 139–155
Filipovic D, Sackin H (1991) A calcium-permeable stretch-activated cation channel in renal proximal tubule. Am J Physiol 260: F119–F129
Finn AL, Reuss L (1975) Effects of changes in the ionic composition of the serosal solution on the electrical properties of the toad urinary bladder epithelium. J Physiol (Lond) 250: 541–548
Fisher RS, Baxendale LM, Helman SI (1986) Sustained increases of apical membrane channel density after inhibition of transport in frog skin. Fed Proc (Abstr) 45: 516

Forte JG, Hanzel DK, Urushidani T, Wolosin JM (1989) Pumps and pathways for gastric HCl secretion. Ann NY Acad Sci 574: 145–158
Foskett JK (1990) $[Ca^{2+}]_i$ modulation of Cl^- content controls cell volume in single salivary acinar cells during fluid secretion. Am J Physiol 259: C998–C1004
Foskett JK, Melvin JE (1989) Activation of salivary secretion: coupling of cell volume and $[Ca^{2+}]_i$ in single cells. Science 244: 1582–1585
Foskett JK, Spring KR (1985) Involvement of calcium and cytoskeleton in gallbladder epithelial cell volume regulation. Am J Physiol 248: C27–C36
Foskett JK, Ussing HH (1986) Localization of chloride conductance to mitochondria-rich cells in frog skin epithelium. J Membr Biol 91: 251–258
Frömter E, Gebler B (1977) Electrical properties of amphibian urinary bladder epithelium. III. The cell membrane resistances and the effect of amiloride. Pflügers Arch Eur J Physiol 371: 99–108
Frömter E, Curci S, Gitter AH (1990) Electrophysiological aspects of gastric ion transport. In: Young JA, Wong PYD (eds) Epithelial secretion of water and electrolytes. Springer, Berlin Heidelberg New York, pp 293–311
Furlong TJ, Spring KR (1990) Mechanisms underlying volume regulatory decrease by *Necturus* gallbladder epithelium. Am J Physiol 258: C1016–C1024
Garcia-Perez A, Burg M (1990) Importance of organic osmolytes for osmoregulation by renal medullary cells. Hypertension 16: 595–602
Garcia-Perez A, Burg M (1991) Role of organic osmolytes in adaptation of renal cells to high osmolality. J Membr Biol 119: 1–13
Garcia-Perez A, Martin B, Murphy HR, Uchida S, Murer H, Cowley BD, Handler JS, Burg M (1989) Molecular cloning of cDNA coding for kidney aldose reductase. J Biol Chem 28: 16815–16821
Garty H (1986) Mechanism of aldosterone action in tight epithelia. J Membr Biol 90: 193–204
Garty H, Furlong TJ, Ellis DE, Spring KR (1991) Sorbitol permease: an apical membrane transporter in cultured renal papillary epithelial cells. Am J Physiol 260: F650–F656
Germann WJ, Ernst SA, Dawson DC (1986) Resting and osmotically induced basolateral K conductances in turtle colon. J Gen Physiol 88: 253–274
Giebisch GH (1987) Cell models of potassium transport in the renal tubule. Curr Top Membr Transp 28: 133–184
Giraldez F, Sepúlveda FV (1987) Changes of the apparent chloride permeability of *Necturus* enterocytes during the sodium-coupled transport of alanine. Biochim Biophys Acta 898: 248–252
Giraldez F, Valverde MA, Sepúlveda FV (1988) Hypotonicity increases apical membrane Cl^- conductance in *Necturus* enterocytes. Biochim Biophys Acta 942: 353–356
Gleeson D, Corasanti JG, Boyer JL (1990) Effects of osmotic stresses on isolated hepatocytes. II. Modulation of intracellular pH. Am J Physiol 258: G299–G307
Graf J (1990) Hepatic electrolyte transport and bile formation. In: Young JA, Wong PYD (eds) Epithelial Secretion of water and electrolytes. Springer, Berlin Heidelberg New York, pp 313–332
Graf J, Korn P, Peterlik M (1972) Choleretic effects of ouabain and ethacrynic acid in the isolated perfused rat liver. Naunyn Schmiedeberg's Arch Pharmacol 272: 230–233
Graf J, Haddad P, Häussinger D, Lang F (1988) Cell volume regulation in liver. Renal Physiol Biochem 11: 202–220
Graf J, Schild L, Boyer JL, Giebisch G (1990) Volumetric analysis of water and solute fluxes in isolated rat hepatocyte couplets. Renal Physiol Biochem 13: 166
Grasset E, Gunter-Smith P, Schultz SG (1983) Effects of Na-coupled alanine transport on intracellular K activities and K conductance of the basolateral membranes of *Necturus* small intestine. J Membr Biol 71: 89–94
Greger R (1985) Ion transport mechanisms in thick ascending limb of Henle's loop of mammalian nephron. Physiol Rev 65: 760–797
Greger R, Schlatter E, Wang F, Forrest JN Jr (1984) Mechanisms of NaCl secretion in rectal gland tubules of spiny dogfish (*Squalus acanthias*). III. Effects of stimulation of secretion by cyclic AMP. Pflügers Arch Eur J Physiol 402: 376–384

Grinstein S, Rothstein A, Sarkadi B, Gelfand EW (1984) Responses of lymphocytes to anisotonic media: volume-regulatory behavior. Am J Physiol 246: C204–C215

Grinstein S, Goetz-Smith JD, Steward D, Beresford BJ, Mellors A (1986) Protein phosphorylation during activation of Na^+/H^+ exchange by phorbol esters and by osmotic shrinking: possible relation to cell pH and volume regulation. J Biol Chem 261: 8009–8016

Gullans S, Blumenfeld JD, Balschi JA, Kaleta M, Brenner RM, Heilig CW, Hebert SC (1988) Accumulation of major organic osmolytes in rat renal inner medulla in dehydration. Am J Physiol 255: R626–R634

Gullans S, Heilig CW Stromski ME, Blumenfeld JD (1989) Methylamines and polyols in kidney, urinary bladder, urine, liver, brain, and plasma. Renal Physiol Biochem 12: 191–201

Gunther-Smith P, Grasset E, Schultz SG (1982) Sodium coupled amino acid and sugar transport by *Necturus* small intestine: an equivalent electrical circuit analysis of a rheogenic co-transport system. J Membr Biol 66: 25–39

Haddad P, Graf J (1989) Volume-regulatory K^+ fluxes in the isolated perfused rat liver: characterization by ion transport inhibitors. Am J Physiol 257: G357–G363

Haddad P, Thalhammer T, Graf J (1989) Effect of hypertonic stress on liver cell volume, bile flow and volume regulatory K^+ fluxes. Am J Physiol 256: G563–G569

Haddad P, Beck JS, Boyer JL, Graf J (1991) Role of chloride ions in liver cell volume regulation. Am J Physiol 261: G340–G348

Häussinger D, Lang F (1990) Exposure of perfused liver to hypotonic conditions modifies cellular nitrogen metabolism. J Cell Biochem 43: 355–361

Häussinger D, Hallbrucker C, vom Dahl S, Lang F, Gerok W (1990a) Cell swelling inhibits proteolysis in perfused rat liver. Biochem J 272: 239–242

Häussinger D, Lang F, Bauers K, Gerok W (1990b) Control of hepatic nitrogen metabolism and glutathione release by cell volume regulatory mechanisms. Eur J Biochem 193: 891–898

Häussinger D, Stehle T, Lang F (1990c) Volume regulation in liver: further characterization by inhibitors and ionic substitutions. Hepatology 11: 243–254

Hazama A, Okada Y (1988) Ca^{2+} sensitivity of volume regulatory K^+ and Cl^- channels in cultured human epithelial cells. J Physiol (Lond) 402: 687–702

Hazama A, Okada Y (1990) Biphasic rises in cytosolic free Ca^{2+} in association with activation of K^+ and Cl^- conductance during regulatory volume decrease in cultured human epithelial cells. Pflügers Arch Eur J Physiol 416: 710–714

He XL, Ship J, Wu XZ, Brown AM, Wellner RB (1989) Beta-adrenergic control of cell volume and chloride transport in an established rat submandibular cell line. J Cell Physiol 138: 527–535

Hebert SC (1986a) Hypertonic cell volume regulation in mouse thick limbs. I. ADH dependency and nephron heterogeneity. Am J Physiol 250: C907–C919

Hebert SC (1986b) Hypertonic cell volume regualtion in mouse thick limbs. II. Na^+-H^+ and Cl^--HCO_3^- exchange in basolateral membranes. Am J Physiol 250: C920–C931

Hebert SC, Sun A (1988) Hypotonic cell volume regulation in mouse medullary thick ascending limb: effects of ADH. Am J Physiol 255: F962–F969

Helander HF, Leth R, Olbe L (1986) Stereological investigations on human gastric mucosa. I. Normal oxyntic mucosa. Anat Rec 216: 373–380

Helman SI, Fisher RS (1977) Microelectrode studies of the active Na transport pathways of frog skin. J Gen Physiol 69: 571–604

Helman SI, Nageö W, Fisher RS (1979) Ouabain on active transepithelial Na transport in frog skin: studies with microelectrodes. J Gen Physiol 74: 105–127

Hermansson K, Spring KR (1986) Potassium induced changes in cell volume of gallbladder epithelium. Pflügers Arch Eur J Physiol 407: S90–S99

Higgins TJ, Gebler B, Frömter E (1977) Electrical properties of amphibian urinary bladder epithelium. I. The cell potential profile of *Necturus maculosus*. Pflügers Arch Eur J Physiol 371: 87–97

Hoffmann EK, Simonsen LO (1989) Membrane mechanisms in volume and pH regulation in vertebrate cells. Physiol Rev 69: 315–382

Hoffmann EK, Schiodt M, Dunham P (1986) The number of chloride-cation cotransport sites on Ehrlich ascites cells measured with [^{3}H] bumetanide. Am J Physiol 250: C688–C693

Howard LD, Wondergem R (1987) Effects of anisosmotic medium on cell volume, transmembrane potential and intracellular K^+ activity in mouse hepatocytes. J Membr Biol 100: 53–61

Hudson RL, Schultz SG (1984) Effects of sodium-coupled sugar transport on intracellular sodium activities and sodium pump activity in *Necturus* small intestine. Science 224: 1237–1239

Hudson RL, Schultz SG (1988) Sodium-glycine uptake by Ehrlich ascites tumor cells results in an increase in cell volume and plasma membrane channel activities. Proc Natl Acad Sci USA 85: 279–283

Hunter M (1990) Stretch-activated channels in the basolateral membrane of single proximal cells of frog kidney. Pflügers Arch Eur J Physiol 416: 448–453

Jensen PK, Fisher RS, Spring KR (1984) Feedback inhibition of NaCl entry in *Necturus* gallbladder epithelial cells. J Membr Biol 82: 95–104

Jorgensen PL (1980) Sodium and potassium ion pump in kidney tubules. Physiol Rev 60: 864–917

Kanno T (1990) The metabolic basis of secretion by the exocrine pancreas. In: Young JA, Wong PYD (eds) Epithelial secretion of water and electrolytes. Springer, Berlin Heidelberg New York, pp 267–275

Kessler M, Semenza AG (1983) The small intestinal Na^+, D-glucose cotransporter: an asymmetric gated channel (or pore) responsive to *Delta Psi*. J Membr Biol 76: 27–56

Kimmich GA, Randles J (1988) Na^+-coupled sugar cotransport: membrane potential-dependent Km and Ki for Na^+. Am J Physiol 255: C485–C494

Kinne R, Hannafin JA, König B (1985) Role of NaCl-KCl cotransport system in active chloride absorption and secretion. Ann NY Acad Sci 456: 198–206

Kirk KL, Halm DR, Dawson DC (1980) Active sodium transport by turtle colon via an electrogenic Na-K exchange pump. Nature 287: 237–239

Kirk KL, Schafer JA, DiBona DR (1987) Cell volume regulation in rabbit proximal straight tubule perfused in vitro. Am J Physiol 252: F922–F932

Kleinzeller A, Mills JW (1989) K^+-induced cell swelling of the dogfish shark (*Squalus acanthias*) rectal gland cells is associated with changes of the cytoskeleton. Biochim Biophys Acta 1014: 40–52

Kleinzeller A, Ziyadeh FN (1990) Cell volume regulation in epithelia – with emphasis on the role of osmolytes and the cytoskeleton. In: Beyenbach KW (ed) Cell volume regulation, vol IV. Comp Physiol. Karger, Basel, pp 59–86

Kleinzeller A, Booz GW, Mills JW, Ziyadeh FN (1990) pCMBS-induced swelling of dogfish (*Squalus acanthias*) rectal gland cells: role of the Na^+-K^+-ATPase and the cytoskeleton. Biochim Biophys Acta 1025: 21–31

Knickelbein R, Aronson PS, Atherton W, Dobbins JW (1983) Sodium and chloride transport across rabbit ileal brush border. I. Evidence for Na-H-exchange. Am J Physiol 245: G504–G510

Knickelbein R, Aronson PS, Schron CM, Seifter J, Dobbins J (1985) Sodium and chloride transport across rabbit ileal brush border. II. Evidence for Cl-HCO_3 exchange and mechanism of coupling. Am J Physiol 249: G236–G245

Koefoed-Johnsen U, Ussing HH (1958) The nature of the frog skin potential. Acta Physiol Scand 42: 298–308

Kristensen LO (1980) Energization of alanine transport in isolated rat hepatocytes. Electrogenic Na^+-alanine cotransport leading to increased K^+ permeability. J Biol Chem 255: 5236–5243

Kristensen LO (1986) Associations between transport of alanine and cations across cell membrane in rat hepatocytes. Am J Physiol 251: G575–G584

Kristensen LO, Folke M (1984) Volume regulatory K^+ efflux during concentrative uptake of alanine in isolated rat hepatocytes. Biochem J 221: 265–268

Lambert IH (1989) Leukotriene-D4 induced cell shrinkage in Ehrlich ascites tumor cells. J Membr Biol 108: 165–176

Lambert IH, Hoffmann EK, Christensen P (1987) Role of prostaglandins and leukotrienes in volume regulation by Ehrlich ascites tumor cells. J Membr Biol 98: 247–256
Lang F, Rehwald W (1992) Potassium channels in renal epithelial transport regulation. Physiol Rev 72: 1–32
Lang F, Messner G, Rehwald W (1986) Electrophysiology of sodium-coupled transport in proximal renal tubules. Am J Physiol 250: F953–F962
Lang F, Macknight ADC, Ballanyi K, Völkl H, Beck F, Reuss L, Graf J, Hoffmannn EK, Lauf PK, Deutsch C, Gilles R (1988) Cell volume regulation. Renal Physiol Biochem 3: 114–288
Lang F, Stehle T, Häussinger D (1989) Water, K^+, H^+, lactate and glucose fluxes during cell volume regulation on perfused rat liver. Pflügers Arch Eur J Physiol 413: 209–216
Lapointe JY, Garneau L, Bell PD, Cardinal J (1990) Membrane cross talk in the proximal tubule during alterations in transepithelial sodium transport. Am J Physiol 258: F339–F345
Larsen EH, Ussing HH, Spring KR (1987) Ion transport in mitochondria-rich cells in toad skin. J Membr Biol 99: 25–40
Larson M, Spring KR (1984) Volume regulation by *Necturus* gallbladder: basolateral KCl exit. J Membr Biol 81: 219–232
Larson M, Spring KR (1987) Volume regulation in epithelia. Curr Top Membr Transp 30: 105–120
Lau KR, Hudson RL, Schultz SG (1984) Cell swelling increases a barium inhibitable potassium conductance in the basolateral membrane of *Necturus* small intestine. Proc Natl Acad Sci USA 81: 3591–3594
Lau KR, Hudson RL, Schultz SG (1986) Effect of hypertonicity on the increase in basolateral conductance of *Necturus* small intestine in response to Na^+-sugar cotransport. Biochim Biophys Acta 855: 193–196
Lau KR, Elliott AC, Brown PD, Case RM (1990) Bicarbonate transport by salivary gland acinar cells. In: Young JA, Wong PYD (eds) Epithelial secretion of water and electrolytes. Springer, Berlin Heidelberg New York, pp 171–187
Leblanc G, Morel F (1975) Na and K movements across the membrane of frog skin epithelium associated with transient current changes. Pflügers Arch Eur J Physiol 358: 159–177
Lechene C (1985) Cell volume and cytoplasmic gel. Biol Cell 55: 177–180
Lee CO, Armstrong WMcD (1972) Activities of sodium and potassium ions in epithelial cells of small intestine. Science 175: 1261–1264
Leibowich S, DeLong J, Civan MM (1988) Apical Na^+ permeability of frog skin during serosal Cl^- replacement. J Membr Biol 102: 121–130
Lewis SA, DeMoura JLC (1982) Incorporation of cytoplasmic vesicles into apical membrane of mammalian urinary bladder epithelium. Nature 297: 675–688
Lewis SA, DeMoura JLC (1984) Apical membrane area of rabbit urinary bladder increases by fusion of intracellular vesicles: an electrophysiological study. J Membr Biol 82: 123–136
Lewis SA, Diamond JM (1976) Na^+ transport by rabbit urinary bladder, a tight epithelium. J Membr Biol 28: 1–40
Lewis SA, Donaldson P (1989) Sodium dependence of cation permeabilities and transport. Curr Top Membr Transp 34: 83–104
Lewis SA, Donaldson P (1990) Ion channels and cell volume regulation: chaos in an organized system. News Physiol Sci 5: 112–119
Lewis SA, Eaton DC, Diamond JM (1976) The mechanism of Na^+ transport by rabbit urinary bladder. J Membr Biol 28: 41–70
Lewis SA, Butt AG, Bowler MJ, Leader JP, Macknight ADC (1985) Effects of anions on cellular volume and transepithelial Na^+ transport across toad urinary bladder. J Membr Biol 83: 119–137
Liedtke CM, Hopfer U (1982) Mechanism of Cl^- translocation across small intestinal brush border membrane. I. Absence of Na-Cl cotransport. Am J Physiol 242: G263–G271

Lienhard GE (1983) Regulation of cellular membrane transport by exocytotic insertion and endocytic retrieval of transporters. Trends Biochem Sci 8: 125–127
Lipton P (1972) Effect of changes in osmolarity on sodium transport accross the isolated toad bladder. Am J Physiol 222: 821–828
Lohr J, Sullivan LP, Cragoe EJ, Grantham JJ (1989) Volume regulation determinants in isolated proximal tubules in hypertonic medium. Am J Physiol 256: F622–F631
Loo DDF, Kaunitz JD (1989) Ca^{2+} and cAMP activated K^+ channels in the basolateral membrane of crypt cells isolated from rabbit distal colon. J Membr Biol 110: 19–28
Lopes AG, Guggino WB (1987) Volume regulation in the early proximal tubule of the *Necturus* kidney. J Membr Biol 97: 117–125
Lopes AG, Amzel LM, Markakis D, Guggino WB (1988) Cell volume regulation by the thin descending limb of Henle's loop. Proc Natl Acad Sci USA 85: 2873–2877
Lote CJ, Haylor J (1989) Eicosanoids in renal function. Prostaglandins Leukot Essent Fatty Acids 36: 203–217
Machen TE, Townsley MC, Paradiso AM, Wenzl E, Negulescu PA (1989) H and HCO_3 transport across the basolateral membrane of the parietal cell. Ann NY Acad Sci 574: 447–462
Macknight ADC (1988) Principles of cell volume regulation. Renal Physiol Biochem 11: 114–141
Macknight ADC, Pigrim JP, Robinson BA (1974) The regulation of cellular volume in liver slices. J Physiol 238: 279–294
MacLeod RJ, Hamilton JR (1991a) Volume regulation initiated by Na^+-nutrient cotransport in isolated mammalian villus enterocytes. Am J Physiol 260: G26–G33
MacLeod RJ, Hamilton JR (1991b) Separate K^+ and Cl^- transport pathways are acitvated for regulatory volume decrease in jejunal villus cells. Am J Physiol 260: G405–G415
MacRobbie EAC, Ussing HH (1961) Osmotic behaviour of the epithelial cells of frog skin. Acta Physiol Scand 53: 348–365
Marin R, Proverbio T, Proverbio F (1989) The effect of K^+ on the volume regulation in rat kidney cortex cells. Biomed Biochim Acta 48: 303–315
Marsh DJ, Spring KR (1985) Polarity of volume regulatory increase in *Necturus* gallbladder epithelium. Am J Physiol 249: C471–C475
Maruyama Y (1989) Control of exocytosis in single cells. NIPS 4: 53–56
McCann JD, Welsh MJ (1990) Regulation of Cl^- and K^+ channels in airway epithelium. Annu Rev Physiol 52: 115–135
McCann JD, Li M, Welsh MJ (1989) Identification and regulation of whole-cell chloride currents in airway epithelium. J Gen Physiol 94: 1015–1036
McCarty NA, O'Neil RG (1990) Dihydropyridine-sensitive cell volume regulation in proximal tubule: the calcium window. Am J Physiol 259: F950–F960
Messner G, Oberleithner H, Lang F (1985) The effect of phenylalanine on the electrical properties of proximal tubule cells in the frog kidney. Pflügers Arch Eur J Physiol 404: 138–144
Mills JW (1987) The cytoskeleton: possible role in volume control. Curr Top Membr Transp 30: 75–101
Mills JW, Lubin M (1986) Effect of adenosine 3′,5′-cyclic monophosphate on volume and cytoskeleton of MDCK cells. Am J Physiol 250: C319–C324
Mills JW, Skiest DL (1985) Role of cyclic AMP and the cytoskeleton in volume control in MDCK cells. Mol Physiol 8: 247–262
Mircheff AK, Bradley ME, Yiu SC, Lambert RW (1989) Rapid stimulation associated changes in exocrine Na^+,K^+-ATPase subcellular distribution. Curr Top Membr Transp 34: 121–142
Montero MC, Ilundáin A (1989) Effects of anisosmotic buffers on K^+ transport in isolated chicken enterocytes. Biochim Biophys Acta 979: 269–271
Montrose-Rafizadeh C, Guggino WB (1990) Cell volume regulation in the nephron. Annu Rev Physiol 52: 761–772
Montrose-Rafizadeh C, Guggino WB (1991) Role of intracellular calcium in volume regulation by rabbit medullary thick ascending limb cells. Am J Physiol 260: F402–F409

Moran WM, Hudson RL, Schultz SG (1986) Transcellular sodium transport and intracellular sodium activities in rabbit gall-bladder. Am J Physiol G155–G159
Moriyama T, Garcia-Perez A, Burg M (1989) Osmotic regulation of aldose reductase protein synthesis in renal medullary cells. J Biol Chem 264: 16810–16814
Moriyama T, Garcia-Perez A, Burg M (1990) Factors affecting the ratio of different organic osmolytes in renal medullary cells. Am J Physiol 259: F847–F858
Muallem S, Loessberg PA (1990) Intracellular pH-regulatory mechanisms in pancreatic acinar cells. II. Regulation of H^+ and HCO_3^- transporters by Ca^{2+}-mobilizing agonists. J Biol Chem 265: 12813–12819
Nakahari T, Murakami M, Yoshida H, Miyamoto M, Sohma Y, Imai Y (1990) Decrease in rat submandibular acinar cell volume during ACh stimulation. Am J Physiol 258: G878–G886
Nakanishi T, Burg MB (1989) Osmoregulation of glycerophosphorylcholine content of mammalian renal cells. Am J Physiol 257: C795–C801
Natke E (1990) Cell volume regulation of rabbit cortical collecting tubule in anisotonic media. Am J Physiol 258: F1657–F1665
Nauntofte B, Poulsen JH (1986) Effects of Ca^{2+} and furosemide on Cl^- transport and O_2 uptake in rat parotid acini. Am J Physiol 251: C175–C185
Ngezahayo A, Kolb HA (1990) Gap junction permeability is affected by cell volume changes and modulates volume regulation. FEBS Lett 276: 6–8
Nielsen R (1979) A 3 to 2 coupling ratio of the Na-K pump responsible for the transepithelial Na transport in frog skin as disclosed by the effect of Ba. Acta Physiol Scand 107: 189–191
Novak I (1990) Electrolyte transport in pancreatic ducts. In: Young JA, Wong PYD (eds) Epithelial secretion of water and electrolytes. Springer, Berlin Heidelberg New York, pp 239–252
Novak I, Greger R (1988a) Electrophysiological study of transport systems in isolated perfused pancreatic ducts: properties of the basolateral membrane. Pflügers Arch Eur J Physiol 411: 58–68
Novak I, Greger R (1988b) Properties of the luminal membrane of isolated perfused rat pancreatic ducts. Effect of cyclic AMP and blockers of chloride transport. Pflügers Arch Eur J Physiol 411: 546–553
O'Grady SM, Musch MW, Field M (1986) Stoichiometry and ion affinities of the Na-K-Cl cotransport system in the intestine of the winter flounder (*Pseudopleuronectes americanus*). J Membr Biol 91: 33–41
O'Grady SM, Palfrey HC, Field M (1987) Characteristics and functions of Na-K-Cl cotransport in epithelial tissues. Am J Physiol 253: C177–C192
Olea RS, Pasantes-Morales H, Lazaro A, Cereijido M (1991) Osmolarity-sensitive release of free amino acids from cultured kidney cells (MDCK). J Membr Biol 121: 1–9
O'Neil R (1989) Modulation of Na^+,K^+-ATPase expression in renal collecting duct. Curr Top Membr Transp 34: 209–228
Ozawa T, Saito Y, Nishiyama A (1988) Mechanism of uphill chloride transport of the mouse lacrimal acinar cells: studies with Cl^- sensitive microelectrode. Pflügers Arch Eur J Physiol 412: 509–515
Palmer LG, Sackin H (1988) Regulation of renal ion channels. FASEB J 2: 3061–3065
Palmer LG, Edelman IS, Lindemann B (1980) Current-voltage analysis of apical sodium transport in toad urinary bladder: effects of inhibitors of transport and metabolism. J Membr Biol 57: 59–71
Palmer LG, Frindt G, Silver RB, Strieter J (1989) Feedback regulation of epithelial sodium channels. Curr Top Membr Transp 34: 45–60
Paradiso AM, Negulescu PA, Machen TE (1986) Na^+-H^+ and Cl^--OH^- (HCO_3^-) exchange in gastric glands. Am J Physiol 250: G524–G534
Persson B-E, Larson M (1986) Carbonic anhydrase inhibition and cell volume regulation in *Necturus* gallbladder. Acta Physiol Scand 128: 501–507
Persson B-E, Spring KR (1982) Gallbladder epithelial cell hydraulic water permeability and volume regulation. J Gen Physiol 79: 481–505

Petersen OH (1986) Calcium-activated potassium channels and fluid secretion by exocrine glands. Am J Physiol 251: G1–G13
Petersen OH, Maruyama Y (1984) Calcium activated potassium channels and their role in secretion. Nature 307: 693–696
Pierce SK, Politis AD (1990) Ca^{2+}-activated cell volume recovery mechanisms. Annu Rev Physiol 52: 27–42
Pirani D, Evans LAR, Cook DI, Young JA (1987) Intracellular pH in the rat mandibular salivary gland: the role of Na-H and Cl-HCO_3 antiports in secretion. Pflügers Arch Eur J Physiol 408: 178–184
Pollack LR, Tate EM, Cook JS (1981) Turnover and regulation of Na,K-ATPase in HeLa cells. Am J Physiol 67: 537–552
Powell DW (1979) Transport in large intestine. In: Giebisch G, Tosteson DC, Ussing HH (eds) Membrane transport in biology, vol IVB. Springer, Berlin Heidelberg New York, pp 781–809
Powell DW (1987) Intestinal water and electrolyte transport. In: Johnson LR (ed) Physiology of the gastrointestinal tract. Raven Press, New York, pp 1267–1305
Proverbio F, Duque JA, Proverbio T, Marin R (1988) Cell volume-sensitive Na^+-ATPase activity in rat kidney cortex cell membranes. Biochim Biophys Acta 941: 107–110
Proverbio F, Marín R, Proverbio T (1989) The "second" sodium pump and cell volume. Curr Top Membr Transp 34: 105–120
Quinton PM, Reddy MM (1989) Cl^- conductance and acid secretion in the human sweat duct. Ann NY Acad Sci 574: 438–446
Reuss L (1985) Independence of apical membrane Na^+ and Cl^- entry in *Necturus* gallbladder epithelium. J Gen Physiol 84: 423–445
Reuss L (1988) Cell volume regulation in nonrenal epithelia. Renal Physiol Biochem 3–5: 187–201
Reuss L (1989) Regulation of transepithelial chloride transport by amphibian gallbladder epithelium. Ann NY Acad Sci 574: 370–384
Richards NW, Dawson DC (1986) Single potassium channels blocked by lidocaine and quinidine in isolated turtle colon epithelial cells. Am J Physiol 251: C85–C89
Rick R, Beck FX, Dörge A, Thurau K (1987) Intracellular ion concentration in the frog cornea epithelium during stimulation and inhibition of Cl secretion. J Membr Biol 95: 229–240
Robertson GL, Berl T (1991) Pathophysiology of water metabolism. In: Brenner BM, Rector FC (eds) The kidney. Saunders, Philadelphia pp 677–736
Roden M, Turnheim K (1988) Sodium pump quantity and turnover in rabbit descending colon at different rates of sodium absorption. Pflügers Arch Eur J Physiol 413: 181–189
Rome L, Grantham J, Savin V, Lohr J, Lechene C (1989) Proximal tubule volume regulation in hyperosmotic media: intracellular K^+, Na^+, and Cl^-. Am J Physiol 257: C1093–C1100
Ronson CW, Nixon BT, Ausbel FM (1987) Conserved domains in bacterial regulatory proteins that respond to environmental stimuli. Cell 49: 579–581
Rossier BC, Geering K, Kraehenbuhl JP (1987) Regulation of the sodium pump: how and why? Trends Biochem Sci 12: 483–487
Roy G, Sauve R (1987) Effect of anisotonic media on volume, ion, and amino-acid content and membrane potential of kidney cells (MDCK) in culture. J Membr Biol 100: 83–96
Sachs F (1987) Baroreceptor mechanisms at the cellular level. Fed Proc 46: 12–16
Sachs G, Spenney JG, Lewin M (1978) H^+ transport: regulation and mechanism in gastric mucosa and membrane vesicles. Physiol Rev 58: 106–173
Sackin H (1987) Stretch-activated potassium channels in renal proximal tubule. Am J Physiol 253: F1253–F1262
Sackin H (1989) A stretch-activated K^+ channel sensitive to cell volume. Proc Natl Acad Sci USA 86: 1731–1735
Sackin H (1990) Regulation of renal proximal tubule basolateral potassium channels. Prog Clin Biol Res 334: 231–249

Sackin H, Palmer LG (1987) Basolateral potassium channels in renal proximal tubule. Am J Physiol 253: F476–F487
Saikia TC (1991) Composition of the renal cortex and medulla of rats during water diuresis and antidiuresis. Q J Exp Physiol 50: 146–157
Saito Y, Wright WM (1987) Regulation of intracellular chloride in bullfrog choroid plexus. Brain Res 417: 267–272
Saito Y, Ozawa T, Suzuki S, Nishiyama A (1988) Intracellular pH regulation in mouse lacrimal gland acinar cells. J Membr Biol 101: 73–81
Saito Y, Ozawa T, Nishiyama A (1990) Electrolyte secretion by the lacrimal gland. In: Young JA, Wong PYD (eds) Epithelial secretion of water and electrolytes. Springer, Berlin Heidelberg New York, pp 229–237
Sánchez Olea R, Pasantes-Morales H, Lázaro A, Cereijido M (1991) Osmolarity-sensitive release of free amino acids from cultured kidney cells (MDCK). J Membr Biol 121: 1–9
Sandle GI, Fraser G, Long S, Warhurst G (1990) A cAMP-activated chloride channel in the plasma membrane of cultured human gastric cells (HGT-1). Pflügers Arch Eur J Physiol 417: 259–263
Sands JM, Knepper MA, Spring KR (1986) Na-K-Cl cotransport in apical membrane of rabbit renal papillary surface epithelium. Am J Physiol 251: F457–F484
Schild L, Aronson PS, Giebisch G (1990) Effect of apical Cl^--formate exchange on cell volume in rabbit proximal tubule. Am J Physiol 258: F530–F536
Schild L, Aronson PS, Giebisch G (1991) Basolateral transport pathways for K^+ and Cl^- in rabbit proximal tubule: effects on cell volume. Am J Physiol 260: F101–F109
Schmidt-Nielson B (1975) Comparative physiology of cellular ion and volume-regulation. J Exp Zool 194: 207–220
Schmolke M, Beck F, Guder WG (1989) Effect of antidiuretic hormone on renal organic osmolytes in brattleboro rats. Am J Physiol 257: F732–F737
Schultz SG (1972) Electrical potential differences and electromotive forces in epithelia. J Gen Physiol 59: 794–798
Schultz SG (1981) Homocellular regulatory mechanisms in sodium-transporting tissues: avoidance of extinction by "flush-through". Am J Physiol 241: F579–F590
Schultz SG (1986) Cellular models of epithelial ion transport. In: Andreoli TF, Hoffman JF, Fanestil DD, Schultz SG (eds) Physiology of membrane disorders, 2nd edn. Plenum Press, New York, pp 519–534
Schultz SG (1989a) Volume preservation: then and now. News Physiol Sci 4: 169–172
Schultz SG (1989b) Intracellular sodium activities and basolateral membrane potassium conductances of sodium absorbing epithelia. Curr Top Membr Transp 34: 21–44
Schultz SG, Hudson RL (1986) How do sodium absorbing cells do their job and survive? News Physiol Sci 1: 185–189
Schultz SG, Zalusky R (1964) Ion transport in isolated rabbit ileum. II. The interaction between active sodium and active sugar transport. J Gen Physiol 47: 1043–1059
Sernka TJ (1990) Direct hyposmotic stimulation of gastric acid secretion. Membr Biochem 9: 1–7
Siebens AW, Kregenow FM (1985) Volume regulatory responses of *Amphiuma* red cells in anisotonic media. J Gen Physiol 86: 527–564
Siebens AW, Spring KR (1989) A novel sorbitol transport mechanism in cultured renal papillary epithelial cells. Am J Physiol 257: F937–F946
Silva P, Myers M, Epstein FH (1986) Stoichiometry of sodium chloride transport by rectal gland of *Squalus acanthias*. Am J Physiol 250: F516–F519
Simmons NL (1984) Epithelial cell volume regulation in hypotonic fluids: studies using a model tissue culture renal epithelial cell system. Q J Exp Physiol 69: 83–95
Singh J (1984) Effects of acetylcholine and caerulein on $^{86}Rb^+$ efflux in the mouse pancreas. Evidence for a sodium-potassium-chloride cotransport system. Biochem Biophys Acta 775: 77–85
Soltoff SP, McMillian MK, Cantley LC, Cragoe EJ Jr, Talamo BR (1989) Effects of muscarinic, alpha-adrenergic, and substance P agonists and ionomycin on ion transport mechanisms in the rat parotid acinar cell. The dependence of ion transport on intracellular calcium. J Gen Physiol 93: 285–319

Soltoff SP, McMillian MK, Cragoe EJ Jr, Cantley LC, Talamo BR (1990) Effects of extracellular ATP on ion transport systems and $[Ca^{2+}]_i$ in rat parotid acinar cells. Comparison with the muscarinic agonist carbachol. J Gen Physiol 95: 319–346
Spring KR, Siebens AW (1988) Solute transport and epithelial cell volume regulation. Comp Biochem Physiol 90A: 557–560
Spring KR, Ussing HH (1986) The volume of mitochondria-rich cells of frog skin epithelium. J Membr Biol 92: 21–26
Stetson DL, Beauwens R, Palmisano J, Mitchell PP, Steinmetz PR (1985) A double-membrane model for urinary bicarbonate secretion. Am J Physiol 249: F546–F552
Stoddard JS, Reuss L (1989) Electrophysiological effects of mucosal Cl-removal in *Necturus* gallbladder epithelium. Am J Physiol 257: C568–C578
Strange K (1989) Volume regulation following Na^+ pump inhibition in CCT principal cells: apical K^+ loss. Am J Physiol 257: C1093–C1100
Stuenkel EL, Machen TE, Williams JA (1988) pH regulatory mechanisms in rat pancreatic ductal cells. Am J Physiol 254: G925–G930
Sudo J-I, Morel F (1984) Na^+ and K^+ cell concentrations in collagenase-treated rat kidney tubules incubated at various temperatures. Am J Physiol 246: C407–C414
Sun A, Hebert SC (1989) Rapid hypertonic cell volume regulation in the perfused inner medullary collecting duct. Kidney Int 36: 831–842
Sun AM, Soldzberg SN, Kikeri D, Hebert SC (1990) Mechanisms of cell volume regulation by the mouse medullary thick ascending limb of Henle. Kidney Int 38: 1019–1029
Suzuki M, Kawahara K, Ogawa A, Morita T, Kawaguchi Y, Kurihara S, Sakai O (1990) $[Ca^{2+}]_i$ rises via G protein during regulatory volume decrease in rabbit proximal tubule cells. Am J Physiol 258: F690–F696
Taniguchi J, Guggino WB (1989) Membrane stretch: a physiological stimulator of Ca^{2+}-activated K^+ channels in thick ascending limb. Am J Physiol 257: F347–F352
Tauc M, Le-Maout S, Poujeol P (1990) Fluorescent video-microscopy study of regulatory volume decrease in primary culture of rabbit proximal convoluted tubule. Biochim Biophys Acta 1052: 278–284
Thomas SR, Suzuki Y, Thompson SM, Schultz SG (1983) Electrophysiology of *Necturus* urinary bladder. I. "Instantaneous" current-voltage relations in the presence of varying mucosal sodium concentrations. J Membr Biol 73: 157–175
Tosteson DC, Hoffmann JF (1960) Regulation of cell volume by active cation transport in high and low potassium sheep red cells. J Gen Physiol 44: 169–194
Tune BM, Burg MB (1971) Glucose transport by proximal renal tubules. Am J Physiol 221: 580–585
Turner RJ, George JN, Baum BJ (1986) Evidence for a Na/K/Cl cotransport system in basolateral membrane vesicles from rabbit parotid. J Membr Biol 94: 143–152
Turnheim K (1991) Intrinsic regulation of apical sodium entry in epithelia. Physiol Rev 71: 429–445
Turnheim K, Frizzell RA, Schultz SG (1978) Interaction between cell sodium and the amiloride-sensitive sodium entry step in rabbit colon. J Membr Biol 39: 233–256
Turnheim K, Thompson SM, Schultz SG (1983) Relation between intracellular sodium and active sodium transport in rabbit colon: current-voltage relations of the apical sodium entry mechanism in the presence of varying luminal sodium concentrations. J Membr Biol 76: 299–309
Turnheim K, Plass H, Grasl M, Krivanek P, Wiener H (1986) Sodium absorption and potassium secretion in rabbit colon during sodium deficiency. Am J Physiol 250: F235–F245
Turnheim K, Hudson RL, Schultz SG (1987) Cell Na^+ activities and transcellular Na^+ absorption by descending colon from normal and Na^+-deprived rabbits. Pflügers Arch Eur J Physiol 410: 279–283
Turnheim K, Costantin J, Chan S, Schultz SG (1989) Reconstitution of a calcium-activated potassium channel in basolateral membranes of rabbit colonocytes into planar lipid bilayers. J Membr Biol 112: 247–254

Ubl J, Murer H, Kolb HA (1988a) Hypotonic shock evokes opening of Ca^{2+} activated K channels in opossum kidney cells. Pflügers Arch Eur J Physiol 412: 551–553
Ubl J, Murer H, Kolb HA (1988b) Ion channels activated by osmotic and mechanical stress in membranes of opossum kidney cells. J Membr Biol 104: 223–232
Uchida S, Garcia-Perez A, Murphy H, Burg MB (1989) Signal for induction of aldose reductase in renal medullary cells by high external NaCl. Am J Physiol 256: C614–C620
Ussing HH (1965) Relationship between osmotic reactions and active sodium transport in frog skin epithelium. Acta Physiol Scand 63: 141–155
Ussing HH (1985) Volume regulation and basolateral co-transport of sodium, potassium and chloride ions in frog skin epithelium. Pflügers Arch Eur J Physiol 405 (Suppl 1): S2–S7
Ussing HH (1986) Epithelial cell volume regulation illustrated by experiments in frog skin. Renal Physiol 9: 38–46
Ussing HH (1987) Role of potassium in epithelial transport illustrated by experiments on frog skin epithelium. Curr Top Membr Transp 28: 3–18
Ussing HH (1990) Volume regulation of frog skin epithelium. In: Beyenbach KW (ed) Comparative physiology, vol 4. Cell volume regulation. Karger, Basel, pp 87–113
Ussing HH, Biber TUL, Bricker NB (1965) Exposure of the isolated frog skin to high potassium concentrations at the internal surfaces. J Gen Physiol 48: 425–433
Valtin H (1966) Sequestration of urea and nonurea solutes in renal tissues of rats with hereditary hypothalamic diabetes insipidus: effect of vasopressin and dehydration on the countercurrent mechanism. J Clin Invest 45: 337–345
van Driessche W, Erlij P (1983) Noise analysis of inward and outward Na^+ currents across the apical borders of ouabain-treated frog skin. Pflügers Arch Eur J Physiol 398: 179–188
van Dyke R, Scharschmidt B (1983) (Na,K)-ATPase mediated cation pumping in cultured rat hepatocytes. Rapid modulation by alanine and taurocholate transport and characterization of its relationship to intracellular sodium concentration. J Biol Chem 285: 12912–12919
van Rossum GDV, Russo MA (1981) Ouabain-resistant mechanism of volume control and the ultrastructural organization of liver slices recovering from swelling in vitro. J Membr Biol 59: 191–209
van Rossum GDV, Russo MA (1984) Requirement of Cl^- and Na^+ for the ouabain resistant control of cell volume in slices of rat liver. J Membr Biol 77: 63–76
van Rossum GDV, Russo MA, Schisselbauer JC (1987) Role of cytoplasmic vesicles in volume maintenance. Curr Top Membr Transp 30: 45–74
Venosa RA (1991) Hypo-osmotic stimulation of active Na^+ transport in frog muscle: apparent upregulation. J Membr Biol 120: 97–104
Völkl H, Lang F (1990) Effect of potassium on cell volume regulation in renal straight proximal tubules. J Membr Biol 117: 113–122
Völkl H, Paulmichl M, Lang F (1988) Cell volume regulation in renal cortical cells. Renal Physiol Biochem 11: 158–173
vom Dahl S, Hallbrucker C, Lang F, Gerok W, Häussinger D (1991) Regulation of liver cell volume and proteolysis by glucagon and insulin. Biochem J 278: 519–521
Wade JB (1986) Role of membrane fusion in hormonal regulation of epithelial transport. Annu Rev Physiol 48: 213–223
Wangemann P, Marcus DC (1990) K^+-induced cell swelling of vestibular dark cells is dependent on Na^+ and Cl^- and inhibited by piretanide. Pflügers Arch Eur J Physiol 416: 262–269
Wartford M (1990) A "swell" way to regulate metabolism. TIBS 15: 329–330
Weinman SA, Reuss L (1982) Na-H exchange at the apical membrane of *Necturus* gallbladder. J Gen Physiol 80: 299–321
Welling PA, Linshaw MA (1988) Importance of anion in hypotonic volume regulation of rabbit proximal straight tubule. Am J Physiol 255: R853–R860
Welling PA, O'Neil RG (1989) Cell swelling activates basolateral Cl and K conductances in rabbit proximal tubule. Am J Physiol 258: F951–F962

Welling PA, Linshaw MA, Sullivan LP (1985) Effect of barium on cell volume regulation in rabbit proximal straight tubules. Am J Physiol 249: F20–F27

White JF, Imon MA (1983) Effect of galactose on intracellular potential and sodium activity in urodele small intestine. Evidence for basolateral electrogenic transport. In: Gilles-Baillien M, Gilles R (eds) Intestinal transport. Springer, Berlin Heidelberg New York, pp 295–312

Wiener H, Turnheim K (1990) Calcium-activated potassium channels in basolateral membranes of colon epithelial cells; reconstitution and functional properties. Wien Klin Wochenschr 26: 622–628

Wills NK, Lewis SA (1980) Intracellular Na activity as a function of Na transport across a tight epithelium. Biophys J 30: 181–186

Wills NK, Millinoff LP, Crowe WE (1991) Na^+ channel activity in cultured renal (A6) epithelium: regulation by solution osmolarity. J Membr Biol 121: 79–90

Wirthensohn G, Lefrank S, Schmolke M, Guder WG (1989) Regulation of organic osmolyte concentrations in tubules from rat renal inner medulla. Am J Physiol 256: R128–R135

Wondergem R, Castillo LB (1988) Quinine decreases hepatocyte transmembrane potential and inhibits amino acid transport. Am J Physiol 254: G795–G801

Wong PYD (1988) Mechanism of adrenergic stimulation of anion secretion in cultured rat epididymal epithelium. Am J Physiol 254: F121–F133

Wong PYD (1990) Electrolyte and fluid transport in the epididymis. In: Young JA, Wong PYD (eds) Epithelial secretion of water and electrolytes. Springer, Berlin Heidelberg New York, pp 333–347

Wong SME, Chase HS Jr (1986) Role of intracellular calcium in cellular volume regulation. Am J Physiol 250: C841–C852

Yamauchi A, Kwon HM, Uchida S, Preston AS, Handler JS (1991) Myo-inositol and betaine transporters regulated by tonicity are basolateral in MDCK cells. Am J Physiol 261: F197–F202

Yancey PH, Burg MB (1989) Distribution of major organic osmolytes in rabbit kidneys in diuresis and antidiruesis. Am J Physiol 257: R602–R607

Yancey PH, Haner G, Feudenberger TH (1990) Effects of an aldose reductase inhibitor on organic osmotic effectors in rat renal medulla. Am J Physiol 259: F733–F738

Yantorno RE, Coca-Prados M, Krupin T, Civan MM (1989) Volume regulation of cultured transformed, non-pigmented epithelial cells from human ciliary body. Exp Eye Res 49: 423–437

Chapter 4

Cell Volume and Stimulus-Secretion Coupling

A. Kurtz and *H. Scholz*[1]

Contents

1 General Aspects of Secretion

In all endocrine and most exocrine cells the secretory process comprises the release of organic material by exocytosis or by constitutive secretion. In addition, the transport of water and electrolytes from the basolateral (capillary) side to the apical lumen (acinus) is an essential characteristic of secretion from exocrine glands.

A common feature of all types of gland cells is that the intracellular concentration of calcium plays a key role in the regulation of secretion (Rubin 1982). In general, calcium exerts a stimulatory effect on secretion. Thus, elevation of the cytosolic level of calcium from its resting level, which is around 100 nM, either initiates, facilitates or maintains secretion in the vast majority of gland cells (Douglas 1968; Rasmussen and Barrett 1984; Knight 1987; Petersen and Findlay 1987; Penner and Neher 1988). At present, only two cell types are known in which calcium appears to play a negative rather than a positive role in secretion. These are the parathyroid gland cells, which secrete the calcium regulating hormone PTH (Brown 1991), and the renal juxtaglomerular epitheloid cells, which release the protease renin (Kurtz 1989).

[1] Institut für Physiologie der Universität Regensburg, Universitätsstraße 31, W-8400 Regensburg, FRG

Advances in Comparative and Environmental Physiology, Vol. 14

Apart from calcium, also cyclic AMP appears to play a relevant role in the second messenger control of secretion (Rubin 1982). Interestingly, cyclic AMP is a powerful stimulator of secretion in those cells, in which calcium appears to act inhibitorily, namely in parathyroid cells (Brown 1991) and in renal juxtaglomerular cells (Kurtz 1989). In cells, in which calcium has a positive effect on secretion, cAMP appears more as a modulator amplifying or attenuating the effect of calcium. In platelets and in mast cells, for instance, cAMP even inhibits secretion.

The stimulus-secretion coupling, which summarizes the events by which extracellular factors such as hormones, neurotransmitters or metabolites exert influence on secretion from cells (Douglas 1968), therefore frequently involves a rise in the cytosolic calcium concentration and/or influence on the activity of the adenylate cyclase. Modulation of adenylate cyclase activity by extracellular signals is achieved by receptors coupling either to the inhibitory GTP-binding protein G_i or to the stimulatory GTP-binding protein G_s, leading to inhibition or stimulation of adenylate cyclase activity, respectively (Casey and Gilman 1988). Intracellular increases in calcium are brought about by release of calcium from intracellular stores and by enhancement of transmembrane calcium influx (Penner and Neher 1988). In secretory cells release of calcium from internal stores is usually induced by a receptor-stimulated and G-protein-mediated activation of a phosphodiesterase (phospholipase C), releasing inositol trisphosphate (IP_3), which in turn causes the release of calcium (Berridge and Irvine 1984). In parallel with IP_3 diacylglycerol appears which is the physiological activator of protein kinase C which may also be involved in secretion (Nishizuka 1984). Enhancement of calcium influx into gland cells is brought about by activation of ion channels permeable to calcium, which are activated by membrane depolarization (potential-operated calcium channels, POCC) and by receptor occupancy either directly (receptor-operated calcium channels, ROCC) or indirectly via second messengers (second messenger-activated calcium channels, SMOC) (Penner and Neher 1988). POCC play an important role in the secretion in excitable cells, while ROCC and SMOC are more relevant for non-excitable cells.

The subcellular mechanisms by which the second messenger then induces secretion are less understood. They may involve direct fusogenic effects of calcium (De Lisle and Williams 1986), other GTP-binding proteins (Barrowman et al. 1986), phosphorylations (De Lisle and Williams 1986), swelling of secretory granules (Finkelstein et al. 1986), and also changes in the cellular volume.

There are three major reasons to consider the cell volume of secretory cells in the context of stimulus-secretion coupling:

1. Secretion from exocrine cells essentially involves directed transport of salts and water from the basolateral side to the gland acinus (apical side), which themselves are the essential determinants of cell volume.

2. Second messenger molecules important for the regulation of secretion such as calcium and cAMP have been found to exert a marked influence on the activity of ion channels, which are of importance for the cell volume, such as potassium and chloride channels.
3. Secretion by exocytosis primarily in endocrine cells has been found to be sensitive to changes in the extracellular osmolarity, which is an important determinant of cell volume.

2 Fluid Secretion in Exocrine Acinar Cells

During the last decade a wealth of information has accumulated suggesting that the principles of secretion are very similar among the different types of exocrine glands such as lacrimal, salivary and pancreatic glands (Petersen 1986; Marty 1987; Petersen and Gallacher 1988). There is evidence emerging that eccrine sweat glands also behave very similarly (Brayden et al. 1988; Krouse et al. 1989).

The general plan of an exocrine acinus is shown in Fig. 1. The acinar cells are polarized, having an apical and a basolateral membrane. Due to the

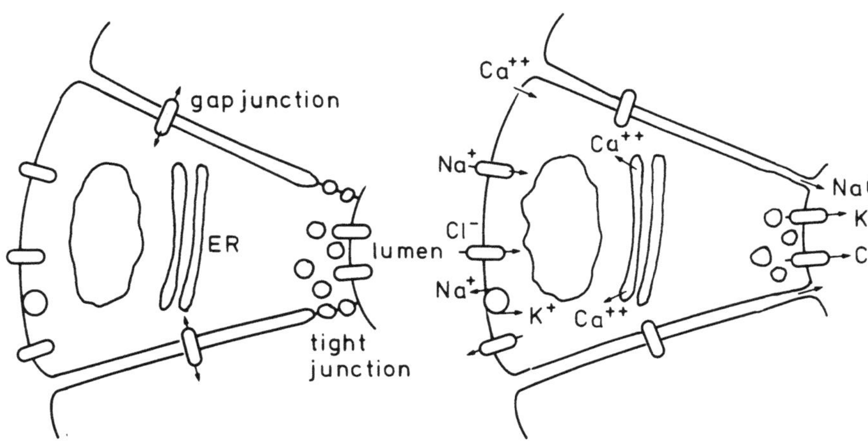

Fig. 1. Organization of an exocrine acinar cell in the resting (*left*) and stimulated state (*right*). The figure is adapted from Marty (1987). Cells are linked by gap junctions, and basolateral and apical domains are separated by tight junctions. The basolateral membrane contains the sodium-potassium ATPase and calcium-activated potassium, chloride and non-selective monovalent cation channels. The apical membrane contains calcium-activated chloride channels and in certain exocrine acinar cells also potassium channels. Stimulation of secretion, for instance by acetylcholine, involves calcium mobilization from intracellular stores, enhancement of transmembrane calcium influx, closure of gap junctions and an increase in the permeability of tight junctions. For clarity, the basolaterally located Na-K-Cl_2 cotransport system has been omitted from the drawing. For more detailed secretion models, the reader is referred to Petersen and Gallacher (1988)

existence of gap junctions acinar cells are capable of communicating electrically and chemically (Iwatsuki and Petersen 1978; Meda et al. 1983). The intercellular space between acinar cells and the acinus lumen is separated by tight junctions, the permeability of which for water and ions can be regulated (Martin and Burgen 1962; Mazariegos et al. 1984). Stimulation of secretion in acinar cells is frequently induced by activation of α_1-adrenergic, muscarinic (M2) or peptidergic receptors (Lundberg and Hökfelt 1983). Receptor occupancy is then followed by a stimulation of the phospholipase C (PLC) activity, which is mediated by a heterotrimeric GTP-binding protein (Cockcroft and Gomperts 1985; Litosch et al. 1985). PLC releases IP_3 from phosphatidylinositolbisphosphate (Berridge and Irvine 1984) and thus causes release of calcium from internal stores (Streb et al. 1983). In parallel, acetylcholine and also cholecystokinin enhance transmembrane calcium influx into acinar cells (Selinger et al. 1973; Kanagasuntheram and Randle 1976; Keryer and Rossignol 1976). As a consequence, the cytosolic levels of calcium rise and begin to oscillate (Wakui et al. 1990). The rise in intracellular calcium induces secretion of organic material by exocytosis through the apical membrane into the acinus lumen. An earlier event related to the rise in calcium is a change in the electrical properties of acinar cells. Frequently, a potassium and a chloride conductance become activated, and it is now generally assumed that they are due to calcium-activated potassium and chloride channels (Marty 1987; Petersen and Gallacher 1988). In certain acinar cells a calcium-activated conductance permeable to monovalent cations has also been found (Maruyama and Petersen 1984a,b). These cation channels appear to be less sensitive to calcium than the potassium and the chloride channels. The calcium-activated potassium channels that are also voltage-sensitive have a high single channel conductance of around 150 pS and they exist at a rather low density of about 100 channels per acinar cell (Marty 1987; Petersen and Gallacher 1988). The calcium-activated chloride channels are less well characterized. They appear to have a comparably small single channel conductance of around 1–2 pS, but they exist at a rather high density of about 10000 channels per acinar cell (Marty 1987; Petersen and Gallacher 1988). The non-specific channels permeable to monovalent cations have single channel conductances of 30–35 pS (Marty 1987; Petersen and Gallacher 1988). Primarily due to the calcium sensitivity of the potassium and chloride channels, the whole cell conductance of acinar cells increases about 1000-fold upon stimulation (Marty 1987). The membrane potential of resting acinar cells has been measured in the range of around −60 mV (Nauntofte and Dissing 1988). It is thus more negative than the chloride equilibrium potential, which has been estimated to be around −20 mV (Lau and Case 1988; Foskett 1990) and it is more positive than the potassium equilibrium potential which is around −90 mV. Due to these electrochemical driving forces, calcium activation of the potassium and the chloride channels should lead to a substantial electroneutral efflux of KCl from the cells. In fact, it has been demonstrated for salivary gland cells that they lose potassium to both saliva (apical side) and blood (basolateral side) upon stimulation. During maximal stimulation, as much as 40% of the total cellular potassium content may be lost (Burgen and Emmelin 1961). For lacrimal acinar cells it has been calculated that the efflux of KCl would deplete the cellular KCl store within 30 s (Marty 1987), provided there is a constant activation by calcium and absence of counterregulatory mechanisms.

3 Volume Changes During Exocrine Secretion

Since the cells lose osmotic active particles with the KCl efflux, the interior of the acinar cell should become hypotonic, leading to a subsequent water efflux and cell shrinkage because of the high water permeability of the plasma membrane. In fact, significant decreases in the cell volume have recently been measured in salivary glands upon stimulation (Foskett and Melvin 1989; Foskett 1990). Thus, the cell volume of salivary gland cells decreases by approximately 20% if the cytosolic concentration of calcium increases from its resting level to concentrations higher than 300 nM (Foskett and Melvin 1989; Fig. 2). The volume changes thereby mirror inversely the intracellular calcium concentration, displaying oscillating volume changes upon intracellular calcium oscillations (Foskett and Melvin 1989). Recently, it has, moreover, been shown that the decrease in the cell volume is followed by a short temporal delay after the decrease in the intracellular chloride concentration (Foskett 1990; Fig. 3).

In regard to the localization of the calcium-activated ion channels, it is assumed that the basolateral membranes contain all three types of calcium-dependent channels, namely potassium channels, non-selective cation

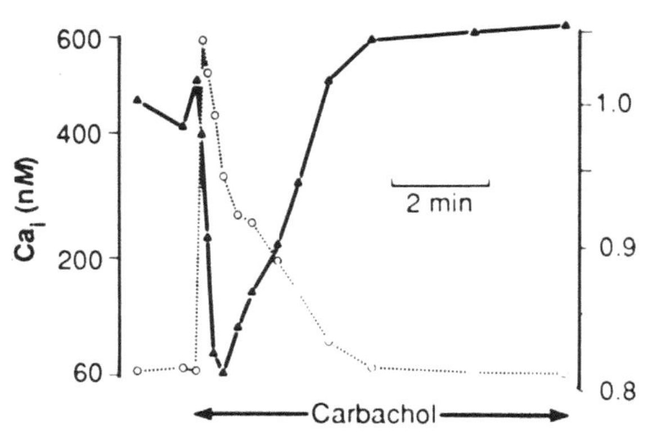

Fig. 2. Time courses of intracellular calcium (○) and cell volume (▲) in a rat salivary gland acinar cell during muscarinic stimulation with carbachol. (Foskett and Melvin 1989)

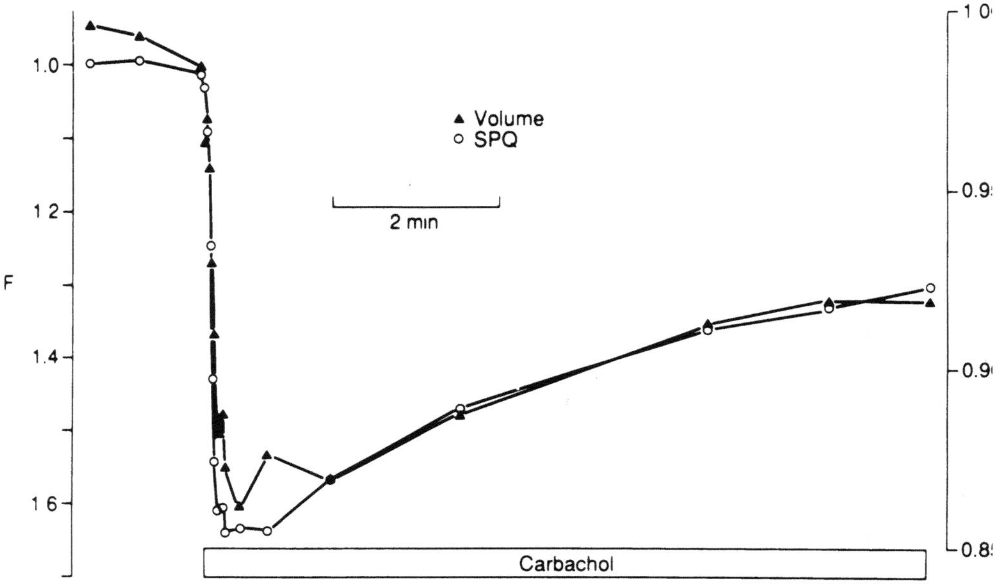

Fig. 3. Time courses of SPQ fluorescence (○), representing intracellular chloride concentration, and cell volume (▲) in a rat salivary gland acinar cell during muscarinic stimulation with carbachol. (Foskett 1990)

channels and chloride channels (Maruyama and Petersen 1984b; Marty et al. 1986). Increase in intracellular calcium will therefore lead to potassium efflux into the basolateral space, including the intercellular spaces, but also into the acinus through the apical membrane, provided potassium channels exist at the apical membrane (Marty 1987). In those acinar cells which also contain non-selective channels permeable to monovalent cations, potassium efflux into the paracellular space will be enhanced and simultaneously sodium will enter the cells through the basolateral membranes due to the favourable electrochemical gradient. Due to the difficult accessibility to luminal membranes, the information about the distribution of ion channels in that particular area is only indirect. Observations that all exocrine acinar cells contain a calcium-activated chloride conductance and the fact that calcium-activated chloride channels could so far only be identified in lacrimal acinar cells (Marty 1987; Petersen and Gallacher 1988) were interpreted that chloride channels were localized in the apical membrane. Such a localization of chloride channels would also fit with a more general organization of secretory epithelia (Frizzel et al. 1986). More direct support for the apical localization of chloride channels was recently provided by Kasai and Augustine (1990) who could show for pancreatic acinar cells that the elevation in the cytosolic calcium concentration at the apical membrane is paralleled by a chloride current, while elevations in calcium at the basolateral membrane activate both chloride and cation conductances. Such a localization of the chloride channels would allow chloride secretion into the acinus lumen und would explain the lumen negativity of the acinus (Lundberg 1958). In turn, the negative charge of the lumen is considered as the driving force for the paracellular movement of sodium ions from the basolateral side into the lumen. Due to osmotic forces built up by the accumulation of NaCl in the lumen, water is thought to follow passively into the lumen through paracellular pathways. The rates of sodium and water flow will also be determined by the permeability of the tight junctions separating the luminal from the basolateral space. Although the mechanisms are not yet clear there is evidence that the permeability of the tight junctions increases upon stimulation of acinar cells (Martin and Burgen 1962; Mazariegos et al. 1984).

4 Volume Counterregulatory Mechanisms During Exocrine Secretion

The enormous increase in membrane ionic conductance raises the question why the acinar cells only shrink to a certain extent during secretion. One explanation is certainly the discontinuous elevation in calcium after stimulation (Foskett and Melvin 1989). Hormonal stimulation of acinar cells causes an initial calcium spike due to IP_3-induced calcium release followed by a more sustained elevation at a lower level due to the enhancement of calcium influx. The contribution of calcium influx to the cytosolic level then becomes superposed by an oscillating, internal calcium-induced calcium release (Wakui et al. 1990).

Since the cellular volume mirrors the intracellular calcium level, decreasing calcium concentrations between the spikes allow a recovery of the cell volume (Foskett and Melvin 1989). A more direct effect attenuating a further decrease in cell volume during secretion is provided by an increased rate in ion influx. This influx is presumably mediated by a basolateral Na-K-Cl_2 cotransport system (Exley et al. 1986; Turner et al. 1986), which is thought to be responsible for the accumulation of intracellular chloride above electrochemical equilibrium, thus providing the driving force for chloride secretion. Loss of KCl through calcium-activated potassium and chloride channels during secretion will increase the inward driving force for the Na-K-Cl_2 cotransport and consequently increase the inward flux rate for sodium, potassium and chloride, which at a certain point will balance KCl efflux (Petersen and Gallacher 1988) and thus stop a further decrease in cellular volume. Sodium influx through the cotransport system will be neutralized by the activity of the Na/K ATPase, which in turn pumps potassium into the cell and thus prevents a further depletion of the cellular potassium stores (Petersen and Gallacher 1988). At the same time, an enhanced influx of chloride through the basolateral membrane could provide the basis for a continuous efflux of chloride through the apical membrane. This more simplifying concept becomes more complicated in pancreatic acinar cells which contain calcium-activated non-selective cation channels in the basolateral membrane, the activation of which during stimulation will tend to attenuate the inward sodium gradient driving the Na-K-Cl_2 cotransport. Moreover, the activation of these channels is responsible for the depolarization of the membrane potentials to values similar to or even higher than the actual chloride equilibrium potential (Petersen 1980). These obvious obstacles for the explanation of a continuous unidirectional flow of chloride from the basolateral to the apical side in pancreatic acinar cells might be solved by a recent observation made by Kasai and Augustine (1990). Utilizing a combination of patch clamp and digital imaging of calcium fluorescence, they provided evidence for a temporal dissociation between the calcium-induced activation of ion channels in the apical and basolateral membranes. They found that upon application of acetylcholine to isolated pancreatic acini the cytosolic calcium concentration rises first at the luminal membrane, causing an activation of the luminal chloride channels which leads to an apical chloride efflux, because the chloride equilibrium potential is more positive than the membrane potential. In a close temporal relation to the decline of the luminal current, which is large enough to release approximately 15% of the total cellular chloride content into the acinus lumen, the calcium elevation spreads out over the acinar cell towards the basolateral membrane. There calcium activates potassium and/or non-selective monovalent cation channels as well as chloride channels. As a consequence, the membrane potential rises over the chloride equilibrium potential leading to chloride influx through calcium-activated chloride channels in the basolateral membrane. This so-called "push and pull" model (Kasai and Augustine 1990) requires discontinuous elevations in calcium, which are in fact provided by the calcium oscillations observed after hormonal stimulation of acinar cells (Pralong et al. 1988; Wakui et al. 1989). One should expect that increasing the number of apical push and basolateral pull cycles by increasing the frequency of the calcium oscillations should enhance the secretory efficacy. This, in fact, has previously been observed by Rapp and Berridge (1981) in salivary glands. Again, with this "push and pull" model a decrease in cell volume due to luminal chloride secretion would be limited by the basolateral influx of sodium through non-selective cationic channels, through Na-K-Cl_2 cotransport, through potassium entry via the cotransport and increased activity of the Na/K ATPase, as well as by chloride influx through basolateral calcium-activated chloride channels. It is as yet unclear whether the changes in the cell volume are essential for secretion from acinar cells or are only passively linked to the ion efflux without physiological relevance. One could imagine that the water flux through the apical and the basolateral membrane upon cell shrinkage directly contributes to fluid secretion.

5 Volume Changes During Endocrine Secretion

While changes in the cellular volume could be demonstrated for exocrine secretion, a direct demonstration for changes in the cell volume during endocrine secretion has so far not been made. Considering the roles of potassium and chloride channels in the regulation of the cell volume, one could by inference assume the occurrence of cell volume changes during the stimulus-secretion coupling in certain endocrine cells. Mast cells, for instance, develop a cyclic AMP-activated chloride conductance upon secretion (Penner et al. 1988) and cyclic AMP has been found to inhibit stimulation of secretion from mast cells (Penner 1988). Because of the rather high resting membrane potential, cAMP-induced activation of chloride channels should cause chloride (and water) influx into the mast cells (Penner et al. 1988). Other secretory cells, for instance parathyroid gland cells (Jia et al. 1988) or pituitary cells (Luini and Brown 1990), possess calcium-activated potassium channels. Since a rise in intracellular calcium is considered a key step in stimulus-secretion coupling, potassium (and water) efflux could be expected during secretion from these cells. Finally, renal juxtaglomerular epitheloid cells possess many calcium-activated chloride channels, which should mediate chloride (and water) efflux upon elevation of the intracellular calcium concentration (Kurtz and Penner 1989). Assuming that such ion and expected water fluxes lead to concomitant changes in the cell volume, the question arises as to the role of the cell volume in endocrine secretion.

Secretion from endocrine cells is either the release of stored material by exocytosis, constitutive secretion or a combination of both. Exocytotic release of material stored in secretory vesicles, such as in mast cells, platelets, leukocytes, pituitary cells, parathyroid gland cells, β-cells of the pancreas, adrenal chromaffin cells or renal juxtaglomerular cells, should lead to a decrease in the cell volume. It is conceivable therefore that regulation of the cell volume in these cells in necessary for the volume substitution of exocytosed material. The cell volume, in turn, will exert influence on the tension of the plasma membrane which could be a determinant for the fusion of the secretory vesicles with the plasma membrane, as it has been found for the tension of the secretory granule membrane (Monck et al. 1991).

The cell volume, moreover, may be relevant for the volume of secretory granules, because it is a well-established observation that secretory granules swell during exocytosis (Finkelstein et al. 1986). It is still a matter of debate whether swelling of the vesicles occurs prior to membrane fusion or after establishment of a permanent fusion pore (Green 1987). In the first case water flux from the cytosol into the secretory granule would be a requirement for secretion, in the second case water flux into the vesicle would occur from the extracellular space. Indirect and morphological evidence suggests that swelling of the secretory granules occurs before membrane fusion, suggesting that the granule's volume, i.e. the water flux into the vesicles, determines fusion and secretion (Yanagimachi and Usui 1974; Pollard et al. 1979; Cohen et al. 1980; Hampton and Holz 1983; Zimmerberg and Whitaker 1985). All of these examinations, however, are hampered by comparably poor time resolutions. Application of a combination of patch-clamp and optical methods on mast cells from beige mice, which contain huge secretory granules, has provided good evidence that the establishment of a fusion pore occurs prior to the granule's swelling and that most likely water flux from the outside then causes swelling of the vesicles (Breckenridge and Almers 1987; Zimmerberg et al. 1987). Explanations for these discrepant findings may be that mutant mast cells are not representative of the general mode of exocytotic secretion, or more likely that small and local swellings of the granules are indeed required for membrane fusion, which are then followed by a strong water flux from the outside (Green 1987).

The evidence that at least a certain swelling of the granules is necessary for fusion is inferred from observations that exocytosis from a variety of cells is dependent on the extracellular osmolarity. Thus, hyperosmolarity is associated with a decreased secretion from a variety of endocrine cells including adenohypophyseal cells (Greer et al. 1985), pancreatic islets (Murakami et al. 1986), adrenal chromaffin cells (Hampton and Holz 1983; Pollard et al. 1984), leukocytes (Sato et al. 1990a) and renal juxtaglomerular cells (Skott 1988). Parathyroid cells in which hyperosmolarity has been found to enhance secretion (Brown et al. 1978; Chen et al. 1987) behave somewhat unusually. Hypoosmolarity, on the other hand, has been found to stimulate secretion in secretory cells including pancreatic islets (Blackard et al. 1975), pituitary cells (Greer et al. 1983; Sato et al. 1990b) and renal juxtaglomerular cells (Skott 1988). For renal juxtaglomerular cells it has been demonstrated that the effectiveness of changes in osmolarity on secretion is dependent on the reflexion coefficient of the particle used to change osmolarity (Frederiksen et al. 1975). From that it appears likely that it is the change in cell or granule volume rather than the change in the intracellular ion concentration that is responsible for the effect on secretion, in such a way that cell and/or granule swelling is favourable for exocytosis, while the opposite holds for cell and granule shrinkage. In keeping with this conclusion is the demonstration that also isotonic swelling of pituitary cells is a powerful stimulus for the secretion of prolactin and thyrotropin from these cells (Sato et al. 1991). Although it is well established that hyperosmolarity causes cell shrinkage, prevents swelling of secretory granules, membrane fusion, dispersal of granule contents after fusion (Pollard et al. 1984; Zimmerberg et al. 1987) and suppresses agonist-induced increases in intracellular calcium (Kazilek et al. 1988), the crucial mechanism by which osmolarity influences exocytosis has not yet been unequivocally identified. Moreover, there appear to be marked differences in the sensitivity of secretion towards changes in osmolality. Thus, osmolality must be increased by 50% with sucrose to inhibit secretion from chromaffin cells half-maximally (Holz 1986) and by 60% to exert a similar effect on insulin secretion from pancreatic islets (Pace and Smith 1983). A 5% rise in osmolality with sucrose, however, has been found to be sufficient to halve renin secretion from renal juxtaglomerular cells in vitro, suggesting that the exocytosis of renin is particularly sensitive to volume changes (Skott 1988). Conversely, a 5% decrease in osmolarity has been found to double renin secretion (Skott 1988), whereas a 15% decrease in osmolarity is required to double, for instance, prolactin secretion from pituitary cells (Sato et al. 1990b).

6 Possible Regulatory Role of the Cell Volume for Renin Secretion

In view of the striking sensitivity of renin secretion from renal juxtaglomerular cells towards extracellular osmolarity, it appears reasonable to further consider a possible regulatory role of the cell or granule volume in secretion in more detail. For juxtaglomerular cells morphological evidence was also produced that the swelling of the granules determines the rate of exocytosis and occurs before membrane fusion (Skott and Taugner 1987). In particular, it has been found that granule swelling is not uniform but restricted to certain areas of the vesicle which then cause a protrusion of the granule's membrane towards the plasma membrane (Taugner et al. 1984). A locally restricted swelling of the granules as a requirement for fusion would also fit recent results, providing evidence for a chemiosmotic control of renin secretion from storage granules (Sigmon and Fray 1991). As mentioned before, the effect of extracellular osmolarity on renin secretion is strongly dependent on the reflexion coefficient of the particle used to change osmolarity (Frederiksen et al. 1975). Thus, an increase in osmolarity by 100 mM sucrose causes a strong inhibition of renin secretion from isolated juxtaglomerular cells, while increasing extracellular NaCl by 100 mM has a stimulatory rather than an inhibitory effect on renin secretion (Fig. 4). Decreasing extracellular osmolality by removal of 50 mM NaCl from the medium, on the other hand, is associated with a marked transient stimulation of renin secretion, followed by a suppression of renin secretion (Fig. 4). Since sodium and chloride are expected to move through the juxtaglomerular cell's plasma membrane in analogy to other vascular smooth muscle cells either via cotransport systems such as the Na-K-Cl_2 cotransport (Owen 1984) or via antiporters such as Na/H exchange (Kurtz et al. 1991) or Cl/HCO_3 exchanger (Skott and Jensen 1989), while sucrose does not enter the cells, it is reasonable to assume that a change in the intracellular ion concentration or ionic strength is not the cause for the effect of extracellular osmolality on renin secretion. Consequently, it appears to be a mechanism related to cell volume and/or secretory granule volume. Also, the transient

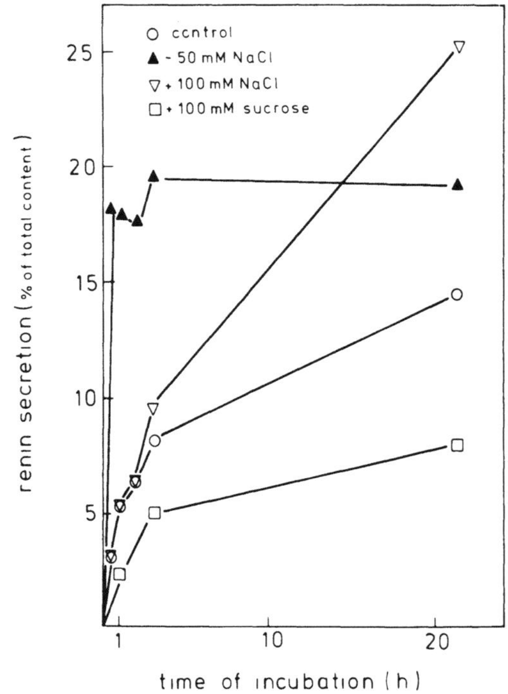

Fig. 4. Renin secretion from cultured mouse renal juxtaglomerular cells after incubation in hypo- and hyperosmotic culture media. Renin secretion is presented as percentage of total cellular renin content. Details of the cell cultures are given in Della Bruna et al. (1991)

stimulation of renin secretion induced by acute hypoosmolarity would be compatible with a relation to the cell volume, because cells stressed by hypoosmotic shock rapidly recover their original cell volume (Wong and Chase 1986). How cell shrinkage inhibits secretion is not yet known. It could be due to an effect on membrane tension which could be of importance for the membrane fusion during exocytosis. A decrease in cell volume could decrease the membrane tension and consequently decrease the effectiveness of secretion. Secondly, a decrease in cell volume could somehow impede water influx into the secretory granules and thus prevent exocytosis. Irregardless of the precise mechanism, it should be recalled at this point that secretion from renal juxtaglomerular cells is an exception of secretory cells in which calcium-mobilizing hormones appear to inhibit rather than to stimulate the secretory process, a phenomenon which is referred to as the "calcium paradox" of renin secretion (Kurtz 1989). It is an as yet unanswered question whether calcium has a rather unique inhibitory effect on secretion in these cells, or whether the inhibitory effect of calcium is indirectly mediated. In fact, the marked sensitivity of renin secretion towards changes in the cell volume could provide an explanation for an indirect mediation.

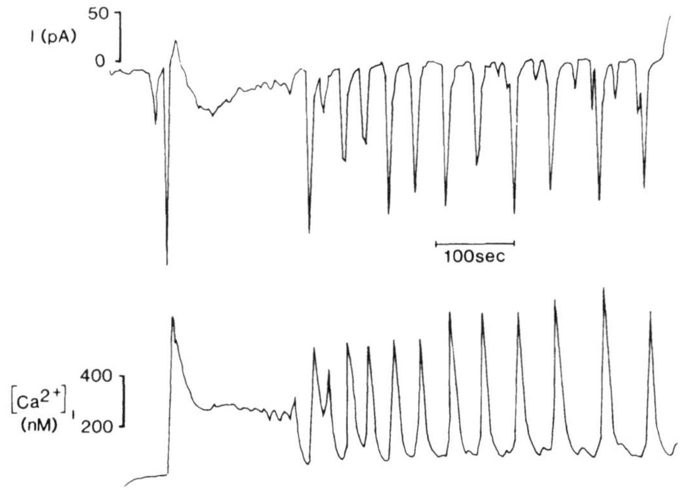

Fig. 5. Simultaneous recordings of chloride current (*upper*) and intracellular calcium concentration (*lower*) in a mouse juxtaglomerular cell during stimulation with internal GTP-g-S (100 μM). Voltage clamp potential was −60 mV (Kurtz 1991). Details of the method are given in Kurtz and Penner (1989)

Recently, it was found that renal juxtaglomerular cells possess calcium-activated chloride channels (Fig. 5, Kurtz and Penner 1989). By ensemble noise analysis a unitary conductance of only a few pS but a high number of about 10000 channels per juxtaglomerular cell has been calculated (Kurtz and Penner, unpubl.), a situation similar to exocrine acinar cells. The whole cell chloride conductance can therefore increase up to 50 nS (Kurtz and Penner 1989) in relation to the intracellular calcium concentration. The resting membrane potential of juxtaglomerular cells is between −70 and −80 mV (Fishman 1976; Bührle et al. 1985) and thus more negative than the chloride equilibrium potential which has been estimated to be around −30 mV in other smooth muscle cells (Aickin and Brading 1985). As a consequence, juxtaglomerular cells have a calcium-regulated chloride efflux. This chloride efflux is probably followed by a potassium efflux through voltage-operated potassium channels which become activated by the membrane depolarization produced by the chloride efflux (Kurtz 1990). The consequence of both events is an efflux of KCl with a substantial loss of intracellular, osmotically active electrolytes. It has been estimated that during one calcium spike about 6% of the intracellular osmolytes leaves the cells (Kurtz 1990). This loss of electrolytes should then be followed by cell shrinkage, as it has been observed in exocrine acinar cells. Since cell shrinkage appears to be a powerful inhibitory mechanism for renin secretion (Skott 1988; Fig. 4), it is not unlikely that the apparent inhibitory effect of calcium on secretion from juxtaglomerular cells could at least in part be mediated via cell volume changes. Similar to other cells calcium-mobilizing hormones induce oscillations of the cytosolic concentration of calcium (Kurtz and Penner 1989). As a consequence, the mean cellular volume of juxtaglomerular cells will be dependent on the frequency of calcium oscillations, which in turn are determined by the hormone concentration (Rink and Jacob 1989). According to this view, the cell volume would be causally involved in the stimulus-secretion coupling in these cells. Interestingly, in vivo investigations have shown that diurnal renin secretion rates are highest at that time when the juxtaglomerular cell volume reaches its maximum during the day (Watanabe et al. 1988).

With regard to the possible role of the cell volume in PTH secretion, the question arises as to whether such an indirect inhibition of secretion by calcium could also provide an explanation for a second "calcium paradox", namely the inhibitory effect of calcium for PTH secretion from parathyroid cells (Brown 1991). Parathyroid cells are considered to have receptors for divalent cations on their surface that are somehow linked to a phospholipase C (Brown et al. 1987; Nemeth and Scarpa 1987). By that mechanism the concentration of calcium could be transduced into the cytosolic concentration of calcium. Evidence that the inhibitory effect of extracellular calcium on PTH secretion could be indirect, was provided by the finding that in electropermeabilized parathyroid cells, in which all transmembrane ion movements are short-circuited, calcium exerts a normal, i.e. positive, effect on PTH secretion (Oetting et al. 1987). A calcium-regulated ion transport system could be given with the calcium-sensitive potassium currents that were recently discovered in parathyroid cells (Jia et al. 1988). These calcium-activated potassium channels were found to possess a rather unusual calcium sensitivity. They become activated at lower calcium concentrations and inactivated at higher cytosolic calcium concentrations. Consistent with these findings would be that parathyroid cells hyperpolarize at lower calcium concentrations (Bruce and Anderson 1979) and depolarize at higher calcium concentrations (Lopez-Barneo and Armstrong 1983). If these calcium-activated potassium channels are involved in the volme regulation of parathyroid cells, one should expect cell shrinkage at lower calcium concentrations because of the potassium efflux and relative cell swelling at higher calcium concentrations due to the inactivation of the potassium channels. PTH secretion from parathyroid cells has been found to be sensitive to changes in extracellular osmolarity (Brown et al. 1978; Chen et al. 1987) and a possible role of the cell volume in the regulation of PTH secretion has therefore, in fact, been considered (Chen et al. 1987). Unlike with other cells, however, an increase in extracellular osmolarity and, consequently, cell shrinkage, appears to be a potent secretogogue in dispersed parathyroid cells (Brown et al. 1978; Chen et al. 1987). Such an unusual influence of the cell volume on PTH secretion would indeed be compatible with a causal role of the calcium-sensitive potassium channels in mediating an inverse dependency of PTH secretion on the extracellular concentration of calcium.

7 Summary

A wealth of mostly indirect evidence suggests that the cell volume of secretory cells is involved in stimulus-secretion coupling. Although a direct demonstration of cell volume changes during secretion has so far only been shown for salivary acinar cells, it is likely that other exocrine acinar cells behave rather similarly. Whether these volume changes occur only passively due to ion efflux or have a physiological meaning cannot be answered as yet.

Changes in the cell volume could be a regulatory step for exocytosis from certain endocrine cells, in particular, those in which calcium exerts an inhibitory rather than a stimulatory effect on secretion. For those cells, however, a direct demonstration for cell volume changes is still lacking.

Acknowledgements. The authors wish to thank Hannelore Trommer for secretarial help and Karl-Heinz Götz for doing the artwork.

References

Aickin CC, Brading AF (1985) Advances in the understanding of transmembrane ionic gradients and permeabilities in smooth muscle obtained by using ion-selective microelectrodes. Experientia 41: 879–887

Barrowman MM, Cockcroft S, Gomperts BD (1986) Two roles for guanine nucleotides in the stimulus-secretion sequence of neutrophils. Nature 319: 504–507

Berridge MJ, Irvine RF (1984) Inositoltrisphosphate, a novel second messenger in cellular signal transduction. Nature 312: 315–321

Blackerd WG, Kikuchi M, Rebinovitch A, Renold AE (1975) An effect of hypoosmolarity on renin release in vitro. Am J Physiol 228: 706–712

Brayden DJ, Cuthbert AW, Lee CM (1988) Human eccrine sweat gland epithelial cultures express ductal characteristics. J Physiol 405: 657–675

Breckenridge LJ, Almers W (1987) Final steps in exocytosis observed in a cell with giant secretory granules. Proc Natl Acad Sci USA 84: 1945–1949

Brown EM (1991) Extracellular Ca^{2+} sensing, regulation of parathyroid cell function, and role of Ca^{2+} and other ions as extracellular (first) messengers. Physiol Rev 71: 371–410

Brown EM, Pazoles CJ, Creutz CE, Aurbach GD, Pollard HB (1978) Role of anions in parathyroid hormone release from dispersed bovine parathyroid cells. Proc Natl Acad Sci USA 75: 876

Brown EM, Enyedi P, Leboff M, Rotberg J, Preston J, Chen C (1987) High extracellular Ca^{2+} and Mg^{2+} stimulate accumulation of inositol phosphates in bovine parathyroid cells. FEBS Lett 218: 113–118

Bruce BR, Anderson NS Jr (1979) Hyperpolarization in mouse parathyroid cells by low calcium. Am J Physiol 236(1): C15–C21

Bührle CP, Nobiling R, Taugner R (1985) Intracellular recordings from renin-positive cells of the afferent arteriole. Am J Physiol 249: F272–F281

Burgen ASV, Emmelin N (1961) Physiology of salivary glands. (Physiol Soc Monogr 8) Arnold, London

Casey PJ, Gilman AG (1988) G protein involvement in receptor-effector coupling. J Biol Chem 263: 2577–2580

Chen CJ, Anast CS, Brown EM (1987) High osmolality: a potent parathyroid hormone secretogogue in dispersed parathyroid cells. Endocrinology 121: 958–964

Cockcroft S, Gomperts BD (1985) Role of guanine nucleotide binding protein in the activation of polyphosphoinositide phosphodiesterase. Nature 314: 534–536
Cohen FS, Zimmerberg J, Finkelstein A (1980) Fusion of phospholipid vesicles with planar phospholipid bilayer membranes. II. Incorporation of vesicular membrane marker into the planar membrane. J Gen Physiol 75: 251–270
De Lisle RC, Williams JA (1986) Regulation of membrane fusion in secretory exocytosis. Annu Rev Physiol 48: 225–238
Della Bruna R, Pinet F, Corvol P, Kurtz A (1991) Regulation of renin secretion and renin synthesis by second messengers in isolated mouse juxtaglomerular cells. Cell Physiol Biochem 1: 98–110
Douglas WW (1968) Stimulus secretion coupling: the concept and clues from chromaffin and other cells. Br J Pharmacol 34: 451–474
Exley PM, Fuller CM, Gallacher DV (1986) Potassium uptake in the mouse submandibular gland is dependent on chloride and sodium and abolished by piretanide. J Physiol 378: 97–108
Finkelstein A, Zimmerberg J, Cohen FS (1986) Osmotic swelling of vesicles: its role in the fusion of vesicles with planar phospholipid bilayer membranes and its possible role in exocytosis. Annu Rev Physiol 48: 163–174
Fishman MC (1976) Membrane potential of juxtaglomerular cells. Nature 260: 542–544
Foskett JK (1990) Ca modulation of Cl^- content controls cell volume in single salivary acinar cells during fluid secretion. Am J Physiol 259: C998–C1004
Foskett JK, Melvin JE (1989) Activation of salivary secretion: coupling of cell volume and Ca in single cells. Science 244: 1582–1585
Frederiksen O, Leyssac PP, Skinner SL (1975) Sensitive osmometer function of juxtaglomerular cells in vitro. J Physiol 252: 669–679
Frizzel RA, Halm DR, Rechkemmer G, Shoemaker RL (1986) Chloride channel regulation in secretory epithelia. Fed Proc 45: 2727–2731
Green DPL (1987) Granule swelling and membrane fusion in exocytosis. J Cell Sci 88: 547–549
Greer MA, Greer SE, Opsahl Z, McCafferty L, Maruta S (1983) Hypoosmolar stimulation of in vitro pituitary secretion of luteinizing hormone: a potential clue to the secretory process. Endocrinology 113: 1531–1533
Greer MA, Greer SE, Opsahl Z, Maruta S (1985) Comparison of hypoosmolar and hyperosmolar effects on in vitro luteinizing hormone secretion by anterior pituitary cells. Proc Soc Exp Biol Med 178: 24–28
Hampton RY, Holz RW (1983) Effects of changes of osmolality on the stability and function of cultured chromaffin cells and the possible role of osmotic forces in exocytosis. J Cell Biol 96: 1082–1088
Holz RW (1986) The role of osmotic forces in exocytosis from adrenal chromaffin cells. Annu Rev Physiol 48: 175–189
Iwatsuki N, Petersen OH (1978) Electrical coupling and uncoupling of exocrine acinar cells. J Cell Biol 79: 533–545
Jia M, Ehrenstein G, Iwasa K (1988) Unusual calcium activated potassium channels of bovine parathyroid cells. Proc Natl Acad Sci USA 85: 7236–7239
Kanagasuntheram P, Randle PJ (1976) Calcium metabolism and amylase release in rat parotid acinar cells. Biochem J 160: 547–564
Kasai H, Augustine GJ (1990) Cytosotic calcium gradients triggering unidirectional fluid secretion from exocrine pancreas. Nature 348: 735–738
Kazilek CJ, Merkle CJ, Chandler DE (1988) Hyperosmotic inhibition of calcium signals and exocytosis in rabbit neutrophils. Am J Physiol 254: C709–C718
Keryer G, Rossignol B (1976) Effect of carbachol on ^{45}Ca uptake and protein secretion in rat lacrimal gland. Am J Physiol 230: 99–104
Knight DE (1987) Calcium and diacylglycerol control of secretion. Biosci Rep 7: 355–367
Krouse ME, Hagiwara G, Chen J, Lewiston NJ, Wine JJ (1989) Ion channels in normal human and cystic fibrosis sweat gland cells. Am J Physiol 257: C129–C140
Kurtz A (1989) Cellular control of renin secretion. Rev Physiol Biochem Pharmacol 113: 1–40

Kurtz A (1990) Do calcium-activated chloride channels control renin secretion? News Physiol Sci 5: 43–46
Kurtz A (1991) Intracellular and membrane events in the activation of afferent arteriolar and granular cells. In: Hatano M (ed) Nephrology. Springer, Berlin Heidelberg New York, pp 323–337
Kurtz A, Penner R (1989) Angiotensin II induces oscillations of intracellular calcium and inhibits inward rectifying potassium current in renal juxtaglomerular cells. Proc Natl Acad Sci USA 86: 3423–3427
Kurtz A, Della Bruna R, Scholz H, Baier W (1991) Amiloride enhances the secretion but not the synthesis of renin in renal juxtaglomerular cells. Pflügers Arch Eur J Physiol 418: 31–37
Lau KR, Case RM (1988) Evidence for apical chloride channels in rabbit submandibular salivary glands. A chloride-selective microelectrode study. Pflügers Arch Eur J Physiol 411: 670–675
Litosch I, Wallis C, Fain JN (1985) 5-Hydroxytryptamine stimulates inositol phosphate production in a cell-free system from blowfly salivary glands. J Biol Chem 260: 5464–5471
Lopez-Barneo J, Armstrong CM (1983) Depolarizing response of rat parathyroid cells to divalent cations. J Gen Physiol 82: 269–294
Luini A, Brown DA (1990) Effects of corticotrophin releasing factor, muscarine and somatostatin on rubidium and potassium efflux from mouse AtT-20 pituitary cells. Eur J Neurosci 2: 126–131
Lundberg A (1958) Electrophysiology of salivary glands. Physiol Rev 38: 21–40
Lundberg JM, Hökfelt T (1983) Coexistence of peptides and classical neurotransmitters. Trends Neurosci 6: 325–333
Martin K, Burgen ASV (1962) Changes in the premeability of the salivary gland caused by sympathetic stimulation and by catecholamines. J Gen Physiol 48: 113–138
Marty A (1987) Control of ionic currents and fluid secretion by muscarinic agonists in exocrine glands. Trends Neurosci 10: 373–377
Marty A, Evans MG, Tan YP, Trautmann A (1986) Muscarine response in rat lacrimal glands. J Exp Biol 124: 15–32
Maruyama Y, Petersen OH (1984a) Calcium activated potassium channels and their role in secretion. Nature 307: 693–696
Maruyama Y, Petersen OH (1984b) Single-channel currents in isolated patches of plasma membrane from basal surface of pancreatic acini. Nature 299: 159–161
Mazariegos MR, Tice LW, Hand AR (1984) Alteration of tight junctional permeability in the rat parotid gland after isoproterenol stimulation. J Cell Biol 98: 1865–1877
Meda P, Findlay I, Kolod E, Orci L, Petersen OH (1983) Short and reversible uncoupling evokes little change in the gap junctions of pancreatic acinar cells. J Ultrastruct Res 83: 69–84
Monck JR, Oberhauser AF, Alvarez de Toledo G, Fernandez JM (1991) Is swelling of the secretory granule matrix the force that dilates the exocytic fusion pore. Biophys J 59: 39–47
Murakami M, Greer SE, Opsahl Z, Greer MA (1986) Effect of osmotic change on insulin secretion by perifused rat pancreatic islets. Program 68th Annu Meet Endocrine Soc, Anaheim CA, p 289 (Abstr)
Nauntofte B, Dissing S (1988) K^+ transport and membrane potentials in isolated rat parotid acini. Am J Physiol 255: C508–C518
Nemeth EF, Scarpa A (1987) Rapid mobilization of cellular Ca^{2+} in bovine parathyroid cells by external divalent cations. J Biol Chem 202: 5188–5196
Nishizuka Y (1984) The role of protein kinase C in cell surface signal transduction. Nature 308: 693–698
Oetting M, Leboff MS, Levy S, Swiston L, Preston J, Chen C, Brown EM (1987) Permeabilization reveals classical stimulus-secretion coupling in bovine parathyroid cells. Endocrinology 121: 1571–1576
Owen N (1984) Regulation of Na/K/Cl cotransport in vascular smooth muscle cells. Biochem Biophys Res Commun 125: 500–508

Pace CS, Smith JS (1983) The role of chemiosmotic lysis in the exocytotic release of insulin. Endocrinology 113: 964–969
Penner R (1988) Multiple signaling pathways control stimulus-secretion coupling in rat peritoneal mast cells. Proc Natl Acad Sci USA 85: 9856–9860
Penner R, Neher E (1988) The role of calcium in stimulus-secretion coupling in excitable and non-excitable cells. J Exp Biol 139: 329–345
Penner R, Matthews G, Neher E (1988) Regulation of calcium influx by second messengers in rat mast cells. Nature 334: 499–504
Peterson OH (1980) The electrophysiology of gland cells. Academic Press, New York
Petersen OH (1986) Potassium channels and fluid secretion. News Physiol Sci 1: 92–95
Petersen OH, Findlay I (1987) Electrophysiology of the pancreas. Physiol Rev 67: 1054–1120
Petersen OH, Gallacher DV (1988) Electrophysiology of pancreatic and salivary acinar cells. Annu Rev Physiol 50: 65–80
Pollard HB, Pazoles CJ, Creutz CE, Zinder O (1979) The chromaffin granule and possible mechanisms of exocytosis. Int Rev Cytol 58: 159–197
Pollard HB, Pazoles CJ, Creutz CE, Scott JH, Zinder O, Hotchkiss A (1984) An osmotic mechanism for exocytosis from dissociated chromaffin cells. J Biol Chem 259: 1114–1121
Pralong WF, Wollheim CB, Bruzzone R (1988) Measurement of cytosolic free Ca^{2+} in individual pancreatic acini. FEBS Lett 242: 79–84
Rapp PE, Berridge MJ (1981) The control of transepithelial potential oscillations in the salivary gland of calliphora erythricephala. J Exp Biol 93: 119–132
Rasmussen H, Barrett PQ (1984) Calcium messenger system: an integrated view. Physiol Rev 64: 938–984
Rink TJ, Jacob R (1989) Calcium oscillations in non-excitable cells. Trends Neurosci 12: 43–46
Rubin RP (1982) Calcium and cellular secretion. Plenum Press, New York
Stao N, Wang X, Greer MA (1990a) Hypoosmolarity stimulates exocytosis from human polymorphonuclear leukocytes. Am J Med Sci 289: 309–312
Sato N, Wang X, Greer MA, Greer SE, McAdams S, Oshima T (1990b) Medium hypoosmolarity stimulates prolactin secretion in GH_4C_1 cells by inducing an increase in cytosolic free calcium. Endocrinology 127: 957–964
Sato N, Wang X, Greer MA (1991) Hormone secretion stimulated by ethanol-induced cell swelling in normal rat adenohypophysial cells. Am J Physiol 260: E946–E950
Selinger Z, Batzri S, Eimerl S, Schramm M (1973) Calcium and energy requirements for K^+ release mediated by the epinephrine α-receptor in rat parotid slices. J Biol Chem 248: 369–372
Sigmon DH, Fray JCS (1991) Chemiosmotic control of renin release from isolated renin granules of rat kidneys. J Physiol 436: 237–256
Skott O (1988) Do osmotic forces play a role in renin secretion? Am J Physiol 255: F1–F10
Skott O, Jensen BL (1989) Influence of bicarbonate on the sensitivity of renin release to sodium chloride. Pflügers Arch Eur J Physiol 414: 651–655
Skott O, Taugner R (1987) Effects of extracellular osmolality on renin release and on the ultrastructure of the juxtaglomerular epitheloid cell granules. Cell Tissue Res 249: 325–329
Streb H, Irvine RF, Berridge MJ, Schulz I (1983) Release of Ca^{2+} from a nonmitochondrial intracellular store in pancreatic acinar cells by inositol-1,4,5-triphosphate. Nature 306: 67–69
Taugner R, Bührle CP, Hackenthal E, Mannek E, Nobiling R (1984) Morphology of the juxtaglomerular apparatus and secretory mechanisms. Contrib Nephrol 43: 76–101
Turner RJ, George NJ, Baum BJ (1986) Evidence for a $Na^+/K^+/Cl^-$ cotransport system in basolateral membrane vesicles from the rabbit parotid. J Membr Biol 94: 143–152
Wakui M, Potter BVL, Petersen OH (1989) Pulsatile intracellular calcium release does not depend on fluctuations in inositol trisphosphate concentration. Nature 339: 317–320

Wakui M, Osipchuk YV, Petersen OH (1990) Receptor-activated cytoplasmic Ca^{2+} spiking mediated by inositol trisphosphate is due to Ca^{2+} induced Ca^{2+} release. Cell 63: 1025–1032
Watanabe T, Matsuba H, Uchiyama Y (1988) Correlation of 24-hour fluctuations in renin granules of juxtaglomerular cells and in renin and angiotensinogen in blood plasma of the rat. Cell Tissue Res 254: 593–598
Wong SME, Chase HS Jr (1986) Role of intracellular calcium in cell volume regulation. Am J Physiol 250: C841–C852
Yanagimachi R, Usui N (1974) Calcium dependence of the acrosome reaction and activation of guinea pig spermatozoa. Exp Cell Res 89: 161–174
Zimmerberg J, Whitaker M (1985) Irreversible swelling of secretory granules during exocytosis caused by calcium. Nature 315: 581–584
Zimmerberg J, Curran M, Cohen FS, Brodwick M (1987) Simultaneous electrical and optical measurements show that membrane fusion precedes secretory granule swelling during exocytosis of beige mouse mast cells. Proc Natl Acad Sci USA 84: 1585–1589

Chapter 5

Cell Volume Regulation and Vascular Endothelial Function

W.C. O'Neill[1]

Contents

1 Introduction

In the vasculature, the endothelium consists of a single monolayer that forms an interface between the blood and the underlying vessel and tissue. Accordingly, it has specialized functions that are critical to normal vascular function. Most apparent is its role as a physical barrier. Endothelial injury with subsequent exposure of subendothelial matrix may be important in the genesis of a variety of thrombotic disorders. Endothelial denudation may be an early event in atherosclerosis, resulting in platelet activation, release of platelet-derived growth factor, and eventual migration and proliferation of smooth muscle cells in the intima (Ross 1986). In addition to being a physical barrier, endothelium is a selective permeability barrier, allowing passage of nutrients and restricting passage of macromolecules. This is particularly important in the cerebral vasculature, where endothelium forms the blood-brain barrier (Janzer and Raff 1987), and in the renal glomerulus, where endothelium contributes to the ultrafiltration barrier. Endothelium also has important paracrine functions in blood vessels that allow it to regulate platelet and leukocyte function, on the one hand, and vascular smooth muscle tone, on the other (Vane et al. 1990). Through these various properties the endothelium plays a critical role in governing thrombosis, inflammation, and hemodynamics.

[1] Renal Division, Department of Medicine, Emory University School of Medicine, Atlanta, Georgia, USA

Advances in Comparative and Environmental Physiology, Vol. 14

Endothelial function is regulated by both neurohumoral and mechanical stimuli through mechanisms that are not fully understood. Changes in ion transporters are important early events in the response to these stimuli. Many of these transporters serve to regulate cell volume and may be responsible for the morphological changes that occur in endothelial cells (Majno et al. 1969; Laposata et al. 1983; Wells et al. 1985; Killackey et al. 1986; Levesque et al. 1989; Shirinsky et al. 1989). Regulation of endothelial cell volume may therefore be an integral component of the response of endothelium to neurohumoral and physical stimuli. Since volume-regulatory pathways have only recently been described in vascular endothelium, their precise role in endothelial function is not fully understood.

2 Regulatory Volume Decrease in Endothelium

Recovery of cell volume after hypotonic swelling has been demonstrated in endothelial cells cultured from rat cerebral microvessels (Kempski et al. 1985), porcine aortae (Shepard et al. 1987), and human umbilical veins (Mazzoni et al. 1989), but the mechanism has not been determined. Endothelial cells cultured from bovine aortae also exhibit a brisk regulatory volume decrease (RVD) in hypotonic medium (Ling and O'Neill 1991), due to loss of K (Fig. 1). Cell K^+ content is restored during subsequent reincubation in isotonic medium, indicating that the original K loss with

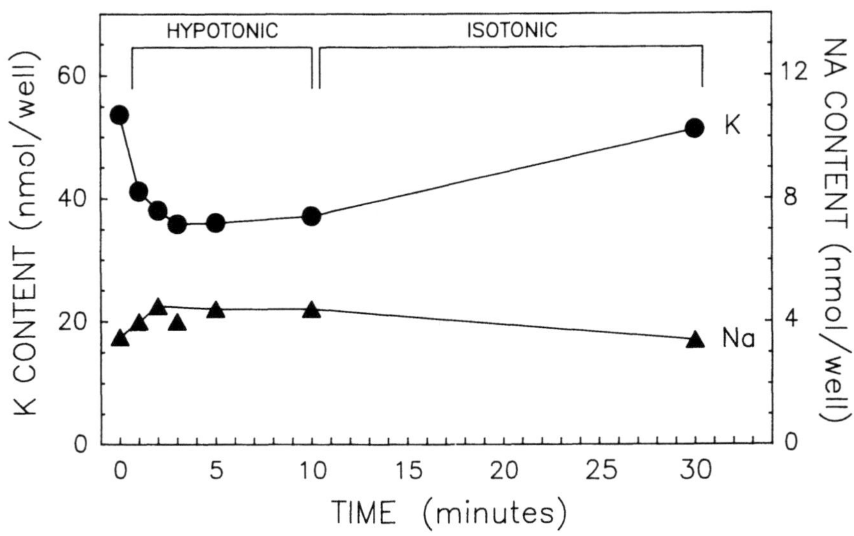

Fig. 1. Regulatory volume decrease in bovine aortic endothelial cells. K content decreased, and Na content increased, with recovery in isotonic medium. Ion content was determined by atomic absorption spectrophotometry

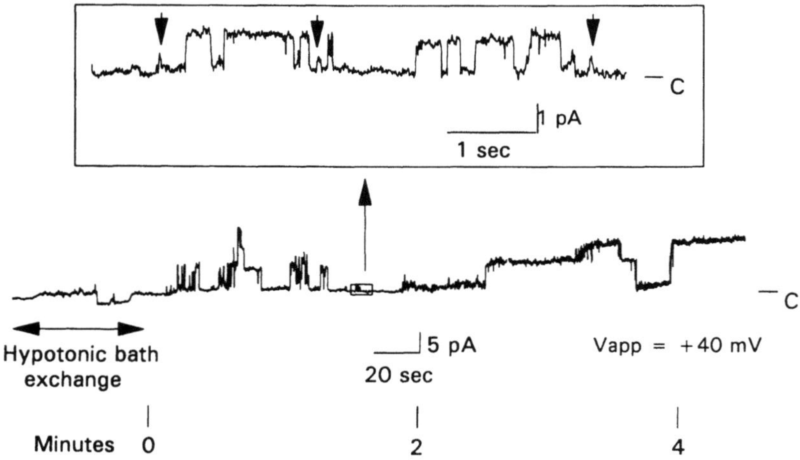

Fig. 2. Cation channels activated by hypotonic swelling of bovine aortic endothelial cells. Large current events are from 165-pS channels. *Inset* is a magnified portion showing 42-pS channels and 28-pS channels (*arrows*)

hypotonic exposure was not due to cell lysis. We have examined the mechanism for this K loss using patch clamp methodology (Ling and O'Neill 1991). Two Ca-dependent, K-selective channels (165 and 42 pS) are activated in cell-attached patches when bovine aortic endothelial cells are swollen (reducing bath osmolality from 285 to 215 mOsmol/kg H_2O), with a time course that corresponds to RVD (Fig. 2). Outward K current through these channels appears to be responsible for the K efflux observed in swollen endothelial cells.

The 165-pS channel was rarely observed under isotonic conditions, but its open probability (P_o) increased dramatically in hypotonic media (Fig. 3). Under isotonic conditions the same 165-pS channel is activated by depolarization or by an increase in cytoplasmic Ca (as determined in excised patches). This channel was highly K-selective, with a P_K/P_{Na} of 25 and may be identical to the 105 pS voltage-sensitive, Ca-activated K channel that has previously been identified in bovine aortic endothelial cells (Fichtner et al. 1987). These characteristics are similar to swelling-activated "maxi-K" channels described in a variety of mammalian cells (Sun et al. 1990; Lang and Rehwald 1991; Ling et al. 1991).

The 42-pS channel is apparent as brief, flickering openings of smaller amplitude, superimposed on the larger conductance K channels (Fig. 2, inset). This channel also differed from the 165-pS channel in that it displayed inward rectification. Studies in isotonic medium revealed a 40-pS channel with inward rectification and high K selectivity (P_K/P_{Na} = 64) that appears to correspond to the 42-pS channel in swollen cells. Open probability for the 42-pS channel was low (<0.05) and independent of applied membrane potential between −80 and +60 mV, but was increased substantially by a

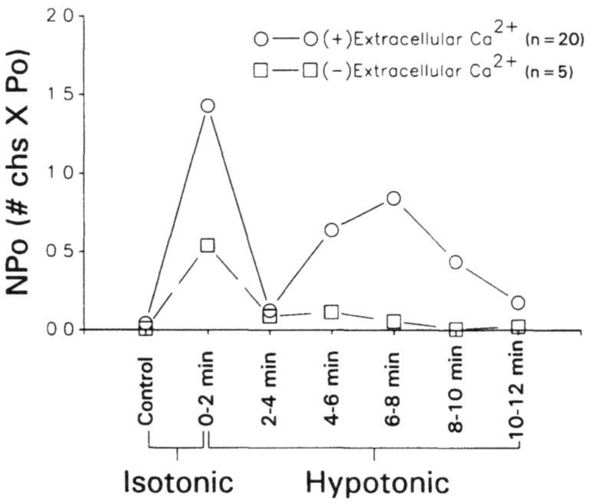

Fig. 3. Activation of 165-pS channels by cell swelling in bovine aortic endothelial cells in the presence and absence of extracellular calcium. NP_o is the number of channels multiplied by the open probability (P_o)

rise in cytoplasmic Ca or by exposure to exogenous bradykinin. Activation of this 40-pS, Ca-dependent K channel by bradykinin and ATP has been previously reported in aortic endothelial cells (Sauve et al. 1988; Colden-Stanfield et al. 1990). Activation of the 40-pS K channel by bradykinin is similar to activation by shrinkage in that it is biphasic, with the initial activation being independent of external Ca.

Activation of both potassium channels in swollen cells was biphasic. An initial, brief (0–2 min) opening occurred within seconds of hypotonic bath exchange and was followed by a delayed, but prolonged opening that lasted over 5 min. Removal of extracellular Ca from the hypotonic bath eliminated the delayed activation of the 165-pS channel but not the initial activation (Fig. 3). Removal of extracellular Ca had identical effects on the activation of the 42-pS channel by cell swelling.

Swelling of aortic endothelial cells also activated a 28-pS channel (Fig. 2, inset) that was nonselective for cations. Of particular importance is the inward conductance of Ca through this channel. When external (intrapipette) cations were replaced with 110 mM $CaCl_2$, inward current (Ca influx) was stimulated by cell swelling. Open probability for this latter 13-pS conductance was increased by membrane hyperpolarization. Activation of this nonselective channel could produce the Ca influx required for the delayed activation of K channels in swollen cells. Nonselective cation channels have been observed in endothelial cells obtained from a variety of sites including human umbilical vein, porcine aorta, bovine aorta, and brain microvessels from rat and pig (Fichtner et al. 1987; Lansman et al. 1987; Adams et al. 1989; Vigne et al. 1989; Nilius 1990). Activation occurs in response to histamine, hyperpolarization, or membrane stretch. Two of these nonselective channels have been shown to conduct inward Ca current (8 and 19 pS) with values for P_{Ca}/P_{Na} of 0.02 and 8.4 (Lansman et al. 1987; Nilius

1990). Given the large inward gradient for Na, activation of the 28-pS nonselective channel by swelling would favor inward Na movement, and this may be responsible for the small increase in Na content observed in swollen cells (Fig. 1). Similar nonselective cation channels that are activated by stretch and conduct calcium have been described in nonendothelial tissues, including renal, lens, and choroid plexus epithelia (Eaton 1991). Calcium influx through stretch-activated, nonselective cation channels has been proposed to be an early event in RVD, leading to subsequent activation of Ca^{2+}-dependent K^+ channels (Christensen 1987; Sun et al. 1990; Eaton 1991; Ling et al. 1991).

RVD requires increased efflux of anions (usually Cl) to accompany the efflux of K ions. Although this could be driven through existing conductance pathways by hyperpolarization from K channel activation, concurrent activation of chloride channels occurs in many swollen cells (Eveloff and Warnock et al. 1987; Hoffmann and Simonsen 1989; Ling et al. 1991). There have been no published descriptions of chloride channels in endothelial cells, which is consistent with the observation that replacement of extracellular Cl^- with impermeant anions has no affect on resting membrane potential of endothelial cells (Johns et al. 1987; Olesen et al. 1988b). Certain nonselective channels in nonendothelial tissues are capable of conducting anions as well as cations (Eaton 1991). However, replacement of patch pipette (external) Cl^- with aspartate had no affect on the unitary conductance or reversal potential of nonselective cation channels in endothelial cells (Lansman et al. 1987; Nilius 1990). We have examined anion conductance with measurements of $^{125}I^-$ efflux (Fig. 4). Cells were swollen

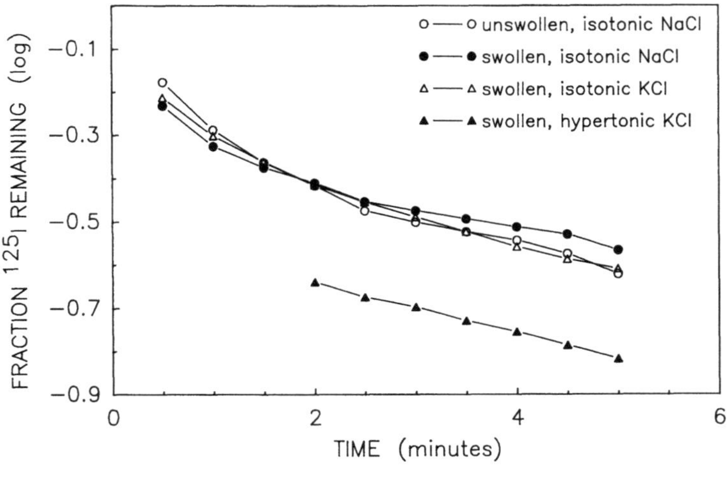

Fig. 4. Iodide efflux in unswollen and swollen bovine aortic endothelial cells. Cells were swollen isotonically by incubation in isotonic KCl medium prior to flux measurements. Influx was measured either into isotonic NaCl medium or into isotonic KCl medium (to prevent volume recovery). *Lower curve* is a continuation of the efflux from swollen cells after restoration of normal cell volume in hypertonic medium

isotonically in a high K medium. Swollen cells were compared to unswollen cells (either control cells never swollen, or isotonically swollen cells placed in hypertonic medium). Efflux was measured either into isotonic NaCl or into isotonic KCl (to prevent RVD). After an initial rapid efflux due to washout of extracellular ^{125}I, the efflux rates (slopes) were the same in swollen and unswollen cells. This indicates that basal anion conductance in aortic endothelial cells is adequate for RVD and that there is no increase in anion conductance in swollen cells.

3 Regulatory Volume Increase

Regulatory volume increase (RVI) in endothelial cells depends on the method of cell shrinkage. No volume recovery was observed in hypertonically

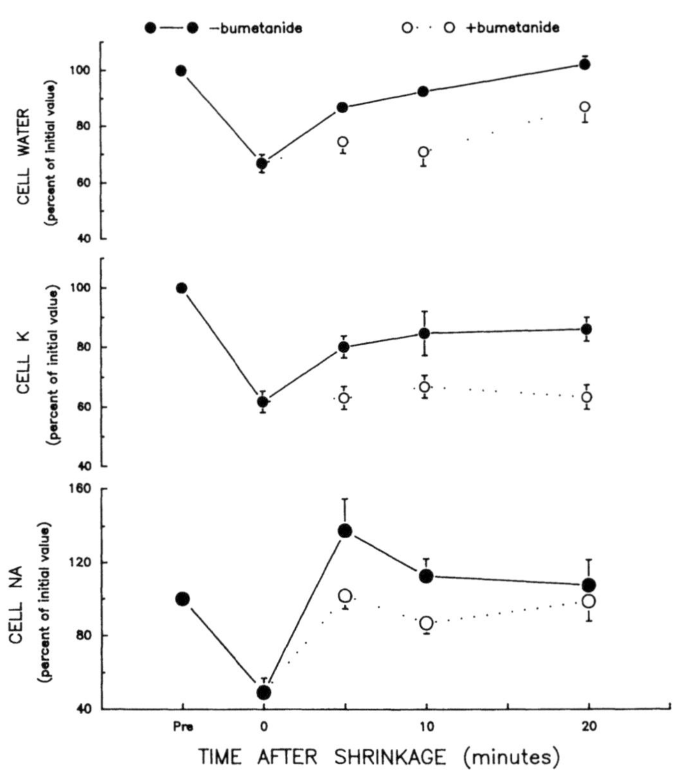

Fig. 5. Regulatory volume increase in bovine aortic endothelial cells. Cell K and Na content (measured by atomic absorption spectrophotometry) and water (measured with [^{14}C]urea) are shown before (*Pre*) and after (0) isotonic shrinkage in Na-free, K-free medium, and then during reexposure to standard isotonic medium in the presence and absence of 50 μM bumetanide

shrunken endothelial cells cultured from porcine aortae (Shepard et al. 1987), or human umbilical veins (Mazzoni et al. 1989). In bovine aortic endothelial cells there is also no volume recovery in hypertonic medium, but a brisk RVI occurs after isotonic shrinkage (O'Neill and Klein 1992), as shown in Fig. 5. Isotonic shrinkage results from the loss of cell K and Cl along with cell water, as opposed to hypertonic shrinkage where only cell water is lost. In Fig. 5 isotonic shrinkage was achieved by preincubating cells in isotonic medium devoid of Na and K (using N-methyl-D-glucamine). Isotonic shrinkage can also be produced by allowing cells to undergo RVD in hypotonic medium and then returning them to isotonic medium, and this also was followed by RVI in aortic endothelial cells (O'Neill and Klein 1992). RVI is inhibited by a low concentration of bumetanide (Fig. 5), a specific inhibitor of Na-K-2Cl cotransport, and is prevented by removing Na or Cl from the medium (Fig. 6). This indicates that RVI in these cells is mediated by Na-K-2Cl cotransport. Some volume recovery occurs in the presence of bumetanide. Since there was no corresponding increase in either Na or K content, this must occur through accumulation of other osmotically active molecules.

As expected, shrinkage of endothelial cells stimulated cotransport (O'Neill and Klein 1992), measured as unidirectional bumetanide-sensitive K^+ influx (Fig. 7). Isotonic shrinkage resulted in greater stimulation of cotransport than did hypertonic shrinkage. Conversely, isotonic swelling (achieved by preincubating cells in isotonic KCl medium) inhibited cotransport (Fig. 8). A 27% increase in endothelial cell volume reduced cotransport to 22% of the baseline activity observed in cells in normal volume. [^{3}H]bumetanide, which binds to Na-K-2Cl cotransporters with high affinity (Haas and Forbush 1986), has been used to determine whether

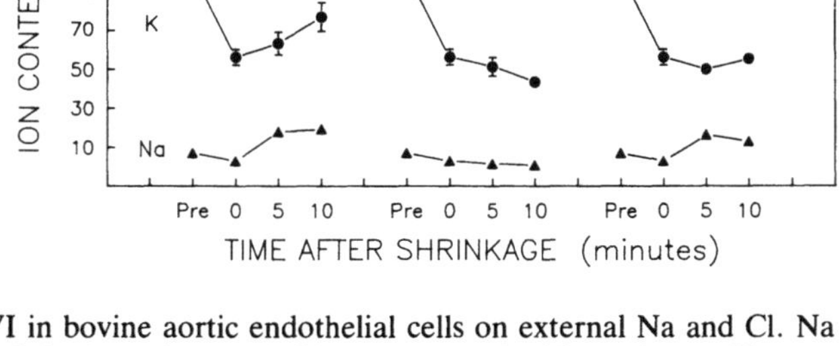

Fig. 6. Dependence of RVI in bovine aortic endothelial cells on external Na and Cl. Na was replaced with N-methyl-D-glucamine, and Cl was replaced with methanesulfonate. Procedure was as described for Fig. 5

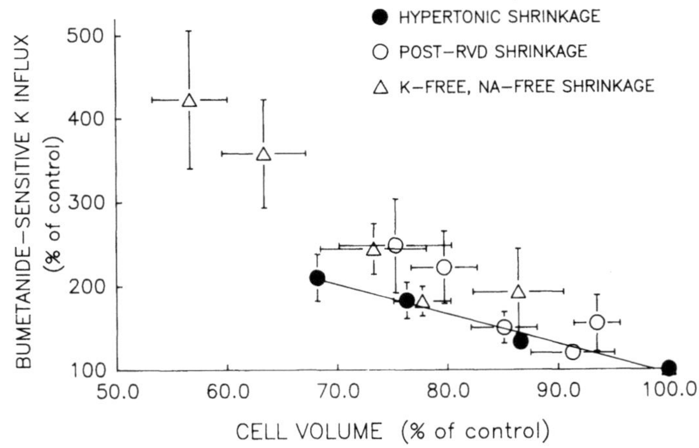

Fig. 7. Activation of Na-K-2Cl cotransport in bovine aortic endothelial cells by cell shrinkage. K influx was calculated from the unidirectional influx of ^{86}Rb in an isotonic, buffered saline solution containing 5.4 mM K. Hypertonic shrinkage was produced by adding sucrose. Cell volume was determined with [^{14}C]urea in hypertonically shrunken cells, and by Na and K content in isotonically shrunken cells

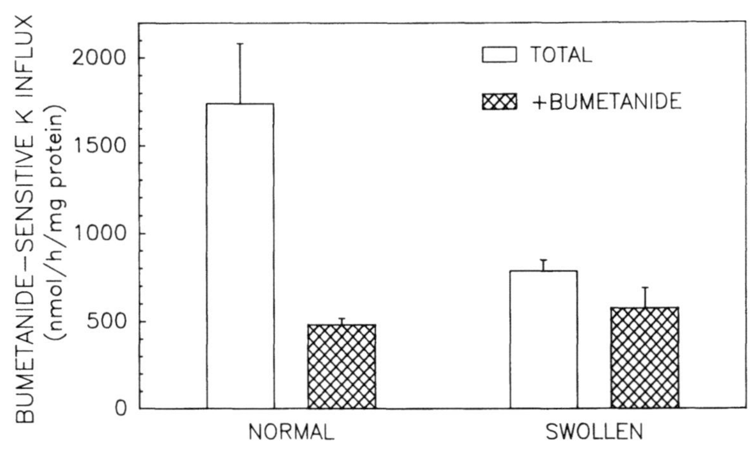

Fig. 8. Inhibition of Na-K-2Cl cotransport in bovine aortic endothelial cells by cell swelling. Cells were swollen 27% by preincubation in isotonic KCl medium, and K influx was measured in isotonic NaCl medium as described for Fig. 7

cotransport is regulated through changes in the density of cotransporters. Values of 230 000 (O'Donnell 1989b) and 122 000 (Klein and O'Neill 1990) specific binding sites per cell have been obtained under basal conditions in aortic endothelial cells. Cell shrinkage increased the number of bumetanide binding sites without altering the transport rate per binding site (O'Neill and Klein 1992). This contrasts with the stimulation of cotransport by bradykinin, which is due to an increase in the transport rate per binding site rather than to an increase in the number of binding sites (Klein and O'Neill 1990).

Na-K-2Cl cotransport has been shown to mediate RVI in a variety of cell types (Eveloff and Warnock 1987). It comprises the majority of K influx in arterial endothelial cells (Brock et al. 1986; O'Donnell 1989a; Klein and O'Neill 1990). Considerably less cotransport activity is observed in endothelial cells cultured from veins (Brock et al. 1986). The activity of cotransport in aortic endothelial cells is modulated by several vasoactive compounds, suggesting that cotransport and cell volume may be important in endothelial function. Cotransport is stimulated by bradykinin (Brock et al. 1986; O'Donnell 1989a; Klein and O'Neill 1990), vasopressin (Brock et al. 1986; O'Donnell 1989a), angiotensin II (O'Donnell 1989a), and thrombin (O'Neill and Klein, unpubl. data). Removal of extracellular calcium or the addition of $LaCl_2$ to inhibit Ca entry, blocks the stimulation of cotransport by bradykinin (Brock et al. 1986). Treatment with bis-(o-aminophenoxy)-ethane-N,N,N',N'-tetraacetic acid (BAPTA) to chelate intracellular Ca inhibits cotransport, while treatment with the Ca ionophores ionomycin or A23187 increases cotransport (Brock et al. 1986; O'Donnell 1991; Kartsonis and O'Neill, unpubl. data). Thus, cotransport is activated by intracellular Ca^{2+} and this is the mechanism of activation by bradykinin. The activation by the other vasoactive compounds has not been investigated but is probably mediated by intracellular Ca^{2+} as well. The addition of 8-Br-cAMP or agents that increase cyclic AMP (i.e., norepinephrine, isoproterenol, forskolin) has been reported to inhibit cotransport (Brock et al. 1986; O'Donnell 1989a). Atrial natriuretic factor (ANF) has been reported to both stimulate (Fujita et al. 1989) and inhibit (O'Donnell 1989a) cotransport via activation of guanylyl cyclase. Other agents reported to inhibit cotransport in endothelial cells include histamine, acetylcholine, and phorbol esters (O'Donnell 1989a).

The mechanism by which cell volume controls Na-K-2Cl cotransport in endothelial cells or any other cell type is not understood. We have found that okadaic acid, an inhibitor of protein phosphatases, stimulates cotransport to the level seen in shrunken cells (Fig. 9) without further stimulation in shrunken cells (O'Neill et al. 1991). This suggests that the regulation of cotransport by cell volume occurs through protein phosphorylation. Similar evidence for protein phosphorylation in the activation of Na-K-2Cl cotransport by cell shrinkage has been obtained in duck red cells (Pewitt et al. 1990). Using activators and inhibitors of known protein kinases (forskolin, 8-Br-cAMP, phorbol esters, staurosporine, and trifluoperazine), we have been unable to identify a kinase that may be involved.

Despite active cotransport, and in the face of intracellular and extracellular ion concentrations that favor net influx, there is no net influx through cotransport in aortic endothelial cells under normal conditions.

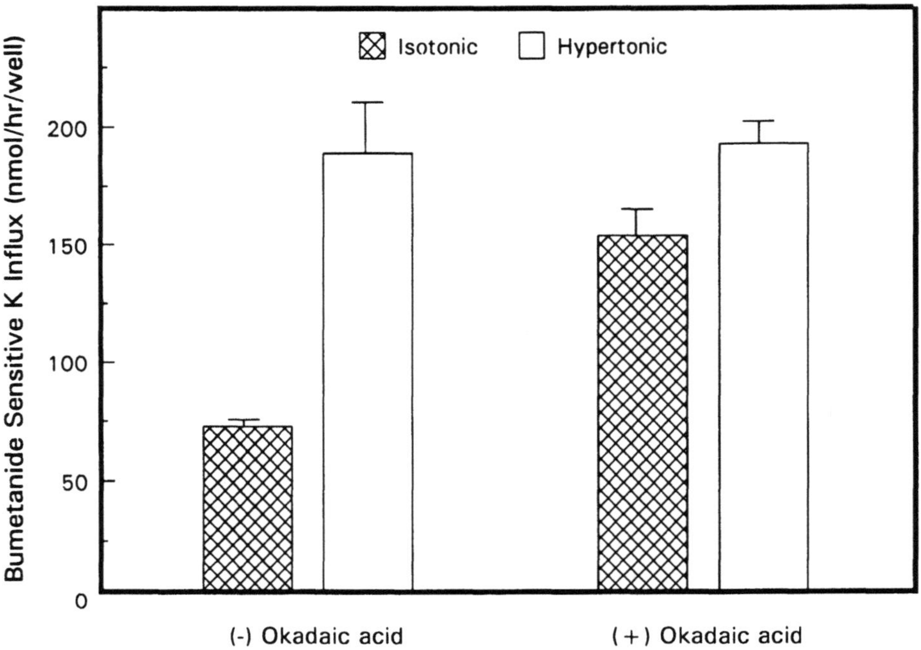

Fig. 9. Activation of Na-K-2Cl cotransport in bovine aortic endothelial cells by okadaic acid. Cells were preincubated with 1 μM okadaic acid prior to measurement of K influx in isotonic and hypertonic (440 mOsmol/l) medium as described for Fig. 7

Cotransport thus functions only in an exchange mode in normal cells. This indicates a dual effect of cell shrinkage on cotransport: an increase in activity that is seen with all forms of shrinkage, and a switch to net influx that is seen only with isotonic shrinkage. The inhibition of net cotransport in normal cells and in hypertonically shrunken cells may be mediated by intracellular Cl^-, which has been shown to inhibit Na-K-2Cl cotransport in squid axons (Breitwieser et al. 1990) and Ehrlich ascites cells (Levinson 1990). In isotonically shrunken cells the intracellular Cl^- concentration is reduced due to KCl loss, allowing net influx by cotransport and recovery of cell K and water (Fig. 5).

Na-H exchange is responsible for RVI in several cell types and has been identified in endothelial cells obtained from a variety of sites (Betz 1983; Kitazono et al. 1988; Gerritsen et al. 1989; Escobales et al. 1990). As in other cells, Na/H exchange is activated by intracellular acidification, hypertonic shrinkage, intracellular Ca, and protein kinase C. Although Na/H exchange in aortic endothelial cells is activated by cell shrinkage (W.C. O'Neill, unpubl. data), it probably does not contribute to cell volume recovery since no regulatory volume increase occurs in hypertonically shrunken endothelial cells (O'Neill and Klein 1992). This may be due to a slow rate of exchange or to limited capacity for Cl/HCO_3 exchange.

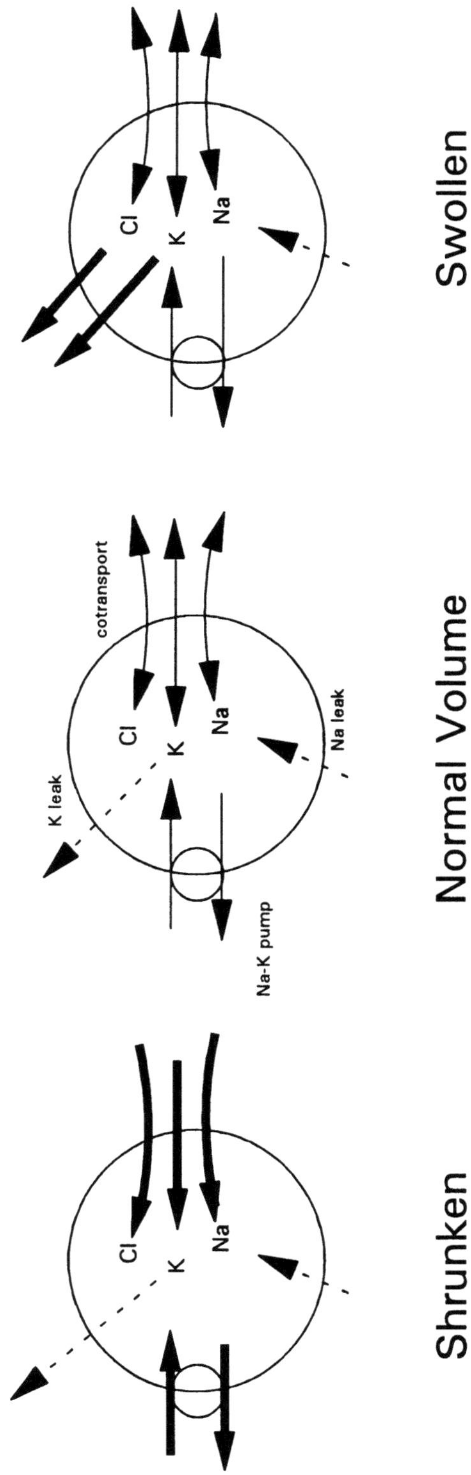

Fig. 10. Cell volume regulation in bovine aortic endothelial cells

4 Cell Volume Regulation and Endothelial Function

The regulation of cell volume in aortic endothelial cells is summarized in Fig. 10. Under normal condition Na-K-2Cl cotransport only mediates exchange, and cell volume is determined by the balance between K and Na leaks and Na-K pump activity. In shrunken cells cotransport is activated and switches to net influx. Na-K pump activity increases to exchange incoming Na with K. A decrease in K leak also contributes to RVI. Cell swelling activates K channels with resulting loss of K and Cl ions.

As is the case with many cells that exhibit volume regulation, vascular endothelial cells are not normally exposed to osmotic stresses in vivo. However, endothelium is exposed to substantial mechanical stresses, including hydrostatic pressure (force perpendicular to the endothelium), shear (force tangential to the endothelium), and stretch, all of which alter endothelial morphology and function (Lansman 1988; Levesque et al. 1989; Shirinsky et al. 1989; Rubanyi et al. 1990; Bevan and Laher 1991). Volume regulatory transporters may be activated by mechanical stresses and produce changes in cell volume and shape that counteract the mechanical stresses. Regulation of the transporters by cell volume may merely be a consequence of the mechanical effects of varying cell volume. An increase in cellular volume of only 1% produces a membrane tension of 0.8 dynes/cm^2 (Sackin 1987). Physiological shear stress on endothelial cells in large arteries ranges upto 30 dynes/cm^2 under steady flow conditions, and can increase under conditions of turbulent/pulsatile flow (Olesen et al. 1988a). In addition to altering cell morphology to conform to mechanical stress, mechanically activated transporters may also serve as transducers that convert physical stimuli into biochemical and/or electrical signals.

There is now substantial evidence that ion channels (including those activated by cell swelling) can be activated by mechanical stimuli (Sachs 1987; Morris 1990; Eaton 1991; Watson 1991), and much of this evidence has been obtained in endothelial cells. Direct stretching of the cell membrane in patch pipettes opens a nonselective cation channel in aortic endothelial cells (Lansman et al. 1987). Like the nonselective channel activated by cell swelling (Ling and O'Neill 1991), this stretch-activated channel conducts Ca ions and may contribute to the rise in [Ca_i] that occurs after deformation of endothelial cells with a blunt micropipette (Goligorsky 1988) or after exposure to flow (Ando et al. 1988). Exposure of aortic endothelial cells in culture to laminar fluid flow (resulting in a shear stress of 0.7–30 dynes/cm^2) induces an inwardly rectifying potassium current (Olesen et al. 1988a). Whether this is related to the K channels activated by cell swelling and is dependent on intracellular Ca is unknown. The membrane tension generated by the degree of cell swelling in our experiments is in the same range known to activate both stretch-activated channels (Sackin 1987) and shear stress-activated potassium current (Olesen et al. 1988a).

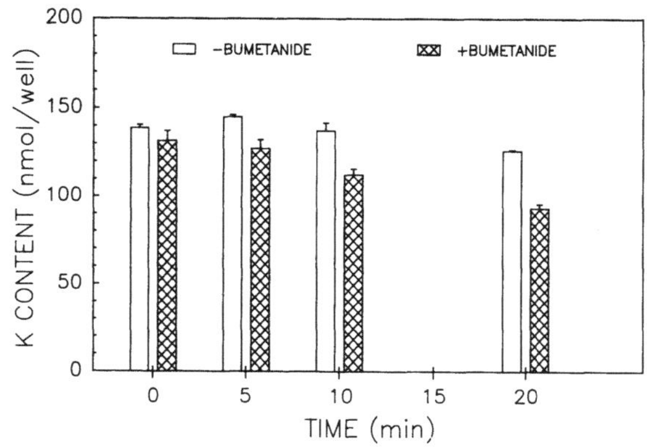

Fig. 11. Effect of bradykinin on the Na and K content of bovine aortic endothelial cells. Bradykinin concentration, 10 μM; bumetanide concentration, 50 μM

In addition to being activated by changes in cell volume, volume-regulatory transporters in endothelial cells are also activated by vasoactive substances. The resulting changes in cell volume may contribute to the morphological responses to vasoactive substances (Majno et al. 1969; Laposata et al. 1983; Wells et al. 1985; Killackey et al. 1986). Agents that increase intracellular Ca concentration would be expected to produce cell skrinkage through activation of K channels. However, the same agents activate Na-K-2Cl cotransport (Brock et al. 1986; Klein and O'Neill 1990; O'Donnell 1991), suggesting that there is compensation for cell shrinkage. This is demonstrated by bradykinin, which induces K loss and cell shrinkage that are exaggerated when Na-K-2Cl cotransport is inhibited with bumetanide (Fig. 11). Thus, the effect of vasoactive substances on Na-K-2Cl cotransport may determine the extent of cell shrinkage and shape change that occurs.

The tight regulation of cell volume exhibited by endothelium may be necessary to maintain the integrity of the endothelial barrier. Agents that increase endothelial permeability in vivo often produce endothelial retraction resulting in intercellular gaps (Majno et al. 1969; Laposata et al. 1983; Killackey et al. 1986). Hypertonic shrinkage of endothelial cells also increases monolayer permeability (Shepard et al. 1987), presumably through the same mechanism. We have confirmed this finding and have demonstrated a protective effect of volume-regulatory transport (Fig. 12). Aortic endothelial cells were grown to confluence on collagen-coated microcarrier beads and permeability was measured by the uptake of albumin-bound Evan's blue dye into the beads (Killackey et al. 1986). Shrinkage of the cells to 80% of their original volume increased permeability (open bar). Permeability was partly restored during a subsequent 20-min period in standard medium to allow volume recovery (Fig. 6; narrow crosshatch). However, inclusion of bumetanide during this period to inhibit Na-K-2Cl

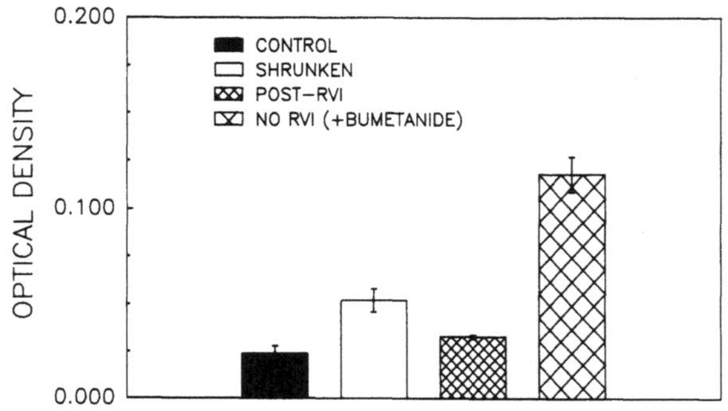

Fig. 12. Effect of cell volume on endothelial monolayer permeability. Bovine aortic endothelial cells were grown on collagen-coated microcarriers until permeability barrier was established. Control cells, isotonically shrunken cells, and isotonically shrunken cells incubated in standard medium without or with bumetanide to inhibit RVI were treated with Evans Blue stain bound to albumin for 2 min and then washed. Permeability was determined as uptake of dye into the beads (O.D.)

cotransport and volume recovery resulted in substantially greater permeability (wide crosshatch).

In summary, cell volume regulation may be an integral component of a multifaceted mechanism that controls the physical state of the endothelium, as shown in Fig. 13. Mechanosensors activate intracellular signals (Ando et al. 1988; Nollert et al. 1989; Nollert et al. 1990) that in turn mediate changes in ion transport (Lansman et al. 1987; Olesen et al. 1988a), cytoskeletal structure (Wong and Pollard 1983; Ives et al. 1986; Sato et al. 1987; Shirinsky et al. 1989), and secretion of vasoactive substances (Grabowski et al. 1985; Yoshizumi et al. 1989; Buga et al. 1991). Changes in ion transport and cytoskeletal structure may counteract mechanical stresses by altering cell shape. Altered secretion of vasoactive substances may counteract mechanical stress on the endothelium by changing vessel diameter (Zarins et al. 1987). The same intracellular signals are activated by endothelial agonists independently of mechanical stimuli, resulting in changes in ion transport (Brock et al. 1986; O'Donnell 1989a; Klein and O'Neill 1990), cytoskeletal structure (Majno et al. 1969; Galdal et al. 1983; Garcia et al. 1986), and secretion of vasoactive substances (McIntyre et al. 1985; Myers et al. 1989) that are required for the action of vasoactive substances.

Acknowledgments. The collaboration and advice of Dr. Janet Klein and Dr. Brian Ling, the expert technical assistance of Ms. Pamela Perry, and the secretarial assistance of Ms. Valerie Panditaratne are gratefully acknowledged. Work from Dr. O'Neill's laboratory was supported by NIH grant DK 01643, National Kidney Foundation of Georgia Affiliate Grant, and American Heart Association Georgia Affiliate Grant-In-Aid.

Address for correspondence: Renal Division, Emory University Hospital, 1364 Clifton Road, N.E., Atlanta, Georgia 30322, USA.

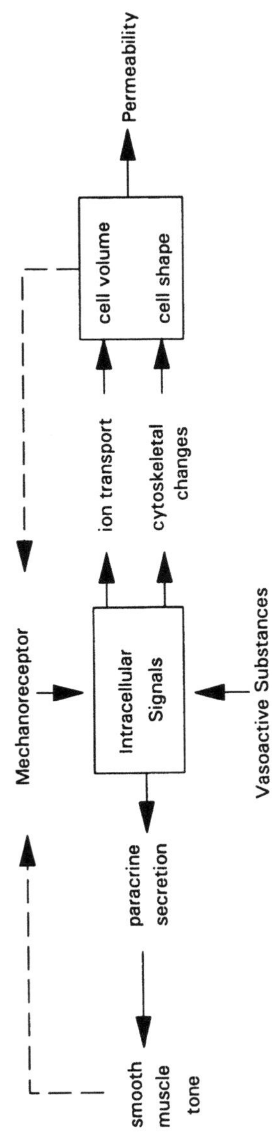

Fig. 13. Role of cell volume regulation in the response of endothelium to mechanical and humoral stimuli

References

Adams DJ, Barakeh J, Laskey R, van Breemen C (1989) Ion channels and regulation of intracellular calcium in vascular endothelial cells. FASEB J 3: 2389–2400

Ando J, Komatsuda T, Kamiya A (1988) Cytoplasmic calcium response to fluid shear stress in cultured vascular endothelial cells. In Vitro. Cell Dev Biol 24: 871–877

Betz AL (1983) Sodium transport in capillaries isolated from rat brain. J Neurochem 41: 1150–1157

Bevan JA, Laher I (1991) Pressure and flow-dependent vascular tone. FASEB J 5: 2267–2273

Breitwieser GE, Altamirano AA, Russell JM (1990) Osmotic stimulation of Na^+-K^+-Cl^- cotransport in squid giant axon is $[Cl^-]_i$ dependent. Am J Physiol 258: C749–C753

Brock TA, Brugnara C, Canessa M, Gimbrone MA Jr (1986) Bradykinin and vasopressin stimulate Na^+-K^+-Cl^- cotransport in cultured endothelial cells. Am J Physiol 250: C888–C895

Buga GH, Gold ME, Fukuto JM, Ipnarro LJ (1991) Shear stress-induced release of nitric oxide from endothelial cells grown on beads. Hypertension 17: 187–193

Christensen O (1987) Mediation of cell volume regulation by Ca^{2+} influx through stretch-activated channels. Nature 330: 66–68

Colden-Stanfield M, Schilling WP, Possani LD, Kunze DL (1990) Bradykinin-induced potassium current in cultured bovine aortic endothelial cells. J Membr Biol 116: 227–238

Eaton DC (1991) Channels in epithelial tissue. In: Schafer J (ed) Membrane transport in biology. Springer, Berlin Heidelberg New York (in press)

Escobales N, Longo E, Cragoe EJ Jr, Danthuluri NR, Brock TA (1990) Osmotic activation of Na^+-H^+ exchange in human endothelial cells. Am J Physiol 259: C640–C646

Eveloff JL, Warnock DG (1987) Activation of ion transport systems during cell volume regulation. Am J Physiol 252: F1–F10

Fichtner H, Forbe U, Busse R, Kohlhardt M (1987) Single nonselective cation channels and Ca^{2+} activated K^+ channels in aortic endothelial cells. J Membr Biol 98: 125–133

Fujita T, Hagiwara H, Ohuchi S, Kozuka M, Ishido M, Hirose S (1989) Stimulation of Na-K-Cl cotransport in cultured vascular endothelial cells by artial natriuretic peptide. Biophys Res Commun 159: 734–740

Galdal KS, Evansan SA, Nilsen E (1983) Thrombin-induced shape changes of cultured endothelial cells: metabolic and functional observations. Thromb Res 32: 57–66

Garcia JGN, Siflinger-Birnboim A, Bizios R, DelVecchio PJ, Fenton JW, Malik AB (1986) Thrombin-induced increase in albumin permeability across the endothelium. J Cell Physiol 128: 96–104

Gerritsen ME, Perry CA, Moatter T, Cragoe EJ Jr, Medow MS (1989) Agonist-specific role for Na^+-H^+ antiport in prostaglandin release from microvessel endothelium. Am J Physiol 256: C831–C839

Goligorsky MS (1988) Mechanical stimulation induces Ca_i^{2+} transients and membrane depolarization in cultured endothelial cells. FEBS Lett 240: 59–64

Grabowski EF, Jaffe EA, Weksler BB (1985) Prostacyclin production by cultured endothelial cell monolayers exposed to step increases in shear stress. J Lab Clin Med 105: 36–43

Haas M, Forbush B (1986) [^{3}H]Bumetanide binding to duck red cells. Correlation with inhibition of (Na+K+2Cl) co-transport. J Biol Chem. 261: 8434–8441

Hoffmann EK, Simonsen LO (1989) Membrane mechanisms in volume and pH regulation in vertebrate cells. Physiol Rev 69: 315–382

Ives SL, Eskin SG, McIntire LV (1986) Mechanical effects on endothelial morphology: in vitro assessment. In Vitro. Cell Dev Biol 22: 500–507

Janzer RC, Raff MC (1987) Astrocytes induce blood-brain barrier properties in endothelial cells. Nature 325: 253–257

Johns A, Lategan TW, Lodge NJ, Ryan US, van Breeman C, Adams DJ (1987) Calcium entry through receptor-operated channels in bovine pulmonary artery endothelial cells. Tissue Cell 19: 733–745

Kempski O, Spatz M, Valet G, Baethmann A (1985) Cell volume regulation of cerebrovascular endothelial in vitro. J Cell Physiol 123: 51–54

Killackey JJF, Johnston MG, Movat HZ (1986) Increased permeability of microcarrier-cultured endothelial monolayers in response to histamine and thrombin. Am J Pathol 122: 50–61

Kitazono T, Takeshige K, Cragoe EJ Jr, Minakami S (1988) Intracellular pH changes of cultured bovine aortic endothelial cells in response to ATP addition. Biochem Biophys Res Commun 152: 1304–1309

Klein JD, O'Neill WC (1990) Effect of bradykinin on Na-K-2Cl cotransport and bumetanide binding in aortic endothelial cells. J Biol Chem 265: 22238–22242

Lang F, Rehwald W (1991) Potassium channels in renal epithelial transport regulation. Physiol Rev 72: 1–32

Lansman JB (1988) Going with the flow. Nature 331: 481–482

Lansman JB, Hallam TJ, Rink TJ (1987) Single stretch-activated ion channels in vascular endothelial cells as mechanotransducers? Nature 235: 811–813

Laposata M, Dovarsky DK, Shin HS (1983) Thrombin-induced gap formation in confluent endothelial cell monolayers in vitro. Blood 62: 549–556

Levesque MJ, Sprague EA, Schwartz CJ, Nerem RM (1989) The influence of shear stress on cultured vascular endothelial cells: the stress response of an anchorage-dependent mammalian cell. Biotechnol Prog 5: 1–8

Levinson C (1990) Regulatory volume increase in Ehrlich ascites tumor cells. Biochim Biophys Acta 1021: 1–8

Ling BN, O'Neill WC (1991) Potassium channel activation by hypotonic shock in bovine aortic endothelial cells. Clin Res 39: 194A

Ling BN, Webster CL, Eaton DC (1991) Eicosanoids modulate apical Ca^{2+}-dependent K^+ in principal cells. Am J Physiol (in press)

Majno G, Shea SM, Leventhal M (1969) Endothelial contraction induced by histamine-type mediators. An electron microscopy study. J Cell Biol 42: 647–672

Mazzoni MC, Lundgren E, Arfors K-E, Intaglietta M (1989) Volume changes of an endothelial cell monolayer on exposure to anisotonic media. J Cell Physiol 140: 272–280

McIntyre TM, Zimmerman GA, Satch K, Prescott SM (1985) Cultured endothelial cells synthesize both platelet activating factor and prostacyclin in response to histamine bradykinin and adenosine triphosphate. J Clin Invest 76: 271–280

Morris CE (1990) Mechanosensitive ion channels. J Membr Biol 113: 93–107

Myers PR, Guerra R, Harrison DG (1989) Release of No and EDRF from cultured bovine aortic endothelial cells. Am J Physiol 256: H1030–H1037

Nilius B (1990) Permeation properties of a non-selective cation channel in human vascular endothelial cells. Pflügers Arch Eur J Physiol 416: 609–611

Nollert MU, Hall ER, Eskin SG, McIntire LV (1989) The effect of shear stress on the uptake and metabolism of archidonic acid by human endothelial cells. Biochem Biophys Acta 1005: 72–78

Nollert MU, Eskin SG, McIntire LV (1990) Shear stress increases inositol triphosphate levels in human endothelial cells. Biochem Biophys Res Commun 170: 281–287

O'Donnell ME (1989a) Regulation of Na-K-Cl cotransport in endothelial cells by atrial natriuretic factor. Am J Physiol 257: C36–C44

O'Donnell ME (1989b) [^{3}H]Bumetanide binding in vascular endothelial cells. Quantitation of Na-K-Cl cotransporters. J Biol Chem 264: 20326–20330

O'Donnell ME (1991) Endothelial cell sodium-potassium-chloride cotransport – evidence of regulation by Ca^{2+} and protein kinase C*. J Biol Chem 266: 11559–11566

Olesen S-P, Clapham DE, Davies PF (1988a) Haemodynamic shear stress activates a K^+ current in vascular endothelial cells. Nature 331: 168–170

Olesen S-P, Davies PF, Clapham DE (1988b) Muscarinic-activated K^+ current in bovine aortic endothelial cells. Circ Res 62: 1059–1064

O'Neill WC, Klein JD (1992) Regulation of vascular endothelial cell volume by Na-K-2Cl cotransport. Am J Physiol 262: C436–C444

O'Neill WC, Kartsonis N, Klein JD (1991) Role of protein phosphorylation in the regulation of Na-K-2Cl cotransport by cell volume in aortic endothelial cells. J Am Soc Nephrol (Abstr) 2: 747

Pewitt EB, Hedge RS, Haas M, Palfrey HC (1990) The regulation of Na/K/2Cl cotransport and bumentanide binding in avian erythrocytes by protein phosphorylation and dephosphorylation. J Biol Chem 19: 20747–20756

Ross R (1986) The pathogenesis of atherosclerosis – an update. New Eng J Med 314: 488–500

Rubanyi GM, Romero JC, Vanhoutte PM (1986) Flow-induced release of endothelium-derived relaxing factor. Am J Physiol 250: H1145–H1149

Rubanyi GM, Freay AD, Kauser K, Johns A, Harder DR (1990) Mechanoreception by the endothelium: mediators and mechanism of pressure- and flow-induced vascular responses. Blood Vessels 27: 246–257

Sachs F (1987) Baroreceptor mechanisms at the cellular level. Fed Proc 46: 12–16

Sackin H (1987) Stretch-activated potassium channels in renal proximal tubule. Am J Physiol 253: F1253–F1262

Sato M, Levesque MJ, Nerem RM (1987) Micropipette aspiration of cultured bovine aortic endothelial cells exposed to shear stress. Arteriosclerosis 7: 276–286

Sauve R, Parent L, Simoneau C, Roy G (1988) External ATP triggers a biphasic activation process of a calcium-dependent K^+ channel in cultured bovine aortic endoethelial cells. Pflügers Arch Eur J Physiol 412: 469–481

Shepard JM, Goderie SK, Brzyski N, Del Vecchio PJ, Malik AB, Kimelberg HK (1987) Effects of alterations in endothelial cell volume on transendothelial albumin permeability. J Cell Physiol 133: 389–394

Shirinsky VP, Antonov AS, Birukov KG, Sobolevsky AV, Romanov YA, Kabaeva NV (1989) Mechano-chemical control of human endothelium orientation and size. J Cell Biol 109: 331–339

Sun AM, Saltzberg SN, Kikeri D, Hebert SC (1990) Mechanisms of cell volume regulation by the mouse medullary thick ascending limb of Henle. Kidney Int 38: 1019–1029

Vane JR, Änggåard EE, Botting RM (1990) Regulatory functions of the vascular endothelium. New Eng J Med 323: 27–36

Vigne P, Campigny G, Marsault R, Barbry P, Frelin C, Lazdunski M (1989) A new type of amiloride-sensitive cationic channel in endothelial cells of brain microvessels. J Biol Chem 264: 7663–7668

Watson PA (1991) Function follows form: generation of intracellular signals by cell deformation. FASEB J 5: 2013–2019

Wells SL, Shepro D, Hechtman HB (1985) Vasoactive amines modulate actin cables (stress fibers) and surface area in cultured bovine endothelium. J Cell Physiol 123: 337–342

Wong AJ, Pollard TD (1983) Actin filament stress fibers in vascular endothelial cells in vivo. Science 219: 867

Yoshizumi M, Kurihara H, Sugiyama T, Takaku F, Yanagisana M, Masaki T, Yazaki Y (1989) Hermodynamic shear stress stimulates endothelin production by cultured endothelial cells. Biochem Biophys Res Commun 161: 859–864

Zarins CK, Zatina MA, Giddend DP, Ku DN, Glagov S (1987) Shear stress regulation of artery lumen diameter in experimental arterogenesis. J Vasc Surg 5: 413–420

Chapter 6

Effects of Swelling on Glial Cell Function

H.K. Kimelberg[1,2], *E.R. O'Connor*[2], and *Helmut Kettenmann*[3]

Contents

1 Astroglial Swelling in Situ – Contribution to Pathology

In the mammalian CNS astrocytic swelling is seen early in numerous pathological states such as ischemia (Jenkins et al. 1982), traumatic brain edema (Gerschenfeld et al. 1959; Castejon, 1980; Bullock et al. 1991), hypoglycemia

[1] Division of Neurosurgery A-60, Albany Medical College, Albany, New York 12208, USA
[2] Department of Pharmacology and Toxicology, Albany Medical College, Albany, New York 12208, USA
[3] Institut für Neurobiologie der Universität Heidelberg, W-6900-Heidelberg, FRG

Advances in Comparative and Environmental Physiology, Vol. 14

and status epilepticus (Siesjo 1981), prolonged hypoxia (Yu et al. 1972), acute hypoxia with hypercapnia (Bakay and Lee 1968), hepatic encephalopathy and Reye's syndrome (Norenberg 1981; Davis et al. 1990) and experimental allergic encephalomyelitis (Eng et al. 1989). Brain damage during these same pathological states is thought in many cases to be partly due to an inappropriate release of endogenous excitatory amino acids (EAAs) such as L-glutamate and L-aspartate. At the same time, extracellular K^+ levels can reach 30–80 mM (Somjen 1979; Sykova 1983). Such K^+ levels are known to produce marked astrocytic swelling (Zadunaisky et al. 1963; van Harreveld 1982; Kimelberg and Ransom 1986) and hypotonic medium-induced (Kimelberg et al. 1990a) or high K^+-induced (Kimelberg and Goderie 1991) cell swelling leads to the release of excitotoxic amino acids. These excitotoxins are thought to cause neuronal injury due to excessive EAA receptor activation (Choi 1988; Choi and Rothman 1990). Thus, we propose the hypothesis that astrocytic swelling in pathological states may be deleterious in part due to the release of excitotoxins, and may be one reason why inhibiting such swelling contributes to the prevention of neurological deficit or death in animal head injury or ischemia models (Kimelberg et al. 1987; Barron et al. 1988; Kohut et al. 1991).

2 Methods for Studying Glial Swelling

In order to further elucidate the mechanisms and consequences of astrocytic swelling it is necessary to turn to cell cultures or complex cell-specific techniques for measurements in brain slices or in vivo such as intracellular ion-specific microelectrodes (ISMs).

2.1 Methods for Measuring Cell Volume in Cell Cultures

Exposure of primary astrocyte cultures to hypoosmotic media leads to rapid cell swelling followed by restoration of cell volume towards normal values in the continued presence of hypotonic media (Kimelberg and Frangakis 1985; Olson et al. 1986), a process termed regulatory volume decrease or RVD (Grinstein et al. 1984; Hoffmann 1987; Gilles et al. 1991). Other glial cell cultures, namely, C_6 glioma cells, also show RVD after exposure to hypotonic medium (Kempski et al. 1983). High K^+ in the medium and glutamate leads to smaller astrocytic swelling that also occurs more slowly (Kimelberg et al. 1992).

The use of cell cultures for in vitro studies on cell volume regulation has allowed researchers to focus on understanding mechanisms at the cellular level. Cell cultures can be classified according to a variety of criteria, including whether the cells are normal or transformed to a malignant state. Both these types can be grown directly from tissue (primary cultures) and can then be passaged up to a maximum of 50 times for normal cells and indefinitely for transformed cells (Hayflick and Moorehead 1961). Usually cells from transformed lines are frozen at an early passage and then thawed and grown for a fixed number of passages as required. Growing substrate-attached cultures in different kinds of plastic vessels makes transport studies quite simple, and cultures of normal cells from the nervous system are invariably of this type. Some cell types can be grown as suspension cultures, e.g., lymphocytes, and such cells are very convenient for studies using Coulter Counter or light-scattering techniques to measure cell volume (Grinstein 1989). This technique has been applied to C_6 glioma cells (Kempski et al. 1983) and primary astrocyte cultures (Olson et al. 1986). A disadvantage of this technique is that these cells have to be removed from culture dishes either by enzymatic treatment, usually with trypsin and a Ca^{2+} chelator, or by mechanical scraping. Detachment of substrate-attached cells may affect the shape, leakiness and physiological properties of cell membranes, introducing uncertainty with regard to physiological function.

For attached cells, intracellular volume has up to now been measured using equilibration with ^{14}C-3-O-methyl glucose (3-OMG), as originally described for hepatocytes (Kletzien et al. 1975), and ^{14}C-urea for glial cells (Martin and Shain 1979). We have used both methods for primary astrocyte cultures (Kimelberg and Frangakis 1985; Kimelberg and Goderie 1988) and, although both these methods report an RVD and RVI response, such indirect radiolabeled tracer studies introduce significant methodological problems. One of the most important drawbacks of this method is its inability to measure rapid changes in cell volume. This technique is limited to collecting data points at probably not less than 30-s intervals because one needs to remove the extracellular tracer by washing. Other problems concerning this technique are whether there is leakage of the tracer when the cell is swollen; what intracellular volume is actively being measured, i.e., are the tracers excluded and/or concentrated in certain areas of the cell; does swelling and volume regulation cause the equilibrium rate of the tracer to change; is the tracer extruded by some additional transport system; are there impurities in the marker that would cause error in the measurement of volume? It is also important that the marker is nonmetabolizable as this would also cause errors in the volume measurement.

Other in vitro methods used for studying cell volume in astrocytes include fixing the cells at differing stages of RVD and imaging them using serial three-dimensional reconstruction on a high voltage electron microscope (Parsons et al. 1989). A method for determining individual cell volumes by optical sectioning using differential interference microscopy in frog urinary bladder cells has been used by Davis and Finn (1987). Briefly, the tissue is viewed under high magnification and the image is monitored with a video camera and the data stored on video tape. An individual cell is optically scanned repeatedly before and after a change in solution. Optical sections of the cell are taken and the cross-sectional area of the cell in each of the sections is measured by computer-aided planimetry. It is not clear whether

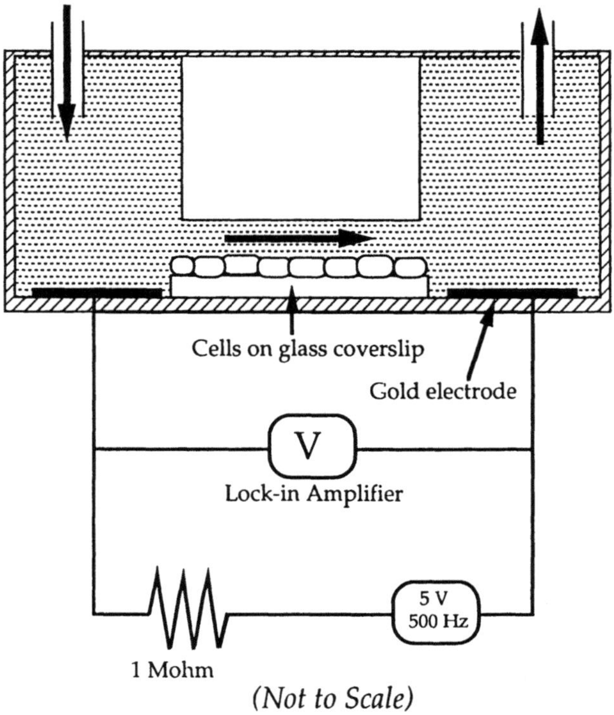

Fig. 1. Design of the electrical impedance system for measure volume of cell monolayers. Astrocytes or other cells are grown on #1 coverslips and placed in a confined channel containing a salt solution; the electrical resistance of the channel is measured using an applied AC field. The gold electrodes are connected through a large resistance (1 MΩ) to a lock-in amplifier that supplies a 500 Hz, 5 V signal to the system. The lock-in amplifier is used to resolve small voltage changes with high noise rejection. Apart from the solution in the channel, the two chambers are insulated from each other. The solution is continuously perfused as indicated.

this method would be applicable for glial cells in culture as they may be too flat.*

An impedance method for measuring dynamic changes in cell volume of substratum-attached endothelial cells has been described recently (Mazzoni et al. 1989; O'Connor et al. 1993). In our version of the method cell cultures are placed in a confined channel containing a salt solution and the impedance or electrical resistance of the channel is measured using an applied AC field (Fig. 1). If the volume of the cells increases, then the volume of the solution within the channel available for current conduction will decrease by the same amount. This allows measurements of volume changes in real time

* Note in proof: Shain et al. (Glia 5: 223–238 (1992)) have shown that confocal microscopy can detect increases in height in flat astroglial cell cultures.

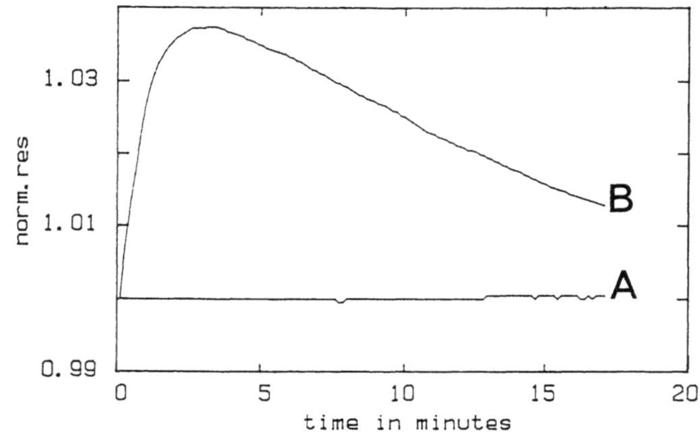

Fig. 2. RVD response of cultured astrocytes using the impedance system (see Fig. 1). *A* Astrocytes in isotonic solution (290 mOsmol), showing the background noise and stability of the system. *B* Typical response of astrocytes to hypotonic solution (−100 mM sucrose). Upon exposure to hypotonic media, astrocytes initially swell and then, in the continued presence of hypotonic media, regulate their volume back to control levels.

During experiments, the chamber is first electronically balanced (the resistance is normalized to 1), with a coverslip coated with a confluent monolayer of cells, in isotonic solution. The perfusing solution can then be switched to a hypotonic solution due to removal of 100 mM sucrose and therefore with the save resistivity.

(see Fig. 2, O'Connor et al. 1993). Volume changes in substratum-attached monolayers can also be made using laser light scattering (Fischbarg et al. 1989; Strange 1991 pers. comm.).

2.2 In Vivo Measurements

In order to correlate the findings in cultures or tissues in vitro with behavior of cells in vivo it is necessary, of course, to have methods to enable such measurements to be made. MacKnight and Leader (1989) have recently discussed such techniques in detail for epithelial cells. Although there are general methods for studying swelling by wet weight/dry weight determinations, decreases in the extracellular space, which can be inferred as being due to cell swelling, and overall mass changes in cell tissue using differential interference contrast, these methods cannot be applied to studying glial cells in a complex multicellular nervous tissue, unless conditions are chosen such that it is known that only this specific cell type has responded. An example of this are measurements of changes in the extracellular space of rat optic nerve to measure changes in glial cell volume (Ransom et al. 1985). In this technique a membrane-impermeant cation TMA^+ (trimethyl ammonium) is injected into the extracellular space and its activity measured by potassium-selective electrodes, which are far more sensitive to TMA^+ than K^+. As the nerve is stimulated the extracellular space decreases and this decrease was proportional to changes in extracellular potassium, which is thought to cause preferential swelling of glial cells. Also, developmental aspects of this study indicated it was likely to be the glial cells that were swelling. The nerve was made axon-free by enucleation.

Ballanyi and Grafe (1988) and Ballanyi et al. (1990) have also used TMA^+, or choline which behaves similarly, but in this case have performed the technically more difficult feat of recording intracellularly from leech neurophil glial cells, which had accumulated these compounds after bath application. It was shown that an elevation in the extracellular potassium concentration caused a depolarization of the cell membrane potential as expected and a decrease in the intracellular choline signal consistent with the fact that the neurophil leech glial cell had swollen. In contrast, application of serotonin to an isolated leech segmental ganglion caused a hyperpolarization of the membrane potential and an increase in the choline signal. These changes are the same as would be expected if the $choline^+$ or TMA^+ membrane gradients were to follow the membrane potential, so a critical requirement of this method is that the large cations are always impermeant. This effect of serotonin was interpreted as due to serotonin causing an increased permeability to potassium and, consequently, an increased efflux of potassium chloride from the cell, causing the cell to shrink. Moreover, the neurophil glial cell is known to have a high resting chloride permeability (Ballanyi et al. 1990). In frog motorneuron, application of taurine caused a depolarization of the membrane potential and an increase in cell volume, i.e., a decrease in the TMA^+ signal. This was interpreted as being due to a taurine-induced increased permeability to chloride.

The impedance method discussed above in relation to cell monolayers was originally applied by van Harreveld (1982) and van Harreveld and Fifkova (1971) to mammalian brain in vivo. In these pioneering studies they showed that conditions such as asphyxia or ischemia caused a marked decrease in the measured impedance of the tissue consistent with cellular swelling and reduction of the ECS. Since these conditions were shown by electronmicroscopy to lead preferentially to astrocytic swelling, it is quite possible that a major contribution to the decrease in the extracellular space was due to astrocytic swelling. However, van Harreveld also found that under the same conditions there was marked dendritic swelling so that the relative contributions of glial and dendritic swelling were unclear. Quantitative morphometry can then be used to estimate the relative contributions of the cell types, as was done by Barron et al. (1988), showing that astrocytic swelling (cytotoxic edema) was significantly increased in the sensorimotor cortex in cats, after closed head injury. However, electronmicroscopic techniques require particular care in regard to rapid fixation otherwise there will be fixation artifacts due to cells essentially becoming ischemic before they are fixed, which will compromise the interpretation. van Harreveld (1966) pioneered the use of rapid freeze substitution to avoid such artifacts.

3 Mechanisms of Astrocytic Swelling

Studies using the above techniques have led to a number of different mechanisms being proposed as leading to astrocytic swelling. Since astrocytes swell in response to diverse conditions, an important question is whether this is due to a number of different mechanisms leading to the same swelling response, or whether it is the result of a number of conditions leading to a single process resulting in astroglial swelling. A variety of possible effectors have been proposed suggesting that there can be a number of different mechanisms. These different mechanisms will now be briefly considered.

3.1 Uptake of Extracellular K^+

1. *Potassium Spatial Buffering*. The classic work of Kuffler et al. (1966) on the optic nerve of the salamander *Necturus maculosa* first showed that the glial cells in this preparation were high intracellular K^+ cells with a negative potential essentially equivalent to the K^+ equilibrium potential, indicating that the cation conductivity of the cell membrane was almost entirely due to K^+. These properties led to the concept of "potassium spatial buffering" as a possible function of glial cells (Orkand et al. 1966). In this model excess potassium, released by active neurons and localized to one region of the extracellular space, is carried away from the neurons through the glial syncytium. The driving force for the flow is the localized increase in extracellular potassium due to neuronal activity. As a result the glial membrane potential in the region of the local increase becomes preferentially depolarized due to its high K^+ selectivity. The difference in the membrane potential between such local and distant regions yields a current flow within the glial cell that carries potassium away from its site of release (see Dietzel et al. 1989 for a recent review). Recent work has shown the existence of a variety of K^+ channels in astrocytes that would have to be fitted to this model (Barres et al. 1990).

2. *Uptake of K^+*. K^+ spatial buffering has been the most studied and discussed aspect of astroglial ion transport, but its involvement in brain function is still to be established (e.g., see Coles 1989; Dietzel et al. 1989). More importantly for the theme of this chapter, it will not necessarily lead to astroglial swelling since it is uniquely designed to generate a current loop carried inward by K^+ at the point of increased $[K^+]_o$ and outward at some distant point, whose location principally depends on the electrical resistance of the astroglial cell membrane (i.e., its length constant). This is true if the cell membrane is indeed only permeable to K^+ and electrical current can only be carried by the same amount of K^+ entering and leaving the cell. In the extracellular space the current is carried by the principal cation, Na^+, or by movement in the opposite direction of Cl^- or HCO_3^-. However, Dietzel

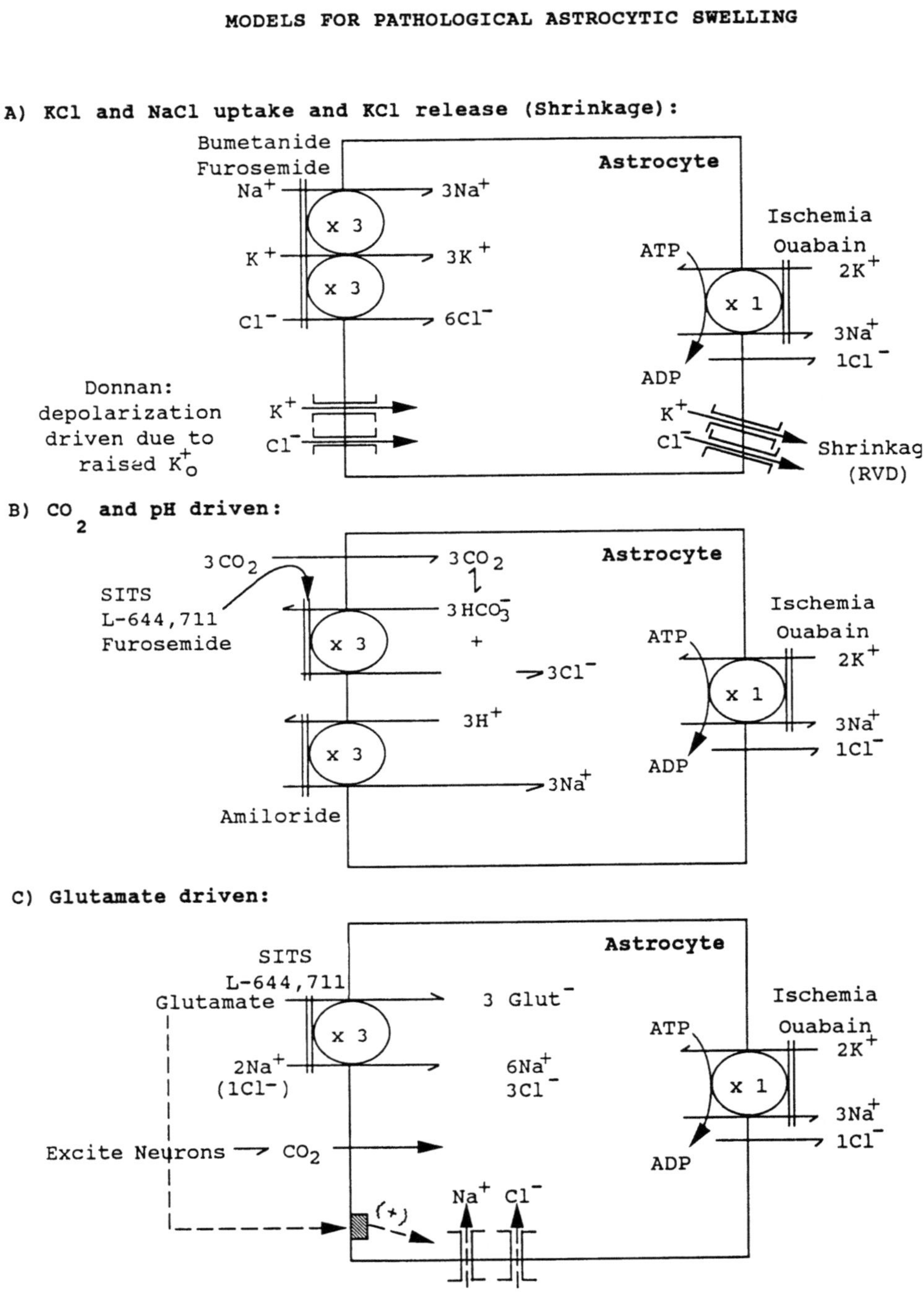

Fig. 3. Models for pathological astrocytic swelling. Transport models for *A*, KCl and NaCl uptake and KCl release; *B*, CO_2 and pH; *C*, glutamate-driven swelling of astrocytes. All models show the accumulation of ions for the number of turns of the inward transport systems that can be balanced by one turn of the ATP-driven ($Na^+ + K^+$) pump.

et al. (1989) proposes that such currents can lead to localized glial swelling. If there is a significant Cl^- conductance, increase in K^+ with reciprocal reduction of Na^+ will also lead to a depolarization-induced net uptake of KCl through diffusional Donnan forces which will lead to swelling (Boyle and Conway 1941; Hodgkin and Horowicz 1959). This has been shown in cultured mouse astrocytes (Walz 1987) and possibly occurs because of the existence voltage-activated Cl^- channels which are activated by membrane depolarization starting around −40 mV (Bevan et al. 1985).

In addition to the channel-mediated processes involved in K^+ spatial buffering or KCl uptake, processes that involve coupled transport of K^+, Na^+ and Cl^-, which will lead to net accumulation of KCl, have been principally found in astrocyte cultures (Kimelberg and Walz 1988). Also, the active exchange transport of Na^+ and K^+ via the (Na^+ + K^+) ATPase will extrude Na^+ and take up K^+ (reviewed in Kimelberg et al. 1982, 1989b; Kimelberg and Ransom 1986). All these different models of K^+ uptake are shown in Fig. 3A.

3.2 Glial Swelling Associated with Acid-Base Changes

It has been shown that the pH of the mammalian CNS is controlled independently of the systemic pH (Katzman and Pappius 1973). Among their other homeostatic functions, glial cells, especially astrocytes, may play a key role in such pH regulation. Thus, recently, significant pH changes have been shown in mammalian astroglial cells during intense neuronal activity which paralleled opposite changes in extracellular pH (Chesler and Kraig 1989; Chesler 1990). In relation to this function of pH homeostasis we originally proposed a model for astrocytic swelling based on experimental observations that a component of the swelling of cat cerebrocortical slices stimulated by

This pump has a stoichiometry of $3Na^+$ ions out for $2K^+$ in, per 1 ATP hydrolyzed to ADP, with presumably $1Cl^-$ (or HCO_3^-) also leaving to accompany the net outward positive charge. Inhibition of the pump can be a result of the specific inhibitor ouabain or ischemia. For the furosemide- and bumetanide-sensitive cotransporter, the uptake is 3KCl + 3NaCl. Normally, $3Na^+$ (+ $1Cl^-$) will be pumped out, $2K^+$ pumped in, thus leading to a net gain of 5KCl, which will also lead to swelling. This is normally compensated by electrochemically driven outward efflux of KCl on channels as shown. Such efflux also effects shrinkage in RVD. It is unknown whether the same channels are involved. Depolarization (Donnan)-driven uptake of KCl through channels is also shown (see Sect. 4.1.3 for a full explanation). *B* This model is explained fully in the text under acid-base changes (see Sect. 3.2) and leads to net accumulation or a pass-through of 3NaCl per ATP. *C* Assuming a stoichiometry of $2Na^+:1$ glutamate$:1Cl^-$, uptake will be, per three turns of the transporter, $6Na^+:3Glut^-:3Cl^-$. Uptake is Cl^--dependent, but it is not clear whether Cl^- enters on the transporter or independently through a channel to effect electrical neutrality. In model *C* stimulation of neuronal metabolism is indicated leading to increased CO_2 and thus swelling via model *B*. At the *bottom* receptor-stimulated increases in Na^+ and Cl^- conductances (see text) are shown high $[K^+]_o$ required bicarbonate ions, and that this same component was specifically inhibited by SITS (Bourke et al. 1979, 1983), an antagonist of the Cl^-/HCO_3^- anion-exchange transport system present in many cell types (Lowe and Lambert 1983). We also found (Kimelberg et al. 1979; Kimelberg 1981) the anion exchange transport system to be present in primary astrocyte cultures, suggesting perhaps that these systems were present on astrocytes in the mammalian CNS, although localization in neurons and other cells, such as endothelial cells, is also possible if not likely (Lowe and Lambert 1983). In this model, the uptake of Na^+ and Cl^- occurs by simultaneous operation of Cl^-/HCO_3^- and Na^+/H^+ transport, with H^+ and HCO_3^- cycling from the intra- to extracellular spaces via membrane-permeant CO_2 (see Fig. 3B). Such mechanisms are thought to underlie swelling in a number of other cell types (Grinstein et al. 1984). Thus, this system, driven by reduced clearance of CO_2 due to ischemia and/or lactic acid-induced acidification of HCO_3^- to H_2CO_3 and then CO_2 and H_2O, coupled with a diminished ability to pump out the intracellularly accumulated Na^+ with Cl^- due to falling energy stores and diminished ATP-driven $(Na^+$-$K^+)$-pump activity, would all lead to astrocytic swelling. Recent work in vitro using different glial cultures has shown that swelling can be caused by activation of Na^+/H^+ exchange subsequent to permeation of the cell by lipid soluble-free acids such as propionate (Jakubovicz et al. 1987) or raised CO_2 in a HCO_3^- buffered medium (Kempski et al. 1988). Walz and Mukerji (1990) also proposed accumulation of lactate as another mechanism that could lead to astrocytic swelling.

3.3 Astroglial Swelling Caused by Glutamate and Other Excitatory Amino Acids

It has been shown that glutamate or kainic acid injected into the CNS causes marked astrocytic swelling. Glutamate-induced swelling of retinal Muller cells and astrocytes from various central nervous system regions has also been observed (Casper et al. 1982; van Harreveld 1982; van Harreveld and Fifkova 1971). The glutamate analog kainic acid has been shown to produce swelling of the astrocytic Bergmann cells of hamster cerebellar cortex (Herndon et al. 1980), and swelling of astrocytes as early as 30 min after direct administration of kainic acid into the corpus striatum of rats has been reported (Matyja 1986).

These effects could also operate by the mechanism shown in Fig. 3C, by glutamate or kainic acid-stimulating neurons and increasing production of CO_2 due to increased metabolism. However, it is also well established that astrocytes take up glutamate by a Na^+-dependent mechanism (Hertz 1979), which may also be partly dependent on Cl^- (Waniewski and Martin 1984). Uptake of glutamate has been reported in astrocyte cultures to occur with a stoichiometry of 1 Na^+:1 glutamate (Drejer et al. 1982), or 2–3 Na^+:1

glutamate (Kimelberg et al. 1989a). This uptake is also depicted in Fig. 3C, and could lead to swelling due to the cells becoming loaded with Na^+ glutamate and the diminished ability of the Na^+-K^+ pump to pump out Na^+. It has also been reported that glutamate interacts with cultured astrocytes to activate a kainic acid receptor (Sontheimer et al. 1988). Some of the observed increases in $^{22}Na^+$ uptake could thus be due to activation of a kainic acid type of glutamate receptor, as depicted at the bottom of Fig. 3C, especially as kainic acid also stimulates uptake of $^{22}Na^+$ and depolarizes the cells and labeled kainic acid is not taken up by these cells (Kimelberg et al. 1989a). We have also observed that glutamate and kainic acid will stimulate ^{36}Cl uptake in astrocytes consistent with uptake of NaCl with cell swelling (unpubl. observ.). A similar scheme for neuronal excitotoxicity via non-NMDA receptors has been recently suggested by McDonald and Johnston (1990), and may lead over a longer time period to astrocyte toxicity (Thakran et al. 1991), which is a possible direct pathological consequence of astrocytic swelling. In astrocytes, the two processes of glutamate uptake and receptor activation of ion channels may clearly work in parallel to effect glial swelling.

3.4 Swelling Due to Fatty Acids and Free Radicals

There has been considerable discussion on the possible role of polyunsaturated fatty acids (PUFA) and free radicals in the initation of brain edema (see Chan and Fishman 1984). A number of the pathological insults which cause edema are associated with release of arachidonic acid and this fatty acid, as well as a number of other PUFAs, have been shown to cause swelling and uptake of Na^+ in brain slices. Chan and Fishman (1984) have suggested that the variety of pathological insults which cause swelling initially increase Ca^{2+} influx into cells leading to activation of phospholipase A, which in turn leads to release of arachidonic acid and other PUFAs. Superoxide ($O_2^{\cdot -}$) and hydroxyl radicals ($^{\cdot}OH$) can also form in brain due to single electron reduction of oxygen, and free radicals can also be generated from PUFAs (Halliwell and Gutteridge 1985). All these species are likely to damage membranes and increase their permeability to ions. This will lead to plasma membrane depolarization, influx of Na^+ and Cl^- and water. Current debate centers on when and how these mechanisms might be involved. However, the effects of these agents might be expected to be nonspecific and it is not clear at present why they would selectively produce swelling of astrocytes.

3.5 Swelling in Hepatic Encephalopathy and Reye's Syndrome

Hepatic encephalopathy and Reye's syndrome have been shown to result in electron-lucent astrocytes when the tissue is examined in the electron microscope, indicating swelling. Such swelling could be related to metabolism of excess ammonia, found in these disorders, by glutamine synthetase which is present at high concentrations in astrocytes (Norenberg 1981). Alternatively, increased ammonia may enter the astrocyte across the blood brain berrier forming ammonium ions intracellularly. Intracellular pH (pH_i) is around 7 in cultured astrocytes (Kimelberg et al. 1982), and direct determination in vivo by ion-specific pH microelectrodes has given a mean value for astrocytes of 7.04 with a range of 6.73–7.38 (Chesler and Kraig 1989). This means that 99% of intracellular ammonia will be in the form of NH_4^+. An increase in NH_3 leading to a greater $[NH_4^+]_i$ would alkalinize the interior of the cell leading to an increase in $[HCO_3^-]_i$ which could then exchange for $[Cl^-]_o$, resulting in an accumulation of NH_4Cl in the astrocyte. It is known that excess CO_2 exacerbates the effects of hepatic encephalopathy (Norenberg 1981), and this could be a source of increased intracellular H^+ and HCO_3^-.

3.6 Swelling in Experimental Allergic Encephalomyelitis (EAE)

Recently, it has been reported that in the early stages of EAE induced in rats by injection of purified guinea-pig myelin there is marked perivascular astrocytic swelling associated with increased immunoreactivity for GFAP. This increased immunoreactivity occurs without an increase in GFAP content suggesting increased exposure of epitopes on GFAP, perhaps due to its depolymerization in the swollen astrocytes (Eng et al. 1989).

3.7 Hypotonic Medium-Induced Swelling

Although it seems likely that astrocytes in situ usually swell by mechanisms other than a hypotonic environment, significant hypoosmolality is seen clinically and is almost always due to lowered plasma Na^+ concentrations (hyponatremia) in children due to kidney failure. Swelling of primary astrocyte cultures due to media made hypotonic by reduction of NaCl is followed by a characteristic regulatory volume decrease (RVD, see Lang et al. and Hoffmann et al., this Vol.), and is associated with release of K^+, Cl^-, and amino acids (Kimelberg et al. 1990a). Work in vivo has shown that astroglial swelling also spontaneously resolves after head injury (Barron et al. 1988). However, the time course of swelling and RVD in vitro is considerably faster than the onset of astroglial swelling and its resolution in vivo. This difference could be due to a balance of slowly developing swelling forces and normally rapid RVD forces in vivo. This slower rate of swelling can be mimicked in primary astrocyte cultures using high extracellular K^+ or L-glutamate (Kimelberg and Goderie 1988; O'Connor et al. 1993).

4 Functional Consequences of Glial Swelling

Some of the effects of astroglial swelling, such as release of amino acids, has been alluded to above. In the rest of this chapter we will principally concentrate on these and other effects of astroglial swelling.

4.1 Swelling and Ion Channels

As mentioned previously, volume regulation can be achieved by the efflux of osmotically active substances across the cell membrane, either by carriers or through membrane channels. Water influx, and thus swelling, can be achieved by the uptake of both K^+ and Cl^- via channels (Boyle and Conway 1941). Conversely, shrinkage is driven by a net loss of anions and cations, usually K^+ and Cl^-. Channels specific (see later, for further analysis of carriers versus channels involved in swelling and shrinkage) for both cations and anions are expressed in glial cells (Barres et al. 1990; Bevan 1990). Moreover, a variety of such channels exist which are differentially expressed in subsets of glial cells and undergo changes in expression during development. As expected, the dominant cation channels are K^+-specific, and the most abundant anion channels are Cl^-, and perhaps HCO_3^-,-specific.

4.1.1 Glial Cells Express Channels Specific for K^+ and Cl^-

As mentioned above, swelling and shrinking can be caused by movements of KCl across the membrane via membrane channels. The dominant channels of the plasma membrane in glial cells are K^+ selective. The different types of glial cells all express a variety of different K^+ channels. Delayed rectifier type channels are found in Schwann cells, astrocytes, and glial precursor cells, inward rectifiers in astrocytes and oligodendrocytes. In addition, A-type and Ca^{2+}-activated K^+ channels are described in glial precursor cells and astrocytes (for review, see Barres et al. 1990; Bevan 1990). Thus, at physiological membrane potentials, there are always active K^+ channels in the membrane of astrocytes, oligodendrocytes, and glial precursor cells.

Thus, the rate-limiting step in the transmembrane translocation of KCl via channels is likely to be the open probability of Cl^- channels. The characterization of Cl^- channels in glial cells is so far much more limited than studies on K^+ channels. Voltage-gated Cl^- channels have been described for cultured rat astrocytes and rabbit Schwann cells. These channels are activated at membrane potentials more positive to $-40\,mV$ (Bevan et al. 1985). The presence of a small Cl^- conductance in oligodendrocytes has been inferred from resistance measurements (Kettenmann 1987). Barres et al. (1988) reported the presence of a Cl^- channel in type 2 astrocytes, which is only active in the inside-out patch configuration, indicating that the open probability of the channel is controlled by a so far unidentified cytoplasmic factor, (Barres et al. 1988). If such a factor exists, it could serve to activate the channel for volume regulation.

4.1.2 Large Conductance Anion Channel

A large conductance ion channel has been described in astrocytes (Bevan et al. 1985; Nowak et al. 1987; Sonnhof 1987; Jalonen et al. 1989, 1991) and Schwann cells (Bevan et al. 1984; Gray and Ritchie 1985), but not in oligodendrocytes. This channel is permeable to anions which include not only Cl^-, but larger anions such as methyl sulfate and isethionate. In addition, it can also pass Na^+ at a greatly reduced rate (Gray et al. 1986). The channel is rarely found in cell-attached patches, but is spontaneously active after isolation of the membrane to form inside-out patches (also see Barres et al. 1988 and 4.1.1 above). There is evidence that it opens when osmotic gradients are applied across inside-out patches or for cell-attached patches when the cells are exposed to hypotonic solutions (Jalonen et al. 1989, 1991). It has been suggested that as yet unidentified internal factors control the activity of the channel, keeping it in the closed configuration under physiological conditions (Gray et al. 1986). Recently, a comparable channel has also been found on the glia limitans of the intact frog optic nerve formed by surface astrocytes (Marrero et al. 1991). This channel type may be responsible for the current increases observed in astrocytes after exposure to hypoosmotic solutions by whole-cell voltage-clamp recordings (Kimelberg et al. 1990c). It is thus likely that activation of this channel leads to an efflux of anions, which with cations will lead to a volume decrease in the cell.

4.1.3 The Cl^- Gradient Determines the Direction of KCl Flux

The K^+ equilibrium potential in glial cells is commonly close to the resting membrane potential (e.g. Kettenmann et al. 1983). Thus, K^+ channels which are commonly open at rest do not lead to an accumulation or loss of K^+ in the glial cells. In contrast, Cl^- in many glial cell cultures, is accumulated above its passive distribution and the Cl^- electrochemical equilibrium is more positive than the membrane potential (for review, see Kettenmann 1990; Kimelberg 1990). Cultured astrocytes from cerebral cortex have the highest level of Cl^- concentration (30 to 50 mM) of all vertebrate glial cells (Kimelberg 1981; Kettenmann et al. 1987). This concentration corresponds to a Cl^- equilibrium potential of −50 to around −30 mV (Kettenmann et al. 1987). Since these cells commonly have membrane potentials of −70 to −90 mV, the difference in potential driving Cl^- through an open channel could be as high as −60 mV. In cultured oligodendrocytes from mouse spinal cord Cl^-_i was closer to, but in many cells still elevated above passive electrochemical distribution (Hoppe and Kettenmann 1989b). In contrast to these findings, glial cells in guinea pig olfactory cortex (Ballanyi et al. 1986) and frog spinal cord (Bührle and Sonnhof 1983) showed a passive Cl^- distribution with the Cl^- equilibrium potential within a few mV of the resting membrane potential. It is not clear whether the differences reflect differences in vivo versus in vitro, species or regional tissue differences, or the contribution of some other unknown factors.

The mechanisms of Cl^- uptake by primary cultures of astrocytes have been studied by Kimelberg and colleagues (Bourke et al. 1979; Kimelberg et al. 1979; Kimelberg 1981). It was found that Cl^- is transported in exchange for HCO_3^-, which is SITS-sensitive, or in combination with K^+ and Na^+ on the furosemide or bumetanide-sensitive cotransporter (Kimelberg and Frangakis 1985). Tas et al. (1986) have also identified a Na^+/K^+ 2Cl cotransporter in primary astrocyte cultures from rat cerebral hemispheres. In oligodendrocytes, a $K^+/Na^+/Cl^-$ cotransport system also seems responsible for the elevation of Cl^- slightly above passive distribution (Hoppe and Kettenmann 1989b).

As a consequence of elevated $[Cl^-]_i$, opening of Cl^- channels will depolarize the membrane potential to a value between the K^+ and Cl^- equilibrium potentials. However, depolarization due to elevated $[K^+]_o$ will only lead to swelling through independent K^+ and Cl^- channels if E_m is more positive than E_{Cl}. For efflux of KCl through independent K^+ and Cl^- channels a membrane potential that is both more positive than E_K and more negative than E_{Cl} is required. In the case of hypotonic media-induced swelling we have shown that there is initially a rapid depolarization, followed by a repolarization (Kimelberg and O'Connor 1988). In the case of astrocytes the depolarization was greater than could, to a first approximation, be explained by dilution of $[K^+]_i$ (see Fig. 4; Kimelberg and Kettenmann 1990). However, $[Cl]_i$ will also be diluted by swelling. Thus, for a twofold swelling E_m depolarizes to approximately $-20\,mV$ (see Fig. 4), presumably due to increased permeability to Na^+, and $[Cl^-]$ is diluted to about 20 mM. Therefore, E_{Cl} is about $-30\,mV$ and an outward driving force on Cl^- can only occur starting several minutes after the initial exposure to hypotonic media when there is a significant repolarization of the membrane potential (Kimelberg and O'Connor 1988). Thus, initially, KCl efflux may occur on the KCl cotransporter, an electrically neutral system which catalyzes the obligatory cotransport of one K^+ with one Cl^- (Lauf 1985). Since it is electrically neutral, this system is driven solely by the transmembrane ion gradients, but formally has a Donnan-type relationship for these gradients (Tosteson 1981);

$$\frac{[K^+]_i}{[K^+]_o} = \frac{[Cl^-]_o}{[Cl^-]_i} . \qquad (1)$$

The ion concentrations likely to be obtained immediately after swelling give values for Eq. (1) of 65/4 = 130/20 (see above). As K^+ is the primary cellular cation $[K]_i$ cannot be reduced, since as K^+ leaves and the cell shrinks, $[K^+]_i$ remains constant. Thus, the relationship in Eq. (1) can only be reached because $Cl^-{}_i$ also leaves with K^+. This occurs when $[Cl^-]_i$ reaches 8 mM, leaving most of the negative counterions for $K^+{}_i$ due to

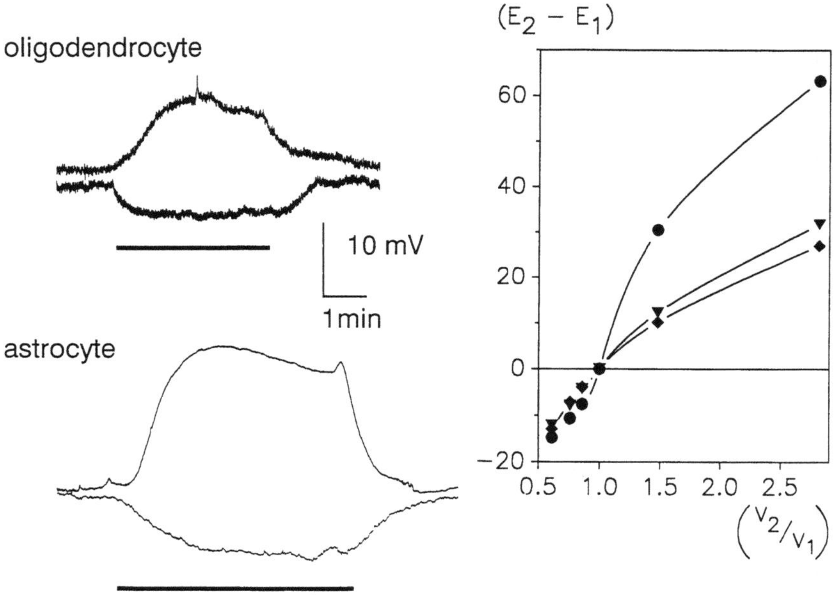

Fig. 4. Comparison of membrane potential changes in cultural oligodendrocytes and astrocytes due to hypo- and hypertonic media changes. The *left traces* show representative membrane potential recordings from oligodendrocytes and astrocytes. Media of different osmolarities were applied as indicated by the *bars*. The control solution for the hypotonic charges had 50 mM of the NaCl replaced with 100 mM sucrose. The depolarizations shown were elicited by media made hypotonic by removal of 100 mM sucrose (*upper traces* in both cases) and the hyperpolarizations (*lower traces*) by media made hypertonic by addition of 100 mM sucrose to standard Ringer's solution. The graph shows the peak potential changes as $E_2 - E_1$ in mV, as a function of the expected changes in volume (V_2/V_1) for the four different conditions based on the cells behaving as perfect osmometers in response to media osmolarity changes. Since peak membrane potential responses were measured, this is reasonable. E_2 is the potential of the altered state and E_1 is the initial potential. The changes are positive for a depolarization and negative for a hyperpolarization; ▼, oligodendrocytes; ●, astrocytes; ◆ shows the predicted changes based on changes in intracellular $[K^+]$ alone due to swelling or shrinkage according to the equation $E_2 - E_1 = 60 \log V_2/V_1$. (Kimelberg and Kettenmann 1990)

impermeant anions. This need for Cl^- for volume regulation may explain why Cl^- is high in astrocytes, but even so there is still a considerable "anion gap" for K^+ efflux. Efflux of HCO_3^- through the Cl^- (anion) channel has not been considered since most of the studies were done in HEPES buffer. However, in situ it is likely that HCO_3^- will also contribute to the efflux of anions. Perhaps efflux of other anions, such as amino acids, which are released when astrocytes swell (Kimelberg et al. 1990a), also contribute. In culture the external concentrations remain unchanged because the solution above the cells represents an infinite reservoir. In vivo ionic concentrations in the extracellular space will be altered by cellular efflux or influx, so that extracellular K^+ and Cl^- will increase as a result of efflux reducing the extent of KCI loss and cell shrinkage (see also Fig. 5).

4.1.4 Rapid K^+ Movements in Situ

Ion movements in glial cells have so far been determined with either ion-selective microelectrodes or by radiotracer studies. While the latter has a time resolution of perhaps 10–30 s when rates are measured, it can only measure levels with a time resolution of 5 min or more, depending on the equilibrium rates. Measurement of ion levels by flame photometry or atomic absorption spectroscopy is simply not sensitive enough for cell monolayers, unless they are grown on large surfaces such as 75 cm^2 T-flasks. In contrast, ion-selective microelectrodes can monitor changes in ion levels within seconds. A recent study on glial cells from the corpus callosum slice has revealed that ion movements, particularly K^+, can occur with unexpected rapidity. In this study, membrane currents in ultrastructurally identified oligodendrocytes were characterized using the patch-clamp technique (Fig. 5; Berger et al. 1991). In contrast to cultured oligodendrocytes, membrane currents of promyelinating oligodendrocytes in the slice decayed during a voltage jump with a time constant of a few milliseconds. This decay was not due to conductance changes, but was due to a marked change in the potassium equilibrium potential within the voltage jump. These changes could be so extensive that intracellular $[K^+]$ equals extracellular $[K^+]$. This finding implies that in the white matter, small membrane polarizations in the physiological range can lead to extensive changes in the K^+ gradient across the glial membrane within few milliseconds and also reverse with a similar time constant. Since there is net loss or gain of K^+, it indicates that volume regulation via channels may occur with a time course in the millisecond range.

4.1.5 Stretch-Sensitive Channels

Ion channels that appear to be activated by swelling or stretch were first identified by Guharay and Sachs (1984) using patch-clamp single channel analysis and have been subsequently described in both prokaryotes and eukaryotes (Kullberg 1987). Many classes of SACs have been described and one important class is the stretch-activated, nonspecific cation channels that allow Na^+ and K^+ to pass as well as divalent cations such as Ca^{2+} and Ba^{2+} (Kullberg 1987; Sachs 1988; Falke and Misler 1989). SACs have been reported to be activated in cell-attached patches when the cells are swollen in hypotonic medium (Christensen 1987; Sachs 1988). Ubl et al. (1988a) reported a Ca^{2+}-activated K^+ channel in opossum kidney cells that was activated by hypotonic shock. A stretch-activated K^+ channel sensitive to cell volume has been described in *Necturus* proximal tubule (Sackin 1989), and a cation-specific Ca^{2+}-permeable channel sensitive to membrane stretch has been described in choroid plexus epithelium (Christensen 1987). A stretch-activated cation channel permeable to calcium was reported in retinal glial cells (Puro 1991), but it was not determined whether an increase in cell volume activated this channel. A preliminary study has shown the opening of K^+-selective SACs in rat primary astrocyte cultures with pressure

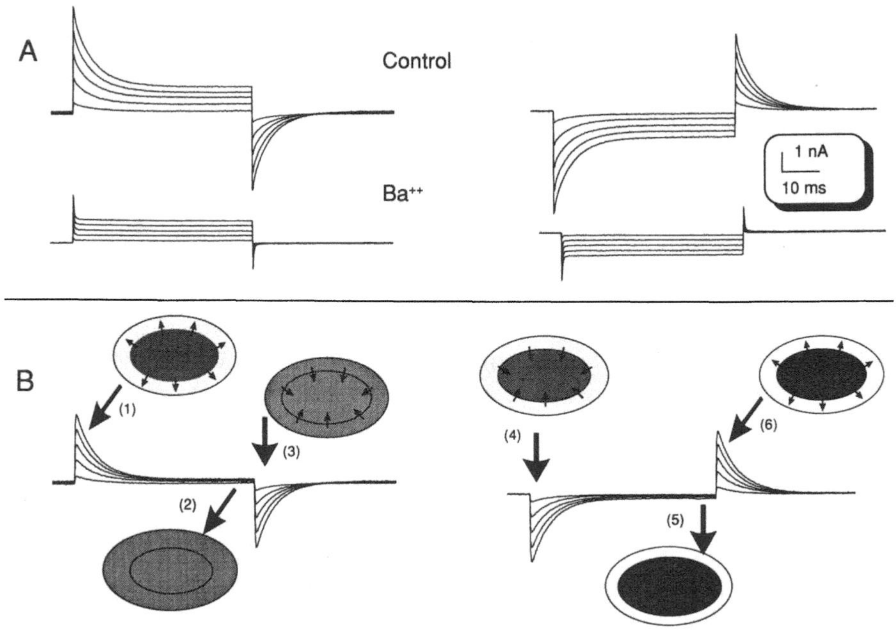

Fig. 5. Rapid movements of K^+ across oligodendrocyte membranes (after Berger et al. 1991). *A* Membrane currents were recorded from oligodendrocytes in slices from the mouse corpus callosum using the patch-clamp technique in the whole cell recording configuration. With depolarizing voltage steps (holding potential, −70 mV), ranging from −50 to 30 mV, outward currents were induced (*upper left trace*). These outward currents decayed rapidly. After the offset of the voltage step, inward tail currents were observed. Conversely, hyperpolarizing voltage steps induced inward currents and outward tail currents (*upper right trace*). The decaying current component and the tail currents were blocked in the presence of Ba^{2+}, a K^+ channel blocker (*lower traces*). Both the decay of the current during the voltage step and the activation of the tail currents were caused by a change in the K^+ gradient across the membrane (Berger et al. 1991). *B* The Ba^{2+}-sensitive current component reflecting the transmembrane K^+ current was isolated by subtracting traces in *A* (*Control* – Ba^{2+}). The schemes illustrate the movements of K^+ from the cell (*inner circle*) to the extracellular space (*outer circle*). During the depolarizing voltage step, K^+ is driven into the extracellular space (*1*). It accumulates and terminates the current (*2*). After the voltage step, a tail current is induced causing the flow of K^+ back into the cell (*3*). The time constants of these K^+ movements range within a few ms. Conversely, during hyperpolarizing voltage steps, K^+ is driven into the cell (*4*) until a new equilibrium is reached (*5*). After the voltage step, the tail current restores the original gradient (*6*) (for further discussion, see Berger et al. 1991)

applied through the cell-attached patch pipette. A variety of different conductance levels, but no ion selectivities, were reported (Ding et al. 1989).* Also, swelling of astrocytes results in a large depolarization of the membrane potential (Kimelberg and O'Connor 1988; Kimelberg and Kettenmann 1990) and exposure of astrocytes in the whole-cell voltage-clamp mode causes increased currents (see Fig. 6; Kimelberg et al. 1990c). This latter observation is consistent with the concept that ion channels are opening.

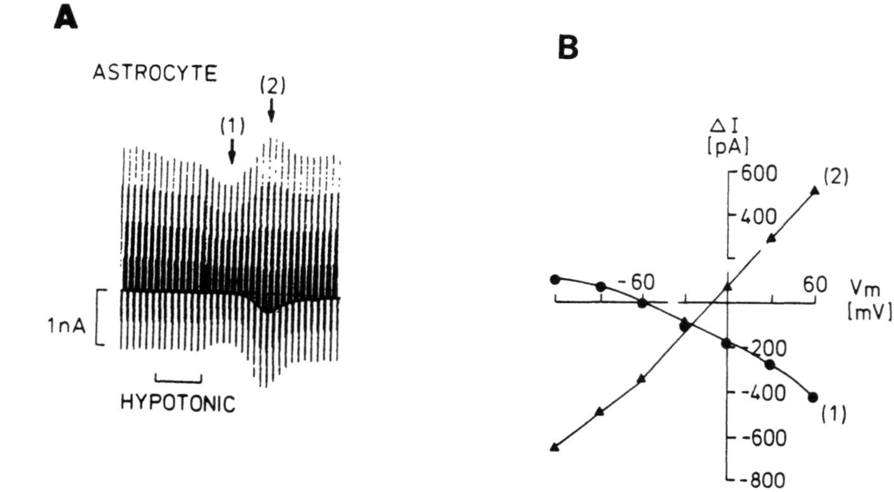

Fig. 6. Swelling-induced currents in a whole-cell voltage-clamped astrocyte. The membrane potential was voltage-clamped near the resting membrane potential (−60 mV) and then brief (200-ms duration) rectangular voltage step commands (−120, −90, −30, 0, +30, and +60 mV) at a frequency of 1 Hz were applied using a computer program; the currents for this are shown. The increased conductance changes induced by a brief (30-s) exposure to hypotonic medium (−100 mM sucrose) are shown in *A* and the current-voltage plot of the swelling-induced changes in current is shown in the ΔI-V plot in *B*. The differences in current were estimated visually by drawing an average horizontal line representing the pre- and posthypotonic current response. In this cell an initial decrease in current conductance preceded the more characteristic conductance increase. The ΔI-V curves labeled (*1*) and (*2*) refer to the regions marked on the current recording. Isotonic control solution was made by replacing 50 mM NaCl with 100 mM sucrose, and hypotonic solution was made by removing this 100 mM sucrose. (Kimelberg et al. 1990c)

4.2 Release of Amino Acids During Swelling

Swelling of primary astrocyte cultures in hypotonic media leads to release of label after cells had been allowed to accumulate ^{3}H-L-glutamate, ^{3}H-D-aspartate, and ^{3}H-taurine, and also release of endogenous glutamate and taurine as measured by high-pressure liquid chromatography (Pasantes-Morales and Schousboe 1989; Kimelberg et al. 1990a). Using microdialysis, Lehmann (1989) demonstrated hypotonic-induced release of taurine, glutamate, and aspartate in hippocampus, dentate gyrus, and piriform cortex in vivo. Application of "Ockham's razor" leads to the proposal that the loss of amino acids was induced by cell swelling in vivo. However, ultrastructural studies were not performed. In vivo, significant hypoosmolality is almost always due to lowered plasma Na^+ concentrations as in hyponatremia. This condition has also been shown to lead to loss of taurine, and to a lesser extent aspartic and glutamic acids and other ninhydrin-positive substances from the brain (Thurston et al. 1987; Wade et al. 1988).

It has been recently reported that swelling of astrocytes in vitro by increasing medium [K^+] with reciprocal reduction of [Na^+] but maintaining isotonicity and [Cl^-], i.e., conditions promoting Donnan swelling as well as depolarization, causes release of [^{3}H] taurine or endogenous taurine (Philibert et al. 1988; Holopainen et al. 1989; Pasantes-Morales and Schousboe 1989). We have also found, using a constant perfusion system, that similar increases in medium K^+ caused increased release of preloaded [^{3}H] D-aspartate, and this release was blocked by the anion transport inhibitor L-644,711 which, in addition to SITS and furosemide, was also found to be effective in blocking hypotonic media-induced release of amino acids (Kimelberg et al. 1990a; Kimelberg and Goderie 1991). L-644,711 also results in improved recovery in an experimental head injury (Kimelberg et al. 1987) and rabbit ischemia models (Kohut et al. 1991).

Pasantes-Morales and Schousboe (1989) have shown that when medium K^+ was increased, but Cl^- was reduced to keep the $K^+ \times Cl^-$ product constant, there was no efflux of [^{3}H]-taurine from cultured astrocytes. Maintaining the $K^+ \times Cl^-$ product constant eliminates Donnan swelling which otherwise occurs when medium K^+ is raised without lowering Cl^- proportionally (Boyle and Conway 1941; Hodgkin and Horowicz 1959). Such effects on volume have been shown in cultured astrocytes, but the raised K^+ will still depolarize the cells (Walz 1987), further supporting the view that the critical factor in K^+-induced release of amino acids is swelling of the cells rather than K^+-induced depolarization.

The hypotonic media-induced release of preloaded [^{3}H] taurine, [^{3}H] L-glutamate, and [^{3}H] D-aspartate is also blocked by L-644,711 (see Fig. 7 for L-644,711 inhibition of hypotonic media-induced [^{3}H] D-aspartate release). Such efflux is also blocked by SITS and furosemide, but 1 mM bumetanide had very little effect (Kimelberg et al. 1990a). At this concentration bumetanide completely inhibits $Na^+ + K^+ + 2Cl^-$ cotransport (Warnock et al. 1983), and at 0.01 mM it maximally inhibits $^{42}K^+$ and $^{36}Cl^-$ uptake by astrocytes (Kimelberg and Frangakis 1985). Threo-B-hydroxyaspartic acid (THBA) is an effective inhibitor of L-glutamate uptake in primary astrocyte cultures (Kimelberg et al. 1989a). It also had no effect on the swelling-induced efflux, suggesting that reversal of the uptake system is not involved in the efflux (Kimelberg et al. 1990a). However, THBA competes with L-glutamate (Kimelberg et al. 1989a), so that THBA would presumably have to enter the cell to block L-glutamate efflux.

Reversal of Na^+-dependent L-glutamate uptake is a possible mechanism for swelling-induced efflux, because we generally make the solution hypoosmotic by omitting 50 mM NaCl and the Na^+-dependent glutamate and aspartate uptake systems can be reversed by altering the transmembrane Na^+ gradient (Erecinska 1987; Sarantis and Attwell 1990). Swelling of astrocytes also leads to depolarization of their membrane potentials (Kimelberg and O'Connor 1988) and this will also reduce the inward driving force. However, we have found that hypotonic media induces release of

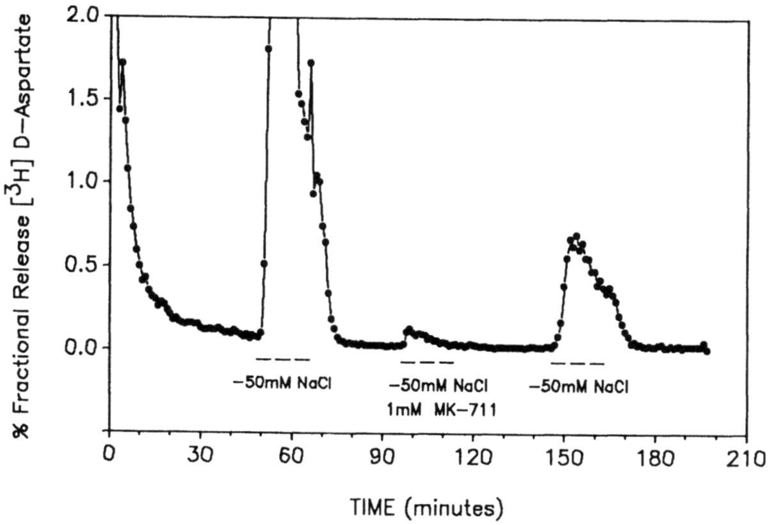

Fig. 7. Release of previously accumulated [^{3}H] D-aspartate in primary astrocyte cultures. Release was measured at 37 °C using a constant perfusion system in which the cells were perfused with HEPES-buffered media containing 145 mM Na^+ (isotonic), or 10-min exposures to hypotonic (removal of 50 mM NaCl) plus or minus the anion transport inhibitor L-644,711 as indicated. Results are expressed as a percentage of the radioactivity present in the cells at each time point released. The initial release is always to varying degrees greater than subsequent releases. The second response is always the same or slightly greater than the third and subsequent responses. As can be seen, 1 mM L-644,711 (MK-711) essentially inhibits the swelling-induced release. Also, note the RVD-like efflux seen clearly in the third response

these amino acids during pre-squilic-bration and exposure to Na^+-free media, conditions which will cause loss of intracellular Na^+ (Kimelberg et al. 1990a).

As noted in the preceding section, L-644,711 is also an effective inhibitor of swelling-induced release of glutamate, aspartate, and taurine from astrocytes in vitro. Nicklas et al. (1987) have shown that addition of kainic acid (KA) to cerebellar or hippocampal slices results in release of glutamate which seemed to come mainly from glia. This release was inhibited by SITS and the authors pointed out that one effect of KA might be to cause the glial cells to swell due to SITS-sensitive increased uptake of Cl^-. It now seems possible that SITS could also be directly inhibiting the release of glutamate from swollen astrocytes.

All the inhibitors capable of inhibiting swelling-induced release of taurine and glutamate also inhibit Cl^-/HCO_3^- anion exchange. However, these data cannot be construed as indicating a correspondence between the swelling-activated amino acid and Cl^-/HCO_3^- exchange transport systems. For one thing these agents also block other Cl^- transport systems such as

Cl^- channels (Landry et al. 1989) and KCl cotransport systems (Warnock et al. 1983; Garay et al. 1986). Possibly, they also have an effect on a unique class of swelling-activated Na^+-independent transport systems for amino acids, which may in turn correspond to the large anion channel (see 4.1.2).

The excitotoxicity of L-glutamate (Olney 1969) and other excitotoxins such as L-aspartate and quinolinic acid (Whetsell et al. 1988; McDonald and Johnston 1990) is thought to be due to activation of NMDA receptors on neurons allowing excessive influx of Ca^{2+} which results in neuronal death, since these effects can be reversed by NMDA antagonists (Simon et al. 1984; Wieloch 1985; Choi 1988; McDonald and Johnston 1990). However, as well as undesirable psychiatric effects, it has been recently shown that a number of these antagonists can themselves cause neuronal toxicity (Olney et al. 1989). Thus, compounds, like L-644,711, that inhibit release of glutamate and other excitotoxins either from swollen astrocytes and perhaps glutamatergic nerve endings may provide a better therapy and indeed may be one of the reasons L-644,711 is effective in animal models of head injury and stroke.

4.3 Role of Calcium in Volume Regulation

There is relatively little known about the signals initiating RVD in glial cells. However, both extra- and intracellular Ca^{2+} have been shown to influence cellular volume regulation in a number of cell types (Foskett and Spring 1985; Cala et al. 1986; Foskett and Melvin 1989; Olson et al. 1990; Pierce and Politis 1990). In some cases removal of extracellular Ca^{2+} blocks volume recovery by preventing osmolyte efflux (Davis and Finn 1987). In other instances, the volume recovery process is not sensitive to $[Ca^{2+}]_o$, but changes in $[Ca^{2+}]_i$ do alter volume recovery following hypoosmotic stress, again by altering the pattern of osmolyte efflux (Grinstein et al. 1982; Cala et al. 1986). In a preliminary report, we have shown that extracellular calcium is required for volume regulation in primary cultures of astrocytes as measured by the impedance method described earlier (O'Connor and Kimelberg 1991).

Most vertebrate cells accomplish regulatory volume decrease (RVD) from hypoosmotic stress by an efflux of K^+ and Cl^-. In some cells this is through separate K^+ and Cl^- channels (Hoffmann et al. 1990), while in others the K^+ movement is totally Cl^--dependent and may be due to coupled KCl cotransport (Pierce and Politis 1990). Most data indicate that Ca^{2+} specifically affects K^+ movement through channels because quinine, a drug that inhibits Ca^{2+}-dependent K^+ transport, blocks or reduces cell volume recovery (Grinstein et al. 1982; Foskett and Spring 1985; Hoffmann et al. 1990; Pierce and Politis 1990). Calcium-activated K^+ channels have been reported in astrocytes (Quandt and MacVicar 1986; Barres et al. 1990) and in retinal Muller cells (Puro et al. 1989). These channels are activated by membrane depolarization and increases in intracellular Ca^{2+} concentration.

One possible source of the increase in free calcium that can be a signal for swelling-induced membrane transport changes (Wong and Chase 1986; Olson et al. 1990; Wong et al. 1990) could be increased calcium influx into the cell via voltage-dependent calcium channels or via stretch-activated channels that are permeable to calcium. Voltage-sensitive L-type Ca^{2+} channels have been described in cortical astrocytes (MacVicar 1984; Barres et al. 1988, 1990; MacVicar and Tse 1988). We have shown in a preliminary report that hypotonic-induced swelling of astrocytes causes a large increase in uptake of extracellular calcium as measured with $^{45}Ca^{2+}$. This uptake was partially blocked (60%) by the voltage-gated calcium channel blocker, nimodipine at 10^{-6}M (O'Connor et al. 1990). Voltage-gated calcium channels could be activated during hypotonic-induced swelling because we have reported that exposing astrocytes to hypotonic media causes a marked depolarization (Kimelberg and O'Connor 1988; Kimelberg and Kettenmann 1990). We have also found that removal of $[Ca^{2+}]_e$ and/or the buffering of $[Ca^{2+}]_i$ did not affect swelling-induced release of preloaded ^{3}H-taurine or ^{3}H-D-aspartate (O'Connor et al. 1990), which has been shown to be stimulated during RVD in astrocytes (Kimelberg et al. 1990a). Thus, we have to assume that swelling-induced efflux of taurine, glutamate, and aspartate does not significantly contribute to cell shrinkage in RVD, which seems to be mainly due to KCl efflux (O'Connor and Kimelberg 1991).

5 Conclusions

Glial cells of all types have a variety of ion channels and carrier-mediated systems which potentially cause them to swell and also allows them to regulate their volume. However, only in astrocytes have these processes been worked out in any detail. Swelling occurs in astrocytes in situ; therefore, the consequences of such swelling for astrocytes are of practical as well as therapeutic importance (Kimelberg 1991b). In vitro, swelling of astrocytes results in loss of K^+ and Cl^-, which could compromise astrocyte regulation of the neuronal ionic microenvironment, and loss of excitatory amino acids which could result in excitotoxically induced neuronal damage. There are numerous potential mechanisms which could cause astroglial swelling, ranging from KCl uptake due to Donnan forces by uptake through separate K^+ and Cl^- channels, to massive uptake of Na^+ and glutamate on the Na^+-dependent glutamate uptake system. Knowledge of the mechanisms of swelling can lead to strategies to inhibit it and knowledge of the mechanisms of swelling-induced release of KCl and excitatory amino acids can lead to strategies to mitigate the potentially deleterious consequences of such release.

Acknowledgments. HKK and EOC acknowledge grant support from NIH (NS 23750) and HK acknowledges grants by the Bundesministerium für Forschung und Technologie, Hermann und Lilly Schilling-Stiftung and Deutsche Forschungsgemeinschaft (Heisenberg-Stipendium, SFB 317). We would like to thank Dr. M. Aschner for assistance with computer graphics.

References

Bakay L, Lee JC (1968) The effect of acute hypoxia and hypercapnia on the ultrastructure of the central nervous system. Brain 91: 697–706

Ballanyi K, Grafe P (1988) Cell volume regulation in the nervous system. Renal Physiol Biochem 11: 142–157

Ballanyi K, Kettenmann H (1990) Intracellular Na^+ regulation in cultured oligodendrocytes. J Neurosci Res 26: 455–460

Ballanyi K, Grafe P, ten Bruggencate G (1986) Ion activities and potassium uptake mechanisms of glial cells in guinea-pig olfactory cortex slices. J Physiol 382: 159–174

Ballanyi K, Grafe P, Serve G, Schlue R (1990) Electrophysiological measurements of volume changes in leech neuropile glial cells. Glia 3: 151–158

Barres BA, Chun LLY, Corey DP (1988) Ion channel expression by white matter glia. I. Type 2 astrocytes and oligodendrocytes. Glia 1: 10–30

Barres BA, Chun LLY, Corey DP (1990) Ion channels in vertebrate glia. Annu Rev Neurosci 13: 441–474

Barron KD, Dentinger MP, Kimelberg HK, Nelson LR, Bourke RS, Keegan S, Mankes R, Cragoe EJ Jr (1988) Ultrastructural features of a brain injury model in cat. I. Vascular and neuroglial changes and the prevention of astroglial swelling by a fluorenyl (aryloxy) alkanoic acid derivative (L-644, 711). Acta Neuropathol Berl 75: 295–307

Berger T, Schnitzer J, Kettenmann H (1991) Developmental changes in the membrane current pattern, K^+ buffer capacity and morphology of glial cells from the corpus callosum slice. J Neurosci 11: 3008–3024

Bevan S (1990) Ion channels and neurotransmitter receptors in glia. Semin Neurosci 2: 467–481

Bevan S, Gray PTA, Ritchie JM (1984) A high conductance anion selective channel in rat Schwann cells. J Physiol 382: 159–174

Bevan S, Chiu SY, Gray PTA, Ritchie JM (1985) The presence of voltage-gated sodium, potassium and chloride channels in rat cultured astrocytes. Proc R Soc Lond Ser B 225: 299–313

Bourke RS, Kimelberg HK, Daze MA, Popp AJ (1979) Studies on the formation of astroglial swelling and its inhibition by clinically useful agents. In: Popp AJ, Nelson LR, Bourke RS, Kimelberg HK (eds) Neural trauma. Raven Press, New York, pp 95–113

Bourke RS, Kimelberg HK, Daze M, Church G (1983) Swelling and ion uptake in cat cerebrocortical slices: control by neurotransmitters and ion transport mechanisms. Neurochem Res 8: 5–24

Boyle PJ, Conway EJ (1941) Potassium accumulation in muscle and associated changes. J Physiol 100: 1–63

Bührle CP, Sonnhof U (1983) Intracellular ion activities and equilibrium potentials in motoneurones and glial cells of the frog spinal cord. Pflügers Arch Eur J Physiol 396: 144–153

Bullock R, Maxwell WL, Graham DI, Teasdale GM, Adams JH (1991) Glial Swelling following human cerebral contusion: an ultrastructural study. J Neurol Neurosurg Psychiat 54: 427–434

Cala PM, Mandel LJ, Murphy E (1986) Volume regulation by *Amphiuma* red blood cells: cytosolic free Ca and alkali metal-H exchange. Am J Physiol 19: C423–C429

Casper DS, Trelstad RL, Reif-Lehrer L (1982) Glutamate-induced cellular injury in isolated chick embryo retina: Muller cell localization of initial effects. J Comp Neurol 209: 79–90
Castejon OJ (1980) Electron microscopic study of capillary wall in human cerebral edema. J Neuropathol Exp Neurol 29: 296–327
Chan PH, Fishman RA (1984) Phospholipid degradation and the early release of polyunsaturated fatty acids in the evolution of brain edema. In: Go TG, Baethmann A (eds) Recent progress in the study and therapy of brain edema. Plenum, New York, pp 193–202
Chesler M (1990) The regulation and modulation of pH in the nervous system. Prog Neurobiol 34: 401–427
Chesler M, Kraig RP (1989) Intracellular pH transients of mammalian astrocytes. J Neurosci 9: 2011–2019
Choi DW (1988) Glutamate neurotoxicity and diseases of the nervous system. Neuron 1: 623–634
Choi DW, Rothman SM (1990) The role of glutamate neurotoxicity in hypoxic-ischemic neuronal death. Annu Rev Neurosci 13: 171–182
Christensen O (1987) Mediation of cell volume regulation by Ca^{2+} influx through stretch-activated channels. Nature 330: 66–68
Coles JA (1989) Functions of glial cells in the retina of the honeybee drone. Glia 2: 1–9
Davis CW, Finn AL (1987) Interactions of sodium transport, cell volume, and calcium in frog urinary bladder. J Gen Physiol 89: 687–702
Davis LE, Blisard KS, Kornfeld M (1990) The influenza B virus mouse model of Reye's syndrome: clinical, virologic and morphologic studies of the encephalopathy. J Neurol Sci 97: 221–231
Dietzel I, Heinenmann U, Lux HK (1989) Relations between slow extracellular potential changes, glial potassium buffering, and electrolyte and cellular volume changes during neuronal hyperactivity in cat brain. Glia 2: 25–44
Ding JP, Bowman CL, Sokabe M, Sachs F (1989) Mechanical transduction of glial cell SACs and SICs. Biophys J 55: 244a
Drejer J, Larsson OM, Schousboe A (1982) Characterization of L-glutamate uptake into and release from astrocytes and neurons cultured from different brain regions. Exp Brain Res 47: 259–269
Eng LF, D'Amelio FE, Smith ME (1989) Dissociation of GFAP intermediate filaments in EAE: observations in the lumbar spinal cord. Glia 2:308–317
Erecinska M (1987) The neurotransmitter amino acid transport system – a fresh outlook on an old problem. Biochem Pharmacol 36: 3547–3555
Falke LC, Misler S (1989) Activity of ion channels during volume regulation by clonal N1E115 neuroblastoma cells. Proc Natl Acad Sci USA 86: 3919–3923
Fischbarg J, Kuang K, Hirsch J, Lecuaona S, Regozinski L, Silverstein SC, Loike J (1989) Evidence that the glucose transporter serves as a water channel in J774 macrophages. Proc Natl Acad Sci USA 86: 8397–8401
Foskett JK, Melvin JE (1989) Activation of salivary secretion: coupling of cell volume and $[Ca^{++}]_i$ in single cells. Science 244: 1582–1585
Foskett JK, Spring KR (1985) Involvement of calcium and cytoskeleton in gallbladder epithelial cell volume regulation. Am J Physiol 248: C27–C36
Garay RP, Hannaert PA, Nazaret C, Cragoe EJ Jr (1986) The significance of the relative effects of loop diuretics and anti-brain edema agents on the Na^+, K^+, Cl^- cotransport system and the Cl^- $NaCO_3^-$ anion exchanger. Naunyn-Schmiedeberg's Arch Pharmacol 334: 202–209
Gerschenfeld HM, Wald F, Zakunaisky JA, DeRobertis EDP (1959) Function of astroglia in the water-ion metabolism of the central nervous system. Neurology 9: 412–425
Gilles R, Hoffmann EK, Bolis L (1991) – Volume and osmolality control in animal cells, Advances in comparative and environmental physiology 9. Springer, Berlin Heidelberg New York

Gray PTA, Ritchie JM (1985) Ion channels in Schwann and glial cells. Trends Neurosci 8: 411–415
Gray PTA, Bevan S, Chiu SY, Shrager P, Ritchie JM (1986) Ionic conductances in mammalian Schwann cells. In: Ritchie RJ, Keynes RD (eds) Ion channels in neural membranes. Liss, New York, pp 145–157
Grinstein S (1989) Measurements of cytoplasmic pH and cellular volume for detection of Na^+/H^+ exchange in lymphocytes. Methods Enzymol 173: 777–790
Grinstein S, Foskett JK (1990) Ionic mechanisms of cell volume regulation in leukocytes. Annu Rev Physiol 52: 399–414
Grinstein S, Dupre A, Rothstein A (1982) Volume regulation by human lymphocytes – role of calcium. J Gen Physiol 79: 849–868
Grinstein S, Rothstein A, Sakardi B, Gelfand EW (1984) Responses of lymphocytes to anisotonic media: volume-regulatory behavior. Am J Physiol 246: C204–C215
Guharay F, Sachs F (1984) Stretch-activated single ion channel currents in tissue-cultured embryonic chick skeletal muscle. J Physiol (Lond) 352: 685–701
Halliwell B, Gutteridge JMC (1985) Oxygen radicals and the nervous system. Trends Neurosci 8: 22–26
Hayflick L, Moorehead PS (1961) The serial cultivation of human diploid cell strains. Exp Cell Res 25: 585–621
Herndon RM, Coyle JT, Addicks E (1980) Ultrastructural analysis of kainic acid lesion to cerebellar cortex. Neuroscience 5: 1015–1026
Hertz L (1979) Functional interactions between neurons and astrocytes. I. Turnover and metabolism of putative amino acid transmitters. Prog Neurobiol 13: 277–323
Hodgkin AL, Horowicz P (1959) The influence of potassium and chloride ions on the membrane potential of single muscle fibres. J Physiol 148: 127–160
Hoffmann EK (1987) Volume regulation in cultured cells. In: Gilles R, Kleinzeller A, Bolis L (eds) Current topics in membranes and transport, vol 30. Academic Press, New York, pp 125–180
Hoffmann EK, Kolb HA (1991) Mechanisms of activation of regulatory volume responses after cell swelling. In: Gilles R, Hoffmann EK, Bolis L (eds) volume and osmolality control in animal cells. Advances in comparative and environmental physiology 9. Springer, Berlin Neidelberg New York, pp 140–185
Hoffmann EK, Simonsen LO, Lambert IH (1990) Separate, Ca^{2+}-Activated K^+ and Cl^- transport pathways in Ehrlich ascites tumor cells. J Membr Biol 91: 227–244
Holopainen I, Kontro P, Oja SS (1989) Release of taurine from cultured cerebellar granule cells and astrocytes: co-release with glutamate. J Neurosci 29: 425–432
Hoppe D, Kettenmann H (1989a) GABA triggers a Cl^- efflux from cultured oligodendrocytes. Neurosci Lett 97: 334–339
Hoppe D, Kettenmann H (1989b) Carrier-mediated Cl^- transport in cultured oligodendrocytes. J Neurosci Res 23: 467–475
Jakubovicz DE, Grinstein S, Klip A (1987) Cell swelling following recovery from acidification in C6 glioma cells: an in vitro model of postischemic brain edema. Brain Res 435: 138–146
Jalonen T, Johansson S, Holopainen I, Oja SS, Arhem P (1989) A high-conductance multistate anion channel in cultured rat astrocytes. Acta Physiol Scand 136: 611–612
Jalonen T, Popp AJ, Kimelberg HK (1991) Potassium and anion channels opened by osmotic gradients across astrocyte plasma membranes. Soc Neurosci Abstr 17(1): 57
Jenkins LW, Becker DP, Coburn TH (1982) A quantitative analysis of glial swelling and ischemic neuronal injury following complete cerebral ischemia. In: Go KG, Baethmann A (eds) Recent progress in the study and therapy of brain edema. Plenum Press, New York, pp 523–537
Katzman R, Pappius HM (1973) Brain electrolytes and fluid metabolism. Williams and Wilkins, Baltimore, pp 224–245
Kempski O, Chaussy L, Gross U, Zimmer M, Baethmann A (1983) Volume regulation and metabolism of suspended C6 glioma cells: an in vitro model to study cytotoxic brain edema. Brain Res 279: 217–228

Kempski O, Staub F, Jansen M, Schodel F, Baethmann A (1988) Glial swelling during extracellular acidosis in vitro. Stroke 19: 385–392
Kettenmann H (1987) K^+ and Cl^+ uptake by cultured glial cells. Can J Physiol Pharmacol 65: 1033–1037
Kettenmann H (1990) Cl^- channels and carriers in cultured glial cells In: Alvarez-Leefmans FJ, Russel J (eds) Chloride channels and carriers in nerve, muscle and glial cells. Plenum Press, New York, pp 193–208
Kettenmann H (1991) Properties of $GABA_A$ receptors in glial cells In: Barnard EA, Costa E (eds) Transmitter amino acid receptors: structures, transduction and models for drug development. Thieme Medical Publishers, New York, pp 37–50
Kettenmann H, Sonnhof U, Schachner M (1983) Exclusive potassium dependence of the membrane potential in cultured mouse oligodendrocytes. J Neurosci 3: 500–505
Kettenmann H, Backus KH, Schachner M (1987) c-Aminobutyric acid opens Cl-channels in cultured astrocytes. Brain Res 404: 1–9
Kimelberg HK (1981) Active accumulation and exchange transport of chloride in astroglial cells in culture. Biochim Biophys Acta 646: 179–184
Kimelberg HK (1990) Chloride transport across gliad membranes. In: Alvarez-Leefmans FJ, Russel J (eds) Chloride channels and carriers in nerve, muscle and glial cells. Plenum Press, New York, pp 193–208
Kimelberg HK (1991a) Swelling and volume control in brain astroglial cells. In: Gilles R, Hoffmann EK, Bolis L (eds) Volume and osmolality control in animal cells. Advances in comparative and environmental physiology 9. Springer, Berlin Heidelberg New York, pp 81–117
Kimelberg HK (1991b) Astrocytic edema in CNS trauma. J Neurotrauma 8: 571–581
Kimelberg HK, Frangakis MV (1985) Furosemide- and bumetanide-sensitive ion transport and volume control in primary astrocyte cultures from rat brain. Brain Res 361: 125–134
Kimelberg HK, Goderie SK (1988) Volume regulation after swelling in primary astrocyte cultures. In: Norenberg MD, Schousboe L, Hertz L (eds) Biochemical pathology of astrocytes. Liss, New York, pp 299–311
Kimelberg HK, Goderie SK (1991) Inhibition of potassium-activated release of D-aspartate from astrocytes by an anion transport inhibitor, Ann NY Acad Sci 633: 619–622
Kimelberg HK, Kettenmann H (1990) Swelling induced changes in electrophysiological properties of cultured astrocytes and oligodendrocytes. I. Effects on membrane potentials, input impedance and cell-cell coupling. Brain Res 529: 255–261
Kimelberg HK, O'Connor E (1988) Swelling of astrocytes causes membrane potential depolarization. Glia 1: 219–224
Kimelberg HK, Ransom BR (1986) Physiological aspects of astrocytic swelling. In: Fedoroff S, Vernadakis A (eds) Astrocytes, vol III. Academic Press, Orlando, pp 129–166
Kimelberg HK, Walz W (1988) Ion transport and volume measurements in cell cultures. In: Bolton AA, Baker GB, Walz W (eds) Neuromethods: neurochemishy, brain electrolytes and water spaces. Humana, clifton, New Jersey, pp 441–492
Kimelberg HK, Biddlecome S, Bourke RS (1979) SITS-inhibitable Cl^- transport and Na^+-dependent H^+ production in primary astroglial cultures. Brain Res 173: 111–124
Kimelberg HK, Bourke RS, Stieg PE et al. (1982) Swelling of astroglia after injury to the central nervous system: mechanisms and consequences. In: Grossman RG, Gildenberg PL (eds) Head injury: basic and clinical aspects. Raven Press, New York, pp 31–44
Kimelberg HK, Cragoe EJ Jr, Nelson LR, Popp AJ, Szarowski D, Rose JW, Woltersdorf OW Jr, Pietruszkiewicz AM (1987) Improved recovery from a traumatic-hypoxic brain injury in cats by intracisternal injection of an anion transport inhibitor. Cent Nerv Syst Trauma 4: 3–14
Kimelberg HK, Pang S, Treble DH, (1989a) Excitatory amino acid- stimulated uptake of $^{22}Na^+$ in primary astrocyte cultures. J Neurosci 9: 1141–1149

Kimelberg HK, Rose JW, Barron KD, Waniewski RA, Cragoe EJ Jr (1989b) Astrocytic swelling in traumatic-hypoxic brain injury: beneficial effects of an inhibitor of anion exchange transport and glutamate uptake in glial cells. Mol Cell Neuropathol 11: 1–31
Kimelberg HK, Goderie SK, Higman S, Pang S, Waniewski RA (1990a) Swelling-induced release of glutamate, aspartate, and taurine from astrocyte cultures. J Neurosci 10: 1583–1591
Kimelberg HK, Goderie SK, O'Connor ER et al. (1990b) Swelling-induced membrane transport changes in astrocytes. In: Seil FJ(ed) Advances in neural regeneration research. Wiley-Liss, New York, pp 199–214
Kimelberg HK, Anderson E, Kettenmann H (1990c) Swelling induced changes in electrophysiological properties of cultured astrocytes and oligodendrocytes. II. Whole cell currents. Brain Res 529: 262–268
Kimelberg HK, Sankar P, O'Connor ER, Jalonen T, Goderie SK (1992) Functional consequences of astrocytic swelling. In: Neuronal-astrocytic interactions: pathological implications. Prog Brain Res (in press)
Kletzien RF, Pariza MW, Becker JE, Potter VR (1975) A method using 3-O-methyl-D-glucose and phloretin for the determination of intracellular water space of cells in monolayer cultures. Anal Biochem 68: 537–544
Kohut JJ, Bednar MM, Kimelberg HK, McAuliffe TL, Gross CE (1991) Reduction in ischemic brain injury in rabbits by the anion transport inhibitor L-644, 711. Stroke 23: 93–97
Kuffler SW, Nicholls JG, Orkand RK (1966) Physiological properties of glial cells in the central nervous system of amphibia. J Neurophysiol 29: 768–787
Kullberg R (1987) Stretch-activated ion channels in bacteria and animal cell membranes. TINS 10: 38–39
Landry DW, Akabas MH, Redhead C, Edelman A, Cragoe EJ Jr., Al-Awqati Q (1989) Purification and reconstitution of chloride channels from kidney and trachea. Science 244: 1469–1472
Lauf PK (1985) $K^+:Cl^-$ cotransport: sulfhydryls, divalent cations, and the mechanism of volume activation in a red cell. J Membr Biol 88: 1–13
Lehmann A (1989) Effects of microdialysis-perfusion with anisoosmotic media on extracellular amino-acids in the rat hippocampus and skeletal muscle. J Neurochem 53: 525–535
Lowe AG, Lambert A (1983) Chloride-bicarbonate exchange and related transport processes. Biochim Biophys Acta 694: 353–374
MacKnight ADC, Leader JP (1989) Volume regulation in epithelia: experimental approaches. Methods Enzymol 171: 744–792
MacVicar BA (1984) Voltage-dependent calcium channels in glial cells. Science 226: 1345–1347
MacVicar BA, Tse FW (1988) Norepinephrine and cyclic adenosine 3':5'-cyclic monophosphate enhance a nifedipine-sensitive calcium current in cultured rat astrocytes. Glia 1: 359–365
Marrero H, Orkand PM, Kettenmann H, Orkand RK (1991) Single channel recording from glial cells on the untreated surface of the frog optic nerve. Eur J Neurosci 3: 813–819
Martin DL, Shain W (1979) High affinity transport of taurine and b-alanine and low affinity transport of c-aminobutyric acid by a single transport system in cultured glioma cells. J Biol Chem 254: 7076–7084
Martin DL, Madelian V, Seligmann B, Shain W (1990) The role of osmotic pressure and membrane potential in K(+)-stimulated taurine release from cultured astrocytes and LRM55 cells. J Neurosci 10: 571–577
Matyja E (1986) Morphologic evidence of a primary response of glia to kainic acid administration into the rat neostriatum: studied in vivo and in vitro. Exp Neurol 92: 609–623
Mazzoni MC, Lundgren E, Arfors KE, Intaglietta M (1989) Volume changes of an endothelial cell monolayer on exposure to anisotonic media. J Cell Physiol 140: 272–280

McDonald JW, Johnston MV (1990) Physiological and pathophysiological roles of excitatory amino acids during central nervous system development. Brain Res 15: 41–70
Nicklas WJ, Zeevalk G, Hyndman A (1987) Interactions between neurons and glia in glutamate/glutamine compartmentation. Biochem Soc Trans 15: 208–210
Norenberg MD (1981) The astrocyte in liver disease. In: Fedoroff S, Hertz L (eds) Advances in cellular neurobiology, vol II. Academic Press, Orlando, pp 304–338
Nowak L, Ascher P, Berwald-Netter Y (1987) Ionic channels in mouse astrocytes. J Neurosci 7: 101–109
O'Connor ER, Kimelberg HK (1991) Effect of calcium on astrocyte volume regulation, ion and amino-acid release. Soc Neurosci Abstr 17(1): 56
O'Connor ER, Schneider A, Kimelberg HK (1990) Effects of astrocytic swelling on Ca^{2+} transport and the release of ^{3}H-D-aspartate. Soc Neurosci Abstr 16(1): 666
Olney JW (1969) Brain lesions, obesity, and other disturbances in mice treated with monosodium glutamate. Science 164: 719–721
Olney JW, Labruyere J, Price MT (1989) Pathological changes induced in cerebrocortical neurons by phencyclidine and related drugs. Science 244: 1360–1362
Olson JE, Sankar R, Holtzman D, James A, Fleischhacker D (1986) Energy-dependent volume regulation in primary cultured cerebral astrocytes. J Cell Physiol 128: 209–215
Olson JE, Fleischhacher D, Murray WB, Holtzmann D (1990) Control of astrocyte volume by intracellular and extracellular Ca^{2+}. Glia 3: 405–412
Orkand RK, Nicholls JG, Kuffler SW (1966) Effect of nerve impulses on the membrane potential of glial cells in the nervous system of amphibia. J Neurophysiol 29: 788–806
Parsons DF, Cole RW, Kimelberg HK (1989) Shape, size, and distribution of cell structures by 3-D graphics reconstruction and stereology. I. The regulatory volume decrease of astroglial cells. Cell Biophys 14: 27–42
Pasantes-Morales H, Schousboe A (1989) Release of taurine from astrocytes during potassium-evoked swelling. Glia 2: 45–50
Philibert RA, Rogers KL, Allen AJ, Dutton GR (1988) Dose- dependent K^+ -stimulated efflux of endogenous taurine from primary astrocyte cultures is Ca^{2+}-dependent. J Neurochem 51: 122–126
Pierce SK, Politis AD (1990) Ca^{2+}-activated cell volume recovery mechanisms. Annu Rev Physiol 52: 27–42
Puro DG (1991) Stretch-activated channels in human retinal Muller cells Glia 4: 456–460
Puro DG, Roberge F, Chan CC (1989) Retinal glial cell proliferation and ion cannels: a possible link. Invest Ophthalmol Visual Sci 30: 521–529
Quandt FN, Mac Vicar BA (1986) Calcium activated potassium channels in cultured astrocytes. Neuroscience 19: 29–41
Ransom BR, Yamate CL, Connors BW (1985) Activity-dependent shrinkage of extracellular space in rat optic nerve: a developmental study. J Neurosci 5: 532–535
Sachs F (1988) Mechanical transduction in biological systems. Crit Rev Biomed Eng 16: 141–169
Sackin H (1989) A stretch-activated K^+ channel sensitive to cell volume. Proc Natl Acad Sci USA 86: 1731–1735
Sarantis M, Attwell D (1990) Glutamate uptake in mammalian retinal glia is voltage- and potassium-dependent. Brain Res 516: 322–325
Serve G, Endres W, Grafe P (1988) Continuous electrophysiological measurements of changes in cell volume of motoneurons in the isolated frog spinal cord. Pflügers Arch Eur J Physiol 411: 410–415
Simon RP, Swan JH, Griffiths T, Meldrum BS (1984) Blockade of N-methyl-D-aspartate receptors may protect against ischemic damage in the brain. Science 226: 850–852
Somjen GG (1979) Extracellular potassium in the mammalian central nervous system. Annu Rev Physiol 41: 159–177
Sonnhof U (1987) Single voltage-dependent K^+ and Cl^- channels in cultured rat astrocytes. Can J Physiol Pharmacol 65: 1043–1050
Sontheimer H, Kettenmann H, Backus KH, Schachner M (1988) Glutamate opens Na^+/K^+ channels in cultured astrocytes. Glia 1(5): 328–336

Sykova E (1983) Extracellular K^+ accumulation in the central nervous system. Prog Biophys Mol Biol 42: 135–189
Tas PWL, Massa PT, Koschel K (1986) Preliminary characterization of an Na^+ K^+ Cl^- cotransport activity in cultured human astrocytes. Neurosci Lett 70: 369–373
Thakran P, Leuschen MP, Chatterjee A (1991) L,644–711 inhibits K^+-stimulated chloride ion uptake and cellular injury in cortical astrocytes. Soc Neurosci Abstr 17(1): 57
Thurston JH, Hauhart RE, Nelson JS (1987) Adaptive decreases in amino acids (taurine in particular), creatine, and electrolytes prevent cerebral edema in chronically hyponatremic mice: rapid correction (experimental model of central pontine myelinolysis) causes dehydration and shrinkage of brain. Metab Brain Dis 2: 223–241
Tosteson DC (1981) Cation countertransport and cotransport in human red cells. Fed Proc 40: 1429–1433
Ubl J, Murer H, Kolb H-A (1988a) Hypotonic shock evokes opening of Ca^{2+}-activated K channels in opossum kidney cells. Pflügers Arch Eur J Physiol 412: 551–553
Ubl J, Murer H, Kolb H-A (1988b) Ion channels activated by osmotic and mechanical stress in membranes of opossum kidney cells. J Membr Biol 104: 223–232
van Harreveld A (1966) Brain tissue electrolytes. Butterworths, London
van Harreveld A (1982) Swelling of the Muller fibers in the chicken retina. J Neurobiol 13: 519–536
van Harreveld A, Fifkova E (1971) Light- and electron-microscopic changes in central nervous tissue after electrophoretic injection of glutamate. Exp Mol Pathol 15: 61–81
Wade JV, Olson JP, Samson FE, Nelson SR, Pazdernik TL (1988) A possible role for taurine in osmoregulation within the brain. J Neurochem 51: 740–745
Walz W (1987) Swelling and potassium uptake in cultured astrocytes. Can J Physiol Pharmacol 65: 1051–1057
Walz W (1988) Physiological consequences of activating seotonin receptors. In: Kimelberg HK (ed) Glial cell receptors. Raven Press, New York, pp 121–130
Walz W, Mukerji S (1990) Simulation of aspects of ischemia in cell culture: changes in lactate compartmentation. Glia 3: 522–528
Walz W, Schlue WR (1982) Ionic mechanism of a hyperpolarizing 5-hydroxytryptamine effect on leech neuropile glial cells. Brain Res 250: 111–121
Waniewski RA, Martin DL (1984) Characterization of L-glutamic acid transport by glioma cells in culture: evidence for sodium- independent, chloride-dependent high affinity influx. J Neurosci 4: 2237–2246
Warnock DG, Greger R, Dunham PB, et al. (1983) Ion transport processes in apical membrane of epithelia. Fed Proc 43: 2473–2487
Whetsell WO, Kohler C, Schwarcz R (1988) Quinolinic acid: a glia- derived excitotoxin in the mammalian central nervous system. In: Norenberg MD, Hertz L, Schousboe A (eds) The biochemical pathology of astrocytes. Liss, New York, pp 191–202
Wieloch T (1985) Hypoglycemia-induced neuronal damage prevented by an N-methyl-D-aspartate antagonist. Science 230: 681–683
Wong SME, Chase HS Jr (1986) Role of intracellular calcium in cellular volume regulation. Am J Physiol 250: C841–C852
Wong SME, DeBell MC, Chase HS Jr (1990) Cell swelling increases intracellular free [Ca] in cultured toad bladder cells. Am J Physiol 258: F292–F296
Yu MC, Bakay L, Lee JC (1972) Ultrastructure of the central nervous system after prolonged hyposia. I. Neuronal alterations. Acta Neuropathol 22: 222–234
Zadunaisky JA, Wald F, De Robertis EDP (1963) Osmotic behavior and ultrastructural modifications in isolated frog brains. Exp Neurol 8: 290–309

Note added in proof: A full paper on this has appeared (Bowman et al., Brain Res. 584:272–286, 1992) as well as a number of reports on swelling activated anion channels (eg. Weiss and Lang, J. Mem. Biol. 126:109–114, 1992; Banderali and Roy, J. Mem. Biol. 126:219–234, 1992; Tseng, Am. J. Physiol. 262:C1056–C1068, 1992; Sorota, Circ. Res. 70:679–687, 1992)

Chapter 7

Cell Volume Regulation: Intracellular Transmission

E.K. Hoffmann[1], *L.O. Simonsen*[2], and *I.H. Lambert*[1]

Contents

1 Introduction

The cell membranes of virtually all animal cells are highly permeable to water, and hence the cell volume will be determined by the cellular content

[1] Institute of Biological Chemistry A, August Krogh Institute, University of Copenhagen, 13 Universitetsparken, 2100 Copenhagen ø, Denmark
[2] Zoophysiological Lab. B, August Krogh Institute, University of Copenhagen, 13 Universitetsparken, 2100 Copenhagen ø, Denmark

Advances in Comparative and Environmental Physiology, Vol. 14

of osmotically active solutes and by the osmolality of the extracellular fluid. It should be noted, however, that in epithelial cells only the membrane on the basolateral side is highly water-permeable. The apical membrane can be quite impermeable to water, as e.g. in the epithelial cells of late distal tubules and collecting ducts of the kidney in the absence of antidiuretic hormone.

Under physiological steady state conditions the cell volume is maintained constant by a "pump-leak steady state", the colloidosmotic pressure being counteracted by the Na^+, K^+ pump (see e.g. Hoffmann and Ussing 1992).

Besides maintaining a constant cell volume under physiological steady state conditions, most cells investigated are also capable of counteracting volume perturbations by volume recovery processes. Swollen cells tend to reduce their volume towards initial values by loss of KCl and concomitant loss of cell water (regulatory volume decrease, RVD), and shrunken cells often show the capacity to increase their volume towards initial values by net uptake of KCl and concomitant uptake of cell water (regulatory volume increase, RVI).

Cell volume perturbations can conceivably be caused either by changes in the osmolarity of the extracellular fluid or by changes in the cell content of osmolytes. The main causes of cell swelling and shrinkage, respectively, in mammalian cells are summarized in Table 1.

Table 1. Cell volume perturbations in mammalian cells

	Cell swelling	Cell shrinkage
Change in tonicity of extracellular fluid	Hyponatraemia (all cells) Excessive water intake (intestinal cells)	Dehydration (all cells) Antidiuresis (kidney medulla)
Change in cellular osmolyte content	Cellular acidosis	Cellular alkalosis
	Salt absorption (epithelial cells)	Secretagogue-induced epithelial secretion (glandular cells)
	Na^+-coupled sugar and amino acid uptake (transporting epithelia) Na^+-coupled amino acid accumulation (liver cells) Glycogenolysis (liver cells) Lipolysis (adipocytes) Excessive lactate production (exercising skeletal muscle) Ischaemia, hypoxaemia (brain cells)	

Under physiological conditions the osmolality of the body fluids is regulated within extremely narrow limits. Under pathophysiological conditions, however, disturbances of body fluid homeostasis can be encountered, and plasma fluid osmolalities ranging between about 220 and 350 mOsmol/kg H_2O have been observed. Under these extreme conditions the cells of the body would – in the absence of volume-regulating mechanisms – swell and shrink by about 30 and 20%, respectively (see Macknight 1983; Macknight and Leaf 1985).

Intestinal epithelial cells and presumably blood cells traversing intestinal capillaries are exposed to hypotonicity after excessive water intake. Moreover, kidney medullary cells and blood cells traversing the capillaries of kidney medulla are exposed to high osmolalities, in particular during antidiuresis, when the osmolality in this region of the mammalian kidney can reach levels higher than 1200 mOsmol (Knepper 1982; Bankir and de Rouffignac 1985).

An increase in the cellular content of osmotically active organic solutes and ions may also represent a challenge to cellular volume regulation and activate membrane transport processes involved in the RVD response. This is presumably the case in epithelial cells in relation to Na^+-coupled uptake of sugars and amino acids (Lau et al. 1984; Schultz et al. 1985; MacLeod 1991), and in liver cells in relation to Na^+-coupled uptake and accumulation of amino acids (Kristensen and Folke 1984; Kristensen 1986).

Intracellular acidosis and alkalosis will affect the net charge of intracellular impermeant macromolecules, and hence the content of intracellular diffusible anions (predominantly chloride), assuming that the cellular K^+ and Na^+ content will remain essentially constant. This means that intracellular pH will affect cell volume. Thus, at low pH_i the negative charge on the impermeant anions decreases, and the cell chloride content and hence cell volume increase. For a detailed discussion of these interrelations, see Hladky and Rink (1977) and Hoffmann and Ussing (1992). Furthermore, an acidification will in most cells activate the Na^+/H^+ exchanger leading to an additional cell swelling (see Hoffmann and Simonsen 1989).

Metabolic events may probably also lead to a rapid increase in the cellular content of organic solutes and hence activate an RVD response (Kristensen 1986). This could conceivably be the case in hepatic glycogenolysis and in adipocyte lipolysis. The metabolic events in exercising skeletal muscle could also lead to accumulation of intracellular organic solutes (predominantly lactate) and activation of volume regulatory responses (see Saltin et al. 1987).

In mammalian brain the glia cells, in particular astrocytes, swell in response to ischaemia and hypoxaemia, mainly caused by uptake of K^+ and Cl^-. $[K^+]_o$ is known to rise to high levels in ischaemia following K^+ loss from the neurons which is partially compensated by uptake into glial cells (see recent reviews by Walz 1989; Kimelberg 1991).

In glandular secretion the primary event in initiation of secretion has been reported to be the simultaneous opening of basal K^+ channels and apical Cl^- channels with resulting K^+ and Cl^- loss (Petersen and Gallacher 1988), and a corresponding cell shrinkage has recently been demonstrated in single acinar cells (Foskett and Melvin 1989) and in perfused glands (Nakahari et al. 1990).

Cell volume regulation has been studied in a wide variety of cells and tissues. For recent reviews, see Chamberlin and Strange (1989), Hoffmann and Simonsen (1989), Okada and Hazama (1989), Schultz (1989a,b), Lewis and Donaldson (1990), Hoffmann and Ussing (1992), Spring and Hoffmann (1992), and a recent series of reviews edited by Gilles et al. (1991).

The cellular mechanisms involved in the activation of volume regulatory responses in various cell types are not clearly understood, and several intracellular signalling systems seem to be involved. Below, we review some recent findings relating to intracellular signalling involved in volume regulation.

2 Principal Ion Transport Systems for Na^+, K^+ and Cl^- Involved in Cell Volume Regulation

Over the last 10–15 years evidence has accumulated to demonstrate that a constant cell volume is maintained not only by adjustment of ion pump rates but also to a large extent by dynamic and controlled changes in a variety of leak pathways. The passive ion leaks play an active role in cell volume control.

In several of the experiments discussed below the Ehrlich ascites tumour cells have been used as an experimental model in order to explore the membrane mechanisms involved in volume regulation and their activation. Table 2 gives an overview of the passive ion transport pathways in Ehrlich cells, their activation, and the more or less specific inhibitors that affect them. The RVD response in Ehrlich cells is mediated mainly via the conductive K^+ and Cl^- channels, and the RVI response mainly via the Na^+, K^+, $2Cl^-$ cotransport system.

Vertebrate cells suspended in hypotonic media initially swell by osmotic water equilibration, but subsequently regulate their volume by a net loss of KCl and concomitant loss of cell water (regulatory volume decrease, RVD) (see reviews by Kregenow 1981; Grinstein et al. 1982b, 1984; Cala 1983; Hoffmann 1983, 1985; Lauf 1985a,b; Siebens 1985; Larson and Spring 1987; Hoffmann and Simonsen 1989; Geck 1990; Lang et al. 1990). The RVD response is illustrated in Fig. 1 in the case of the Ehrlich ascites tumour cells. It can be seen that partial volume recovery is attained in 5–10 min.

Following shrinkage in hypertonic medium, a regulatory volume increase (RVI) can be observed in some tissues and cell types (e.g. *Amphiuma* red

Table 2. Passive ion transport pathways in Ehrlich ascites tumour cells

Transport pathway	Activated by	Inhibited by
K^+ channel	Cell swelling, increase in cytosolic Ca^{2+}, LTD_4, LTD_5	Quinine, barium, anti-calmodulin drugs, unsaturated fatty acids
Cl^- channel	Cell swelling, indirectly activated by Ca^{2+}, LTD_4, LTD_5	Indacrinone, diphenylamine-2-carboxylate, anti-calmodulin drugs, unsaturated fatty acids
SA channel	Cell swelling, mechanical stress	
Na^+, K^+, $2Cl^-$ cotransport and/or Na^+, Cl^- cotransport	Cell shrinkage, phorbol esters	Loop diuretics, e.g. furosemide, bumetanide
K^+, Cl^- cotransport	Cell swelling at low pH or after Ca^{2+} depletion, NEM, $HgCl_2$	Furosemide
Na^+/H^+ exchange	Cellular acidification, phorbol esters, $CuSO_4$	Amiloride
Anion exchange	Alkaline pH	DIDS

Sources: Aabin and Hoffmann (1986); Christensen and Hoffmann (1992); Doppler et al. (1986); Harbak and Simonsen (1991); Hoffmann et al. (1983); Hoffmann et al. (1984); Hoffmann et al. (1986); Jensen et al. (1993); Jessen et al. (1986); Jessen et al. (1989b); Kramhøft et al. (1984); Kramhøft et al. (1986); Kramhøft et al. (1989); Kramhøft et al. (1988); Lambert (1987a); Lambert 1989; Lambert et al. (1984); Lambert et al. (1987); Lambert et al. (1989); Lauritzen et al. (1993); Wiener et al. (1986).

blood cells (Cala 1980) and *Necturus* gallbladder epithelial cells (Larson and Spring 1987)). Other cells, such as Ehrlich ascites tumour cells, simply shrink as osmometers with no sign of volume recovery (see Hoffmann and Simonsen 1989). The same kind of behaviour is shown by lymphocytes (Grinstein et al. 1984), frog skin epithelial cells (Ussing 1982a), and frog urinary bladder epithelial cells (Davis and Finn 1985). Nearly all cells, however, show a volume recovery when hypotonic pretreatment, resulting in a net loss of KCl (the RVD response), is followed by restoration of tonicity. This is illustrated in Fig. 2 in the case of the Ehrlich ascites tumour cell. After the initial osmotic shrinkage the cells recover their volume with an associated KCl uptake. It is not clear why some cells show an RVI response only when the "RVI-after-RVD protocol" is applied. This conditioned RVI response has been called by Lewis and Donaldson (1990) "pseudo-regulatory volume increase".

Table 3 summarizes the main features of the RVD and the RVI response in the Ehrlich ascites tumour cell. For details, see recent reviews by Hoffmann et al. (1988) and Hoffmann and Simonsen (1989).

During RVD, about 70% of the total loss of cellular osmolytes is accounted for by net KCl loss, and the remaining 30% by net loss of amino

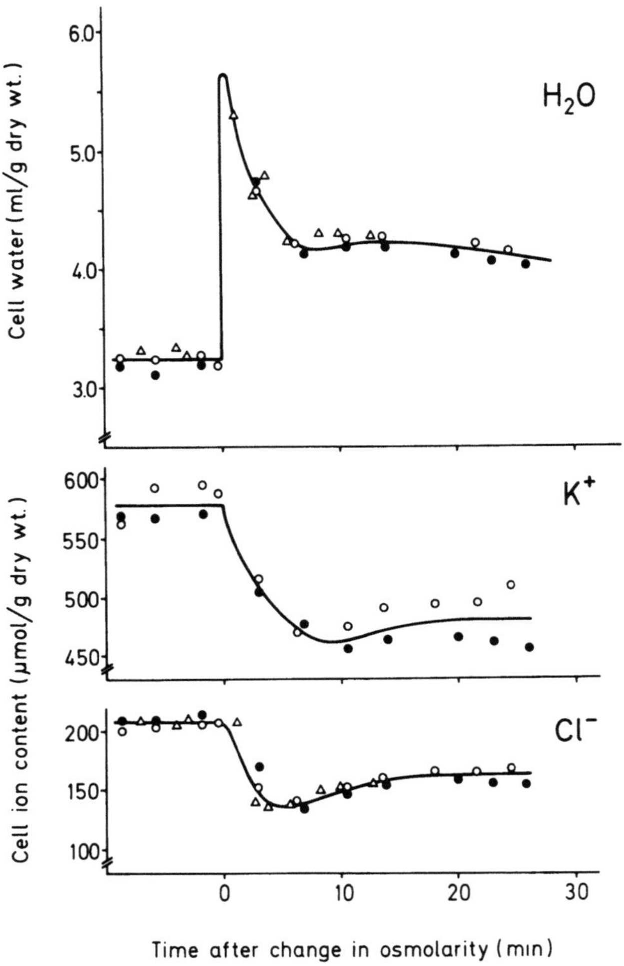

Fig. 1. Regulatory volume decrease (RVD) in Ehrlich ascites tumour cells after reduction of external osmolarity. The cells were pre-incubated in 300 mOsmol standard NaCl medium for 40–50 min, and at time zero resuspended in 150 mOsmol NaCl medium. Cell water, K^+ and Cl^- content were followed with time after reduction of external osmolarity. Measurements from three independent experiments marked individually (○, △, ●) are shown. (After Hoffmann et al. 1984)

acids, predominantly taurine (Hoffmann and Hendil 1976). The volume-induced taurine efflux pathway is a simple, pH-dependent leak pathway which seems to be independent of the saturable Na^+-, and Cl^--dependent uptake system (Lambert 1984, 1985; Lambert and Hoffmann 1993).

A net loss of taurine during RVD is also found in red cells, heart muscle, and brain of the little skate (King and Goldstein 1983), in red cells from the euryhaline fish species flounder and eel (Fugelli and Rohrs 1980; Fugelli and Thoroed 1986; Fincham et al. 1987), in fish myocardium (Vislie

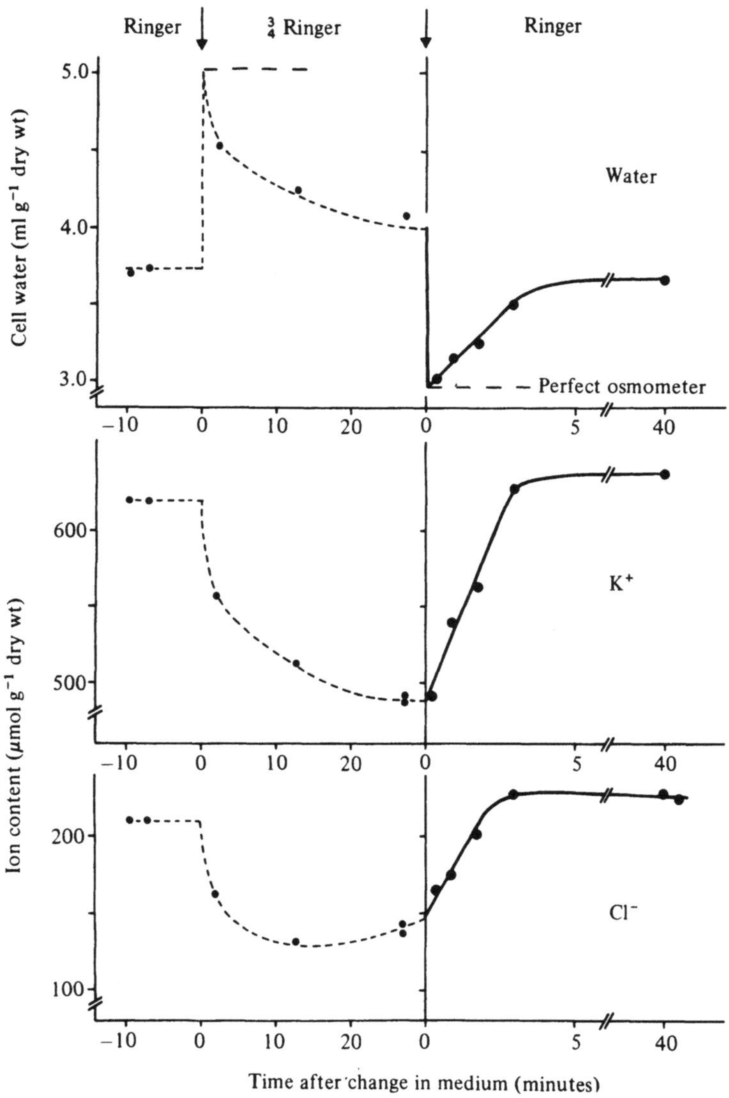

Fig. 2. Regulatory volume increase (RVI) in Ehrlich ascites tumour cells exposed to the "RVI-after-RVD protocol". Cell volume, K^+ and Cl^- content were followed with time after changes in medium osmolarity. The cells were pre-incubated in 300 mOsmol standard NaCl medium. Osmolarity was changed at the first t = 0 from 300 to 225 mOsmol by dilution with distilled water, and at the second t = 0 from 225 to 300 mOsmol by addition of the salts present in normal Ringer's solution. (Hoffmann 1983)

1983), in mammalian heart (Thurston et al. 1981), in cultured Madin-Darby canine kidney (MDCK) cells (Olea et al. 1991) and in mammalian astrocytes (Pasantes-Morales and Schousboe 1988). It is suggested that taurine plays an important role as a volume-protective osmolyte in the brain during hypoxic

Table 3. Characterization of the regulatory volume decrease (RVD) and regulatory volume increase (RVI) response in Ehrlich ascites tumour cells

	RVD	RVI
Transport systems activated	K^+ channels Cl^- channels SA channels Taurine channels	Na^+, K^+, $2Cl^-$ cotransporter Na^+, Cl^- cotransporter (?)
Specific features	Anion-unselective, (accepts Cl^-, SCN^-, NO_3^-) Depolarization of the cell membrane Cellular acidification Increased leukotriene synthesis Reduced prostaglandin synthesis	Cl^--dependent Na^+-dependent K^+-dependent Unchanged membrane potential
Inhibitors	Quinine Barium Anti-calmodulin drugs Indacrinone Diphenylamine-2-carboxylate Cytochalasine B Unsaturated fatty acids	Loop diuretics, e.g. bumetanide, furosemide
"Second messengers"	Ca^{2+}/calmodulin Inositol phosphates, Leukotrienes	Ca^{2+}/calmodulin Protein kinase C

Sources: Aabin and Hoffmann (1986); Christensen and Hoffmann (1992); Christensen et al. (1988) Hoffmann et al. (1983); Hoffmann et al. (1984); Hoffmann et al. (1986); Jensen et al. (1993); Jessen et al. (1989b); Kramhøft et al. (1984); Lambert (1987); Lambert et al. (1987a,b); Lauritzen et al. (1993); see Sect. 4.

ischaemia (Solis et al. 1990). In trout red cells, on the other hand, amino acids make no significant contribution to the volume-induced net loss of osmotically active solutes after advenergic swelling (Borgese et al. 1987).

The use of amino acids for RVD requires that their concentration within the cells is high. The taurine concentration of teleost red cells is high, 25–45 mmol/l cell water (Fincham et al. 1987). This is also the case in a mammalian cell like the Ehrlich ascites tumour cell, where the taurine concentration is measured at 53 mmol/l cell water (Hoffmann and Lambert 1983).

The net KCl loss during RVD is mediated by activation of K^+ channels and Cl^- channels. The K^+ channels can be blocked by barium (Lambert et al. 1984) and by quinine (Hoffmann et al. 1984). The Cl^- transport pathway is strongly inhibited by diphenylamine-2-carboxylate (DPC) and indacrinone (MK-196) (Aabin and Hoffmann 1986), drugs known to inhibit Cl^- channels in Cl^--transporting epithelia (DiStefano et al. 1985) and the potential-gated

Cl^- conductance of toad skin (Dürr and Hviid Larsen 1986). In mouse proximal tubule epithelial cells and in MDCK cells, the Cl^- channel activated by cell swelling was not blocked by the chloride channel blocker anthracene-9-COOH (Lang et al. 1987). In addition, unsaturated fatty acids (e.g. arachidonic acid, oleic acid, linoleic acid, linolenic acid and palmitoleic acid) seem to have a non-specific inhibitory effect on the Cl^- channels in Ehrlich cells (Lambert 1987a).

The volume-induced Cl^- channel is anion-unselective in Ehrlich cells (Hoffmann et al. 1984, 1986) as well as in lymphocytes (Grinstein et al. 1982a,b). In the Ehrlich cell the chloride conductance is increased to a larger extent than that of K^+, and it is concluded that the K^+ conductance limits the rate of volume change during RVD since volume changes are more rapid when valinomycin (Hoffmann et al. 1984) or gramicidin (Hoffmann et al. 1986) is used to provide a parallel pathway for K^+.

In Ehrlich cells a depolarization (about 20 mV in hypotonic solution with half-osmolarity) is observed during RVD, measured by conventional microelectrodes (Lang et al. 1987) or using a fluorescent probe (Lambert et al. 1989). This is consistent with a larger increase in the Cl^- conductance than in the K^+ conductance. The increase in the Cl^- conductance is estimated at 68-fold when Ehrlich cells are transferred to a medium with half the osmolarity of the standard medium, whereas the concomitant increase in the K^+ conductance is only two-fold (Lambert et al. 1989). It may be noted that the depolarization will increase the driving force for K^+ efflux, and hence account for at least part of the net K^+ loss (see Lang et al. 1987). A depolarization during the RVD response has been demonstrated in several other cell types, e.g. human lymphocytes (Grinstein et al. 1982a; Sarkadi et al. 1984a), astrocytes (Kimelberg and O'Connor 1988), cultured opossum kidney cells (Ubl et al. 1988) and neuroblastoma cells (Falke and Misler 1989). In cultured human epithelial cells biphasic membrane potential changes during RVD have been demonstrated at low external Cl^-, with an initial hyperpolarization (sensitive to the K^+ channel blockers Ba^{2+} and quinine) followed by a transient depolarization (sensitive to the stilbene derivative Cl^- channel blocker SITS) (Hazama and Okada 1988, 1990).

The consequences of a loss of K^+ and Cl^- via separate, conductive transport pathways have been analyzed in detail by Lew and Bookchin (1986). For a cell with a Cl^-/HCO_3^- exchanger in its membrane, the K^+ loss can be predicted to be partially balanced by HCO_3^- loss, producing a proton uptake via the Jacobs-Stewart cycle (Jacobs and Stewart 1942) with resulting intracellular acidification, and hence the K^+ loss will exceed the Cl^- loss to an extent which depends upon the cellular pH buffering capacity (see Lew and Bookchin 1986; Freemann et al. 1987). In the case of the Ehrlich ascites tumour cell, the K^+ loss during RVD exceeds the Cl^- loss by a factor about 1.6 (Hendil and Hoffmann 1974), and a cytoplasmic acidification during RVD can be demonstrated using an intracellular pH-sensitive fluorescent probe (Livne and Hoffmann 1990).

The RVI response in Ehrlich cells is mediated by activation of a quiescent, electroneutral, Na^+- and Cl^--dependent cotransport system (Hoffmann et al. 1983; Levinson 1991). The K^+ dependence is discussed below (see Sect. 2.2). The cotransporter is electroneutral and the membrane potential is unchanged during RVI (unpubl. observ.). The cotransporter can be blocked by loop diuretics (Kramhøft et al. 1984). The involvement of cotransport systems in RVI was first described in avian cells (Kregenow 1973; see Kregenow 1981).

A role for Na^+, Cl^- or Na^+, K^+, $2Cl^-$ cotransport in cell volume regulation has also been established in frog skin epithelial cells (see Ussing 1986), red blood cells from rat (Duhm and Göbel 1984) and from ferret (Mercer and Hoffman 1985); and, moreover, in L cells (Gargus and Slayman 1980), in simian virus-transformed 3T3 cells (Bakker-Grunwald et al. 1982), in chick cardiac cells (Frelin et al. 1986), in rabbit cradiac cells (Drewnowska and Baumgarten 1991), in astrocytes from rat cerebral cortex (Kimelberg and Frangakis 1986), in retinal pigment epithelial cells (Adorante and Miller 1990), in the apical membrane of rabbit kidney medullary thick ascending limb of Henle's loop (MTAL; see review by Eveloff and Warnock 1987), in jejunal villus cells (MacLeod and Hamilton 1990) and in tracheal epithelial cells (Musch and Field 1989). However, the Na^+, K^+, $2Cl^-$ cotransport system in red blood cells from other species, including man (Duhm and Göbel 1984), is not affected by changes in cell volume.

In several cell types the principal ion transport systems activated during RVI is parallel Na^+/H^+ and Cl^-/HCO_3 exchange. These transport systems are not discussed here. For references see e.g. Rotin et al. (1991), Hoffmann and Ussing (1992) and Spring and Hoffmann (1992) as well as Lang et al. (this Vol.).

The findings related to the activation mechanisms during RVD and RVI and the second messengers involved are discussed in Sections 3 and 4, respectively.

2.1 K^+ and Cl^- Conductive Pathways Involved in Regulatory Volume Decrease (RVD)

A swelling-induced increase in the conductive Cl^- permeability of the basolateral membrane of frog skin epithelial cells was intimated already in 1961 by MacRobbie and Ussing (1961; see Ussing 1982a, 1985; Kristensen and Ussing 1985). Activation by cell swelling of separate, conductive fluxes of both K^+ and Cl^- was first proposed for Ehrlich ascites tumour cells (Hoffmann 1978).

Separate, conductive K^+ and Cl^- transport pathways have been reported to be activated during RVD in human lymphocytes (Cheung et al. 1982; Grinstein et al. 1982a,b,c, 1984; Sarkadi et al. 1984a,b, 1985), in Ehrlich ascites tumour cells (Hoffmann 1978, 1985; Hoffmann et al. 1986), in human platelets (Livne et al. 1987), in human fibroblasts (Rugolo et al. 1989), in rat astrocytes (Kimelberg and O'Connor 1988), in jejunal villus cells (MacLeod and Hamilton 1991), in the basolateral membrane of rabbit proximal tubule (Welling and O'Neil 1990) and in human intestinal epithelial cells (Hazama and Okada 1988). A conductive K^+ transport pathway activated by cell swelling has been reported in the basolateral membrane of frog and toad urinary bladder epithelial cells (Davis and Finn 1982, 1985), in mouse and rabbit kidney epithelial cells (Simmons 1984; Welling et al. 1985; Eveloff and Warnock 1987; Kirk et al. 1987a,b; Rugolo et al. 1989), in enterocytes (Lau et al. 1984; Schultz et al. 1985), and in liver cells (Kristensen and Folke 1984; Kristensen 1986).

Under physiological steady-state conditions the membrane conductances in Ehrlich cells are estimated at 10.4 $\mu S/cm^2$ for K^+, 3 $\mu S/cm^2$ for Na^+, and 0.6 $\mu S/cm^2$ for Cl^- (Lambert et al. 1989). During the RVD response, the Na^+ conductance is reduced (Hoffmann 1978), whereas the K^+ conductance is increased about 2-fold, and the Cl^- conductance is increased about 60-fold (Lambert et al. 1989), reflecting activation of conductive K^+ and Cl^- channels.

In hypotonically swollen Ehrlich cells a 23-pS channel was reported in patch-clamp studies by Hudson and Schultz (1988). This channel was proposed to be a volume-induced Cl^- channel, but showed similarities to the stretch-activated, non-selective cation channel later reported in Ehrlich cells by Christensen and Hoffmann (1992; see Sect. 3.2).

The volume-induced K^+ and Cl^- channels in Ehrlich cells have recently been directly studied by Christensen and Hoffmann (1992), using the patch-clamp technique. Single channel currents were recorded in excised patches, and in cell-attached patches with the cell exposed to isotonic or hypotonic medium.

2.1.1 Properties of the Volume-Activated K^+ Channel

In isolated inside-out patches a Ca^{2+}-dependent, inwardly rectifying K^+ channel was demonstrated. The single channel conductance for inward current was estimated at 40 pS and for outward current at 15 pS. This K^+ channel shows properties similar to those of small K^+ channels which have previously been characterized in several cell types: human red cells (Grygorczyk et al. 1984, see also Christophersen 1991), Hela cells (Sauvé et al. 1986), aortic endothelial cells (Sauvé et al. 1988), MDCK cells (Friedrich et al. 1988) and rabbit proximal tubule cells (Parent et al. 1988). Previous studies of the Ca^{2+}-activated K^+ channel in Ehrlich cells have suggested single-file behaviour (Hoffmann et al. 1986) similar to that demonstrated in human red cells (Vestergaard-Bogind et al. 1985).

Patch-clamp studies in a cell-attached mode showed K^+ channels with spontaneous activity and with characteristics similar to those of the K^+ channel seen in excised patches. Figure 3A shows typical single channel currents from a K^+ channel in a cell-attached patch with KCl in the electrode

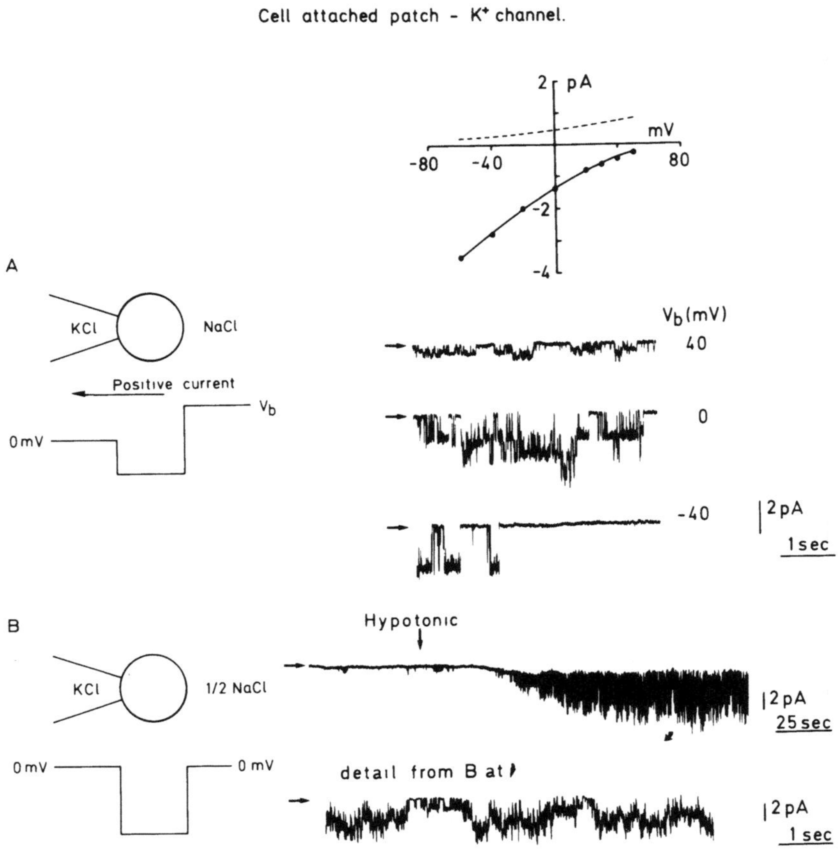

Fig. 3. Activation of the K^+ channels in the Ehrlich ascites tumour cell by cell swelling. *A* The relationship between bath potential and patch current in cell-attached patches showing spontaneous activity of the inward rectifier K^+ channel. Data are from a patch with KCl in the pipette. The conductance was under these circumstances 37 pS at negative bath potentials and 22 pS at positive potentials. The *dotted line* indicates the I/V curve obtained with NMDGCl in the pipette. Downward deflexions indicate negative current. *B* The activation of current through K^+ channels in cell-attached patches when the outside medium was reduced from normal to half-tonicity. The electrode contains KCl and the bath potential was 0 mV. The *lower trace* shows an expanded view of the channel activity. Results are from Christensen and Hoffmann (1992)

and with varying bath potential. The I/V-relationship is also presented. The dotted line shows the I/V-curve with an impermeant cation ($NMDG^+$) in the electrode. From this curve the single channel conductance at 5 mM external K^+ was estimated at about 7 pS (see Table 4). A K^+ channel with similar properties can be activated by addition of Ca^{2+} plus ionophore A23187 (see Sect. 3.3).

Typical recordings from a cell-attached patch with the cell exposed to hypotonic medium is shown in Fig. 3B. It can be seen that a channel

Table 4. Characteristics of membrane channels activated in Ehrlich ascites tumour cells after cell swelling[a]

	K^+ channel	Cl^- channel	SA channel
Single channel conductance (pS)	7[b]	7	23
Activation time (s)[c]	65 ± 11 (8)	100 ± 15 (17)	80 ± 7.5 (13)
Estimated channel activation during RVD			
Open channel density (per cm^2)	3×10^6	6×10^6	
No. of open channels per cell	40	80	

[a] Values from Christensen and Hoffmann (1992).
[b] Estimated value for K^+ efflux under physiological conditions (low K^+ concentration in pipette, zero bath potential, see Fig. 3, dotted line in I/V curve).
[c] Time delay following hypotonic exposure in cell-attached mode, mean ± SD (n).

becomes activated with a time lag following hypotonic exposure. The average time delay after the onset of cell swelling was 65 s (see Table 4). Activation by cell swelling was observed in 9 out of a total of 13 patches containing a K^+ channel. The channel activity typically occurs in waves (see the expanded trace).

From the available evidence the same K^+ channel was observed in excised patches and in cell-attached patches with spontaneous activity, with Ca^{2+}-induced activity and with activity induced by cell swelling. No evidence was found of voltage-gating, or channel activation by membrane-stretch (suction).

From the single channel conductance estimated at 7 pS (see above) and a K^+ conductance at 19 $\mu S/cm^2$ during RVD (Lambert et al. 1989), the time-averaged number of open channels during RVD can be estimated at $3 \times 10^6 cm^{-2}$ or about 40 open channels per cell. The number of open K^+ channels following addition of Ca^{2+} plus ionophore A23187 was estimated at 250 per cell (Christensen and Hoffmann 1992).

It is tempting to suggest that the volume-induced conductive K^+ transport seen in many cells is always mediated by this type Ca^{2+}-activated channels; this may, however, be an oversimplification. In human red blood cell "inside-out" vesicles exposed to hypotonicity, a high K^+ permeability was demonstrated which showed a large Ca^{2+}-independent component (Rossi and Schatzmann 1982). In frog red blood cells two classes of K^+ channels were demonstrated in patch-clamp studies. In osmotically swollen frog red cells an increased K^+ permeability was demonstrated which was mediated predominantly by a small conductance channel which did not display the same strong Ca^{2+} sensitivity as the larger conductance channel (Hamill 1983). In human peripheral blood lymphocytes the available evidence indicates that the Ca^{2+}-activated K^+ channel is not involved in RVD, whereas a voltage-activated, charybdotoxin-sensitive K^+ channel seems to account for the largest portion of the RVD response (see Rotin et al. 1991). In these cells, however, the volume-induced K^+ channel has not been characterized in patch-clamp studies. In rat thymocytes the voltage-activated charybdotoxin-sensitive K^+ channel cannot account for the RVD response which is unaffected by charybdotoxin (see Rotin et al. 1991).

2.1.2 Properties of the Volume-Activated Cl^- Channel

In excised inside-out patches (with NMDG chloride on both sides) a small 5-pS chloride channel with low spontaneous activity was observed. The channel activity was not dependent on internal Ca^{2+}, and could not be activated by membrane-stretch (suction).

Recordings from a cell-attached patch under isotonic conditions is shown in Fig. 4A. Single channel currents with characteristics similar to those seen in isolated patches are seen. The single channel conductance of the fully activated channel, which is only rarely seen under isotonic conditions, was estimated at 7 pS (see Table 4). In addition, single channel conductances of about 3 pS were regularly recorded. These channels were proposed to represent a partly activated state of the 7-pS channel. In contrast to the channels seen in isolated patches, the channels in the cell-attached mode could be activated (with a single channel conductance of 7 pS), by addition of Ca^{2+} plus ionophore A23187, suggesting an indirect effect, initiated by internal Ca^{2+} (see Sect. 3.3).

In cell-attached mode the channel is in most patches (20 patches out of a total of 28 patches presenting a Cl^- channel) activated by hypotonic exposure with a single channel conductance at 7 pS and with a time delay of about 1 min (see Fig. 4B and Table 4).

The analysis of the chloride channel is difficult because of the extremely low single channel conductances. It is proposed that the same channel is seen in excised inside-out patches, in cell-attached mode with spontaneous channel activity, and in cell-attached mode following channel activation by calcium or by cell swelling, but that the fully activated channel with a 7-pS single channel conductance is only rarely seen except after activation by Ca^{2+} or by cell swelling (Christensen and Hoffmann 1992).

From the single channel conductance of the fully activated channel at 7 pS and the chloride conductance at 40 $\mu S/cm^2$ during RVD (Lambert et al. 1989), the number of open channels is estimated at $6 \times 10^6\,cm^{-2}$ or 80 open channels per cell (see Table 4).

A "mini-chloride" channel has also been found in human and murine T-cells and proposed to be involved in the RVD response in these cells. The channel was difficult to resolve at single channel level and the activation of the channel by cell swelling was in both cell types demonstrated in whole-cell recordings (see Rotin et al. 1991).

Christensen et al. (1989) have reported a "mini-chloride" channel in *Necturus* choroid plexus epithelial cells which was activated during volume regulation after hypotonic shock. In rat glands a mini-chloride channel has

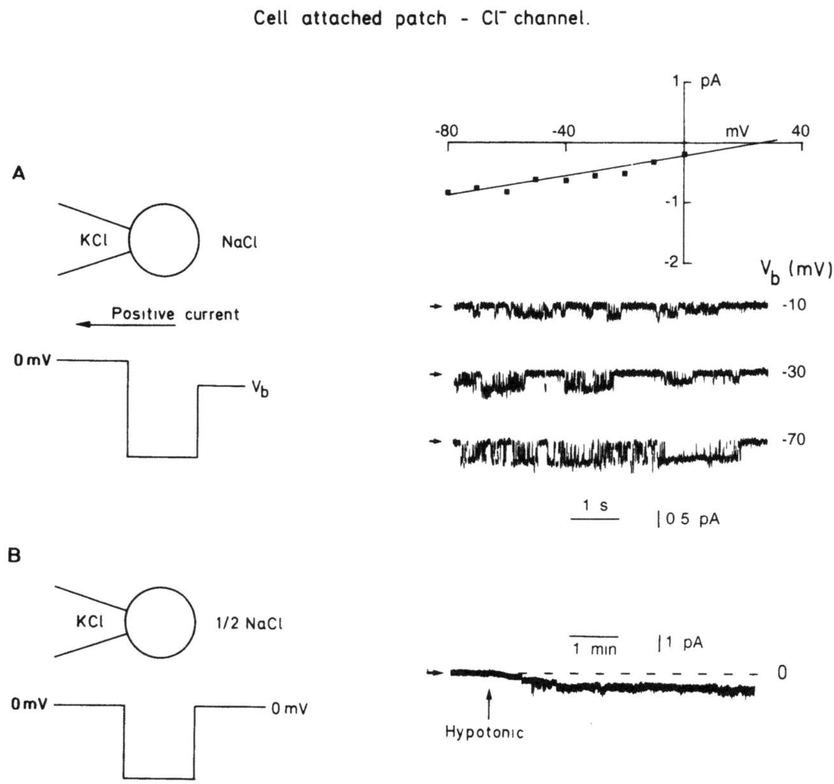

Fig. 4. Activation of a (3–7)-pS Cl^- channel in hypotonically swollen Ehrlich cells. Current recordings for a Cl^- channel from a cell-attached patch with KCl in the electrode. *A* Single-channel current versus bath potential for the Cl^- channel (I/V curve). *B* Activation of the Cl^- channel after cell swelling in hypotonic medium. At the time indicated by the *arrow* the fluid bathing the cell was reduced from a normal Ringer's solution to a solution with half the tonicity. Results are from Christensen and Hoffmann (1992)

been reported which in inside-out patches were directly activated by Ca^{2+} (Marty et al. 1984), in contrast to the findings for the mini-chloride channels in Ehrlich cells (Christensen and Hoffmann 1992).

In patch-clamp studies on Ehrlich cells two other types of chloride channels were recorded, a 400- and a 34-pS chloride channel. A voltage-activated 400-pS channel was regularly observed in excised inside-out patches. This channel has properties similar to the "maxi-chloride" channel reported in many cell types: cultured rat muscle (Blatz and Magleby 1983), macrophages (Schwarze and Kolb 1984; Kolb and Ubl 1987), and a mouse B-lymphocyte hybridoma cell line (Bosma 1989). In the Ehrlich cell this channel was not directly activated by Ca^{2+}. The 34-pS chloride channel could only be detected in excised inside-out patches which failed to present the 400-pS chloride channel. This channel showed properties similar to the chloride channel described in the apical membrane in human airway epithelia cells by Frizzell et al. (1986) and by Welsh (1986), which is activated by cAMP-dependent protein kinase C (Li et al. 1989), and affected in cystic fibrosis. The 400- and 34-pS chloride channels were never observed in cell-attached patches, neither in isotonic medium nor after hypotonic exposure, and there is no evidence for a role in the RVD response of either channel. These two types of chloride channels have again been observed in human B- and T-cell lines, and again without evidence of involvement of either channel in volume regulation (Rotin et al. 1991).

2.2 Na^+, K^+, $2Cl^-$ Cotransport and/or Na^+, Cl^- Cotransport Involved in Regulatory Volume Increase (RVI)

Electroneutral Na^+, K^+, $2Cl^-$ cotransport systems have been found in a wide variety of cell types. For reviews on the Na^+, K^+, $2Cl^-$ cotransport system, see Chipperfield (1986), Lauf et al. (1987), O'Grady et al. (1987) and Haas (1989).

In epithelial cells, Na^+, Cl^- cotransport systems, as originally proposed by Nellans et al. (1973) for the mucosal membrane of intestinal epithelial cells, have been widely demonstrated (Gerencser and Lee 1983) and recognized as playing an important role in epithelial transport (see review by Frizzell et al. 1979). In several epithelia, however, the Cl^--dependent cation cotransport can be attributed to a Na^+, K^+, $2Cl^-$ cotransport system (see reviews by Warnock et al. 1984; Greger 1985; O'Grady et al. 1987).

Activation of Na^+, K^+, $2Cl^-$ cotransport or Na^+, Cl^- cotransport during the RVI response has been demonstrated in several cells and tissues (see Sect. 2). In most cells the RVI response has been ascribed to the Na^+, K^+, $2Cl^-$ cotransport system.

In rabbit kidney medullary thick ascending limb of Henle's loop (MTAL) both Na^+, Cl^- and Na^+, K^+, $2Cl^-$ cotransport systems have been identified, with the Na^+, K^+, $2Cl^-$ system being proposed to predominate during the RVI response (Eveloff and Warnock 1987). In tracheal epithelial cells a K^+-independent Na^+, Cl^- cotransport is activated by osmotic shrinkage (Musch and Field 1989). In Ehrlich ascites tumour cells the RVI response was previously reported to involve activation of a K^+-independent Na^+, Cl^- cotransport system (Hoffmann et al. 1983), but recent evidence demonstrates that activation of the Na^+, K^+, $2Cl^-$ cotransport system predominates during RVI (Jensen and Hoffmann 1993; Levinson 1991; see below).

2.2.1 Ehrlich Ascites Tumour Cells

Activation of the Na^+, K^+, $2Cl^-$ cotransport system has been demonstrated in Ehrlich cells depleted of K^+ by inhibition of the Na^+, K^+ pump by ouabain (Geck et al. 1980) or by pre-incubation at low temperature (Levinson 1987, 1990).

Geck and Pfeiffer (1985) have reported the cotransport rate at steady state to increase with external Na^+ concentration up to an optimal concentration, above which the cotransport rate decreases again. The results are explained in terms of a feedback control system, the cotransport system being switched off when cell volume exceeds a critical value. This is consistent with previous findings of Hoffmann et al. (1983) who reported the cotransport system to be activated in shrunken cells, but silent in control cells at steady state. On the other hand, Levinson (1987) has presented evidence that the cell volume is not a major factor regulating the activity of the cotransport system in Ehrlich cells.

It was previously demonstrated that the primary process during the RVI response in Ehrlich cells is an activation of an otherwise quiescent bumetanide-sensitive Na^+, Cl^- and/or Na^+, K^+, 2Cl cotransport system, with subsequent replacement of Na^+ by K^+ via the Na^+, K^+ pump, stimulated by the Na^+ influx (Hoffmann et al. 1983).

The available evidence suggested involvement of Na^+, Cl^- cotransport rather than Na^+, K^+, $2Cl^-$ cotransport. This conclusion was based on the findings:

1. Under conditions where the Na^+, K^+ pump was inhibited with ouabain, an uptake of Na^+ and Cl^- in about equimolar amounts could be demonstrated, whereas the uptake of K^+ was negligible. The uptake of Na^+ and Cl^- was inhibited by bumetanide. In these experiments, however, a recycling of K^+ could conceivably have masked a K^+ uptake via the cotransport system.
2. Flux ratio analysis of the bumetanide-sensitive Cl^- flux during RVI indicated a Na^+, Cl^- cotransport, since the ratio of the driving forces for Na^+, K^+, $2Cl^-$ cotransport would hardly be sufficient to account for the observed magnitude of the ratio of influx/efflux (Hoffmann et al. 1983). Moreover, a net uptake of KCl could be demonstrated even at low external K^+, under conditions where the calculated driving force for a Na^+, K^+, $2Cl^-$ cotransport system is in the outward direction. In this calculation it was assumed, however, that the activity coefficients for the ions in the medium and in the cytosol are equal. Whether differences in activity coefficients could be large enough to make the energy deficit vanish, as suggested by Geck (1990), is still an open question.

We have recently found that a Na^+, K^+, $2Cl^-$ cotransporter is involved in RVI in Ehrlich cells (Jensen and Hoffmann 1991, 1993). The evidence is that both net Cl^- uptake and unidirectional ^{36}Cl influx during RVI require external K^+. The half-maximal rate of bumetanide-sensitive ^{36}Cl uptake is attained at about 2.5 mM external K^+. In addition, there was a bumetanide-inhibitable K^+ influx during RVI which was dependent on Na^+ and Cl^-. Table 5 shows the activation of bumetanide-sensitive unidirectional chloride and potassium influx during RVI. These findings are at variance with the findings discussed above. Recently, Levinson (1991) also presented evidence

Table 5. Activation of bumetanide-sensitive unidirectional potassium and chloride influx during regulatory volume increase in Ehrlich ascites tumour cells; RVI after RVD protocol (see Fig. 2)

	Bumetanide-sensitive influx ($\mu mol \cdot g$ cell dry $wt^{-1} \cdot min^{-1}$)[a]	
	Physiological steady state	RVI
Cl^- influx[b]	-0.8 ± 2.3 (3)	39 ± 3 (3)
K^+ influx[c]	0.7 ± 0.2 (3)	24 ± 2 (11)[d]

[a] Mean ± SEM (n).
[b] Values from Hoffmann et al. (1983).
[c] Values from Jensen and Hoffmann (1993).
[d] The total bumetanide-sensitive K^+ influx during RVI is a composite figure, comprising K^+ influx via the Na^+, K^+, $2Cl^-$ cotransport system plus some K^+ influx via the Na^+, K^+ pump corresponding to the Na^+ efflux via the Na^+, K^+ pump, in exchange for Na^+ taken up via the cotransport system. In three experiments in the presence of ouabain the bumetanide-sensitive K^+ influx was reduced by about 15%.

that the RVI response in Ehrlich cells is mediated by the Na^+, K^+, $2Cl^-$ cotransport system. The precise relation between the Na^+, Cl^- and the Na^+, K^+, $2Cl^-$ cotransport systems during RVI is an open question. The two transport systems are distinguishable, but may represent alternate modes of operation of the same transporter. Recently, Sun et al. (1991) reported that Na^+, Cl^- cotransport systems may be converted into Na^+, K^+, Cl^- cotransport systems by hormonal control. The addition of antidiuretic hormone (ADH) to mouse medullary thick ascending limb cells converts the K^+-independent Na^+, Cl^- cotransporter into a K^+-dependent Na^+, K^+, Cl^- cotransporter.

3 Activation Mechanisms for RVD

The mechanisms by which cells sense osmotically induced changes in turgor pressure, volume and/or intracellular composition, and transduce those signals into a regulatory response are only partly understood, and several factors have been assigned a regulatory function, e.g. Ca^{2+}, calmodulin, cAMP, eicosanoids, polyphosphoinositide metabolism, protein kinases and the microfilament network. Factors, which have been proposed to play a role in the RVD response in Ehrlich cells, are shown in Fig. 25 in a schematic form.

3.1 The Cytoskeleton

In isolated membrane vesicles from lymphocytes (rabbit thymocytes) the expected increase in K^+ permeability in response to either cytoplasmic Ca^{2+}

or hypotonic swelling was absent, suggesting that an intact cytoarchitecture is required for the RVD response (Grinstein et al. 1983). Moreover, cytochalasin B, which is known to disrupt microfilaments, inhibits RVD in rat liver slices (van Rossum and Russo 1981), in *Necturus* gallbladder epithelial cells (Foskett and Spring 1985), in Ehrlich cells (Hoffmann and Kolb 1991) and in various cultured mammalian cells (Cornet et al. 1987), suggesting that an intact microfilament network is a prerequisite for a normal RVD response. An example of this is seen in Fig. 5 which shows that cytochalasin B inhibits RVD in Ehrlich cells. A possible role of micro-

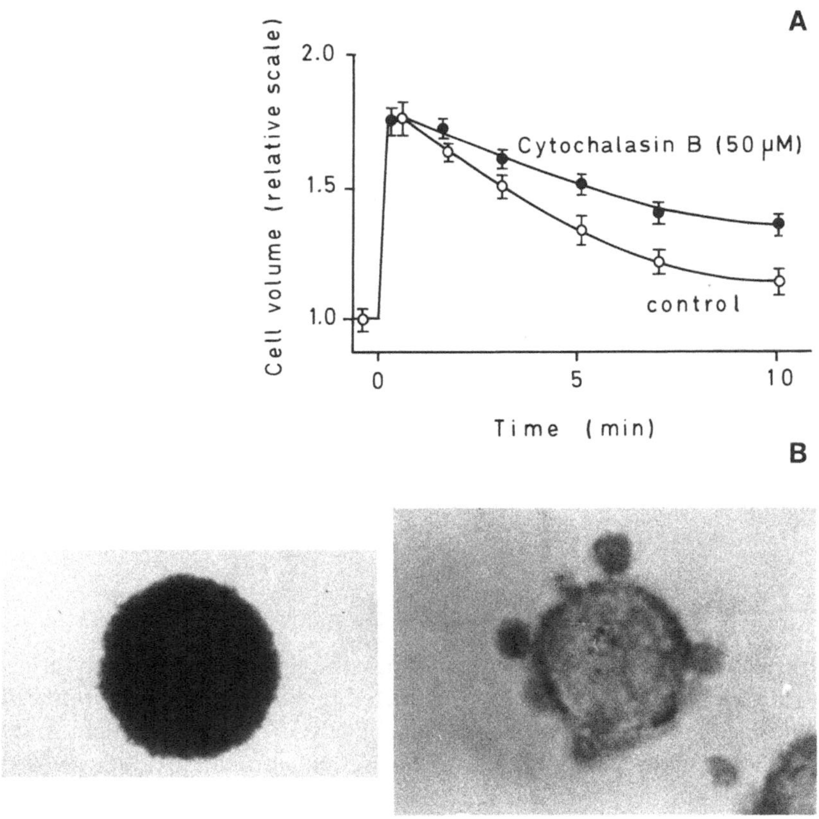

Fig. 5. Role of microfilaments in volume regulation in Ehrlich ascites tumour cells. *A* Effect of cytochalasin B on regulatory volume decrease in Ehrlich cells exposed to hypotonic (150 mOsmol) NaCl medium. At zero time a sample of the cell suspension was diluted 500-fold in hypotonic (150 mOsmol) NaCl medium. The cell volume was followed with time using a Coulter counter and is given relative to the initial cell volume. Cytochalasin B (50 μM, *closed symbols*) was present during a 30-min pre-incubation period before hyposmotic exposure. *B* Microfilament network in Ehrlich cells, stained by an IgSS method using anti-actin antibodies. The *black colour* indicates intact actin filaments. Cells maintained in isotonic medium (*left*) and cells treated over 30 min in isotonic medium with 50 μM cytochalasin B (*right*). Results are from Cornet et al. (1993)

tubules and the microfilament network during volume regulation has also been suggested in cultured rat pheochromocytoma cells (Delpire et al. 1985). Mills and Skiest (1985) and Mills and Lubin (1986) have analyzed the role of the cytoskeleton in cell volume regulation using MDCK cells. They propose that the state of organization of F-actin, which can be influenced by changes in the levels of cAMP, could have an effect on membrane elements that play a role in volume control processes. A coupling of the cytoskeleton with stretch-activated channels has been proposed by Sachs (1987). Changes in the organization of F-actin induced by hypotonic stress have been demonstrated in a cultured mammalian cell (Cornet et al. 1987). In Ehrlich cells the volume-induced depolymerization of F-actin is only observed in the presence of external Ca^{2+}, in contrast to the RVD response (Cornet, et al., 1993). Moreover, F-actin disappeared in the time interval 1 to 10 min after hypotonic exposure, whereas the ion channels are fully activated after about 1 min (see Table 4). These findings suggest that the changes in F-actin may be parallel phenomena rather than causally related to the volume regulatory processes. The possible role of the cytoskeleton in the mechanism of cell volume control has been reviewed by Mills (1987).

Three possible mechanisms by which microfilaments can regulate ion channels are (1) by the insertion of channels from a cytoplasmic store into the membrane, as proposed by Foskett and Spring (1985) and Lewis and de Moura (1982); (2) by acting as the mechanical transducer by which membrane stretch causes channel activation (see below); and (3) by activating a second messenger cascade, the products of which either activate or modulate membrane ion channels. At present, it is not known which of the above mechanisms is subserved by the cytoskeleton in the regulation of ion channels.

3.2 Stretch-Activated (SA) Channels

Ion channels activated by membrane stretch, and presumably attached to cytoskeletal strands, have been detected in a variety of vertebrate cells, e.g. chick skeletal muscle (Guharay and Sachs 1984), mammalian endothelial cells (Lansman et al. 1987), *Necturus* renal proximal tubule cells (Sackin 1987; Filipovic and Sackin 1992), choroid plexus from salamander (Christensen 1987), opossum kidney cell line (Ubl et al. 1988) and human fibroblast cell line (Stockbridge and French 1988; for reviews, see Sachs 1988; Rugolo et al. 1989; Morris 1990). This prevalence of stretch-activated (SA) channels suggests that these channels might constitute part of the volume-sensing signal transduction mechanisms.

The membrane stretch-activated ion channels are in general nonspecific, cation-selective channels which are permeable to Ca^{2+} and may allow sufficient Ca^{2+} to enter to serve a second messenger function (Christensen 1987; Lansman et al. 1987). Christensen (1987) proposed that

SA channels were activated by the swelling-induced increase in membrane tension and that an increase in cytoplasmic Ca^{2+} was produced by influx of Ca^{2+} through these channels. This was based on the findings that (1) amphibian choroid plexus cells contain SA channels (i.e. channels that could be activated by suction) that could bring Ca^{2+} into the cell, and that (2) Ca^{2+}-activated K^+ channels are activated after cell swelling. Swelling-induced activation of the SA channel, however, could not be demonstrated in choroid plexus cells. Recently, single channel reports have appeared, however, demonstrating swelling-induced activation of SA cation (SACat) channels:

1. Opossum kidney cells were exposed to 60% of normal osmolarity while the channel activity in a cell-attached patch was monitored (Ubl et al. 1988). In a given patch, an identical number of channels (up to six per patch) could be activated either by swelling or by direct mechanical stress (i.e. suction). The characteristics of the swelling- and suction-activated channels were identical: a non-selective channel of small conductance. This is illustrated in Fig. 6.

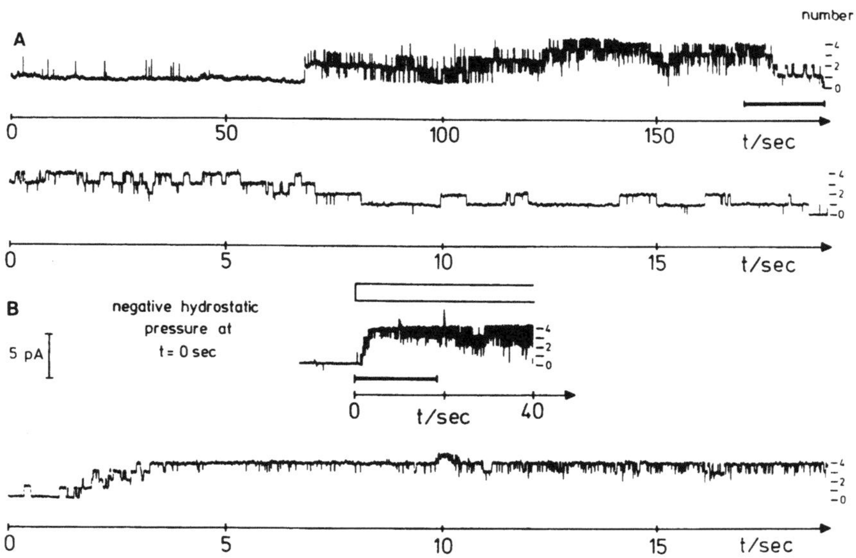

Fig. 6. Activation of a channel population in a cell-attached membrane patch of an opossum kidney cell by two different activation mechanisms. *A* 25 s prior to time zero, the isotonic (310 mOsmol NaCl) bath medium was replaced by a hypotonic (190 mOsmol NaCl) medium. The current trace marked by a *solid bar* is shown on an expanded scale below, indicating a transient closure of the osmotically activated channels. *B* After the transient closure of the osmotically activated channels a negative pressure of 2.5 cmHg is constantly applied to the silent membrane patch, as indicated by the *horizontal, open bar*. The current trace marked by a *solid bar* is shown on an expanded scale below. The pipette solution was 310 mOsmol NaCl and the pipette potential 0 mV. The number of simultaneously open channels is given on the *right-hand scale*. Upward deflections correspond to inward current. Results are from Ubl et al. (1988)

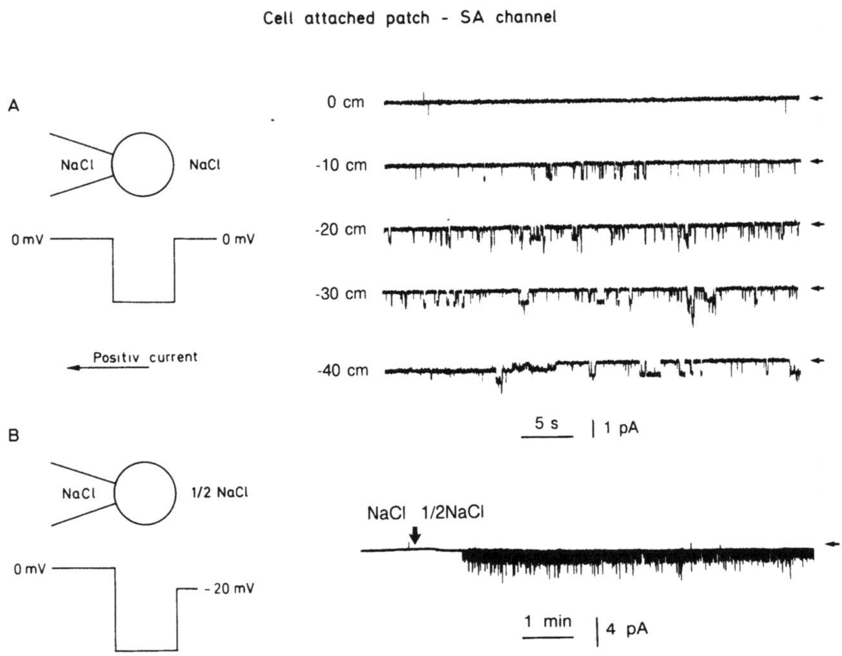

Fig. 7. Stretch activation of the cation channel in Ehrlich acites tumour cells produced either by pipette suction (A) or by osmotic cell swelling (*B*). *A* Current recordings from a cell-attached patch. The figures to the *left* of the traces indicate the negative electrode pressure (in cmH_2O) induced by pipette suction. Recordings are obtained with isotonic NaCl in the electrode at 0 mV bath potential. Inward current through the SA channels is shown as negative deflections. At −40 cm pressure a channel substrate may be noticed. *B* The influence of hypotonicity on the activity of the SA channel in cell-attached patches. Isotonic NaCl in the electrode and −20 mV bath potential (hyperpolarization of the patch). The external NaCl Ringer's solution is reduced from normal tonicity to half-tonicity at the time indicated by the *arrow*. Results are from Christensen and Hoffmann (1992)

2. Falke and Misler (1989), using 65–79% hypotonic shock on *neuroblastoma* cells, observed that SACat channels were activated with a time course similar to cell swelling and that, as expected if the stimulus is mechanical, activity subsided during the regulatory volume decrease phase.
3. Ehrlich ascites tumour cells were exposed to 50% of normal osmolarity and the channel activity in a cell-attached patch was monitored. A typical result is seen in Fig. 7B. After the onset of osmotically induced cell swelling, non-selective cation channels became activated. Channels with identical characteristics could be activated by suction (Fig. 7A), thus identifying the channels as SA channels. It should be noted that the channel activation does not occur instantaneously, but only after about 1 min (see Table 4). At that time the cells will be about 80% swollen. Volume-activated channels have previously been demonstrated in Ehrlich cells after a volume increase of

only 5% (Hudson and Schultz 1988). The 1-min delay in SA channel activation (see above) may perhaps reflect the involvement of other factors in the channel activation than simply the degree of cell swelling. In rabbit proximal tubule cells Lohr and Grantham (1986) have demonstrated an exquisite sensitivity of the cellular volume sensor. This implies that in order to play a physiological role in cell volume regulation the SA channels should be activated by only minute cell swelling. The use of non-physiologically large cell swelling in all published studies leaves open the crucial question of whether a small-scale cell swelling would actually activate SA channels.

Most SA channels have been demonstrated to be unselective cation channels. Two typical examples are illustrated in Table 6. SA channels with similar conductance and ion selectivity have been demonstrated in frog oocytes (Tanglietti and Teselli 1988), in *Xenopus* oocytes (Yang and Sachs 1989) and in renal proximal tubuli (Filipovic and Sackin 1992). Their poor selectivity seems in most cells to contradict an involvement of SA channels as a pathway for the K^+ and Cl^- efflux during RVD. An exception to this is, however, the stretch-activated, volume-sensitive K^+ channels in the basolateral membrane of *Necturus* proximal tubules (Sackin 1989). This channel is illustrated in Fig. 8. Efflux of potassium through these channels may partially mediate renal cell volume regulation (Filipovic and Sackin 1992). The volume-activated K^+ and Cl^- channels described above for Ehrlich cells *cannot* be activated by mechanical stress (suction) (Christensen and Hoffmann 1992).

It seems likely that the function of the unselective stretch-activated channels is correlated with their observed Ca^{2+} permeability (see Christensen 1987; Lansman et al. 1987; Ubl et al. 1988; Table 6). A resulting increase in $[Ca^{2+}]_i$ could be part of the volume-activated signal system. Ca^{2+} current via gadolinium-sensitive ion channels activated by osmotic cell swelling has recently been demonstrated in whole-cell recordings in intestine 407 cells (Okada et al. 1990). This Ca^{2+} entry seems to account for the major part

Table 6. Single channel conductance for SA channels, activated by pipette suction in opossum kidney cells and in Ehrlich ascites tumour cells; patch-clamp analysis in cell-attached mode

Cation in the pipette solution	Single channel conductance (pS)	
	Opossum kidney cells[a]	Ehrlich cells[b]
Na^+	30–40	23
K^+	22	23
Ca^{2+}	17	–
Ba^{2+}	17	16
$NMDG^+$	10	≈0

[a] See Hoffmann and Kolb (1991).
[b] Data from Christensen and Hoffmann (1992).

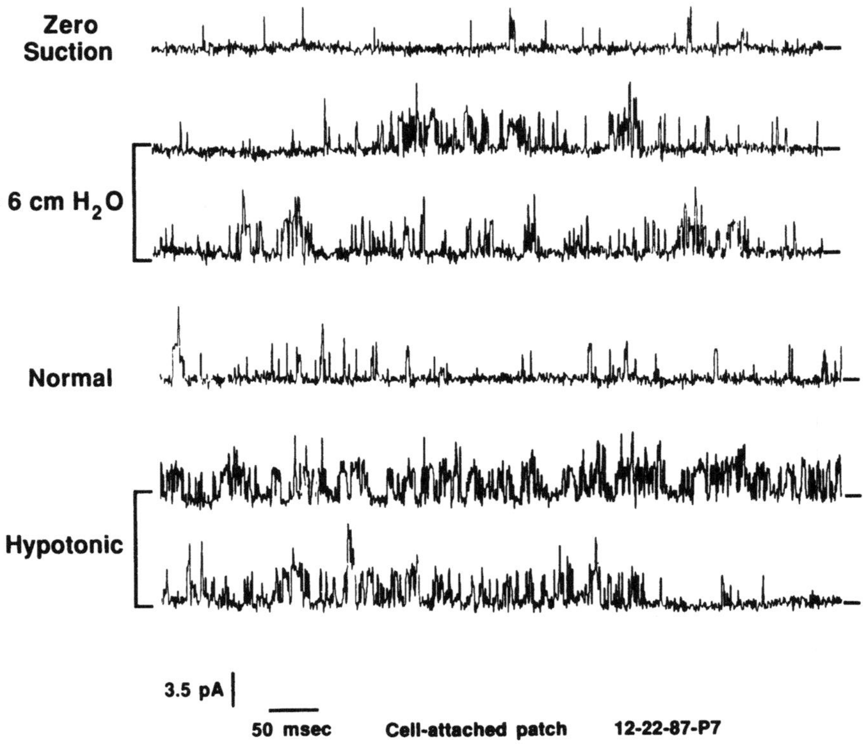

Fig. 8. Effect of pipette suction and hypotonic medium on a short open-time K^+ channel in a cell-attached patch from the basolateral membrane of *Necturus* proximal tubule. All records in this figure were obtained from the same patch. The patch pipette contained predominantly Na-gluconate, the bathing solution predominantly K-gluconate + sucrose (normal), or − sucrose (hypotonic). Upward deflections from closed state (*horizontal bars*) denote outward currents from cell to pipette. (Sackin 1989)

of the swelling-induced increase in cytosolic free Ca^{2+} (see Sect. 3.3) in intestine 407 cells, since the swelling-induced Ca^{2+} response is virtually abolished by gadolinium which is a specific blocker of stretch-activated ion channels (Okada et al. 1990).

A significant cell depolarization during RVD has been observed for several cell systems and may play a role in K^+ channel activation (for refs., see Hoffmann and Simonsen 1989; Rotin et al. 1991); such depolarization could partly be caused by activation of cation unselective stretch-activated channels. Hypothetically, such membrane depolarization might also result in Ca^{2+} entry via activation of voltage-gated Ca^{2+} channels.

3.3 Ca/Calmodulin

In most cells investigated Ca^{2+} seems to be involved in the RVD response following cell swelling, but the nature of the role played by Ca^{2+} is far from clear. For a recent review of the role of Ca^{2+} in cell volume regulation, see Pierce and Politis (1990).

In several cell types extracellular Ca^{2+} and Ca^{2+} entry across the cell membrane is a prerequisite for the swelling-induced RVD response. In these cells the volume recovery is blocked by removal of external Ca^{2+}, and the volume response can be inhibited by addition of Ca^{2+} channel blockers. In other cell types the volume response is unaffected by removal of external Ca^{2+}, and in these cells swelling-induced release of Ca^{2+} from internal stores has been proposed to be involved in the RVD response. A number of studies on epithelial cells have demonstrated that extracellular Ca^{2+} is a prerequisite for RVD (Davis and Finn 1985; Wong and Chase 1986; Hazama and Okada 1988; McCarty and O'Neil 1990; Okada et al. 1990). In isolated toad bladder cells the cellular ^{45}Ca uptake increased three-fold during RVD (Wong and Chase 1986).

A sizeable increase in cytosolic, free Ca^{2+} during RVD has been reported in isolated toad bladder cells (Wong and Chase 1986), in cultured intestinal epithelial cells (Hazama and Okada 1988, 1990) and in rabbit medullary

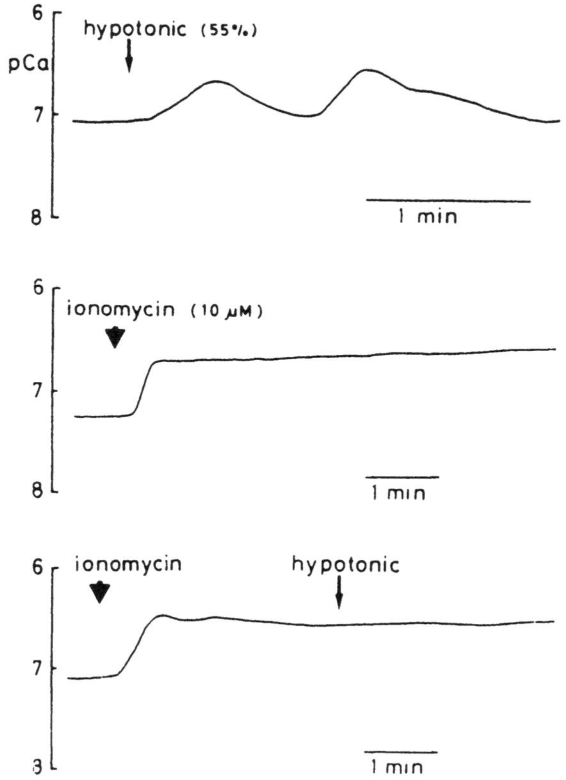

Fig. 9. Cytosolic Ca^{2+} measurements with Fura-2 in single monolayer intestine 407 cells, showing Ca^{2+} increase in response to hypotonicity (55%) and ionomycin (10 µM). (Hazama and Okada 1990)

thick ascending limb cells (Montrose-Rafizadeh and Guggino 1991). This is illustrated in Fig. 9 which shows a biphasic Ca^{2+} increase induced by hypotonicity in the case of intestine 407 cells. Recently, it has been demonstrated that swelling of hepatocytes induced by Na^+-dependent amino acid accumulation also increases the concentration of free cytosolic Ca^{2+} (Baquet et al. 1991).

Ca^{2+} entry across the cell membrane, however, is not necessary to activate RVD in Ehrlich cells (Hendil and Hoffmann 1974; Hoffmann et al. 1984) or in lymphocytes (Grinstein et al. 1982c, 1984). In these cell types release of Ca^{2+} from internal stores has been proposed to be involved in the RVD response. In Ehrlich cells prolonged pre-incubation in Ca^{2+}-free media containing EGTA reduces the RVD response only slightly (Hoffmann et al. 1984). It can be demonstrated, however, that the activation of K^+ and Cl^- channels is substantially reduced, but that this is partly compensated by an activation of a KCl cotransport (Kramhøft et al. 1986). No change in cytosolic, free Ca^{2+} concentration could be detected in human lymphocytes during RVD using Quin-2 (Rink et al. 1983) or indo-1 (Grinstein and Smith 1989). Moreover, in lymphocytes the RVD response was found to be unaffected by chelation of cytosolic Ca^{2+} (Grinstein and Smith 1989) or by depletion of intracellular Ca^{2+} (Grinstein and Smith 1990). A modulation of the Ca^{2+} sensitivity of the transport pathways has been suggested to play a role during RVD (Hoffmann et al. 1984).

In Ehrlich cells the RVD response in hypotonic medium can be accelerated by maneuvers which are presumed to increase intracellular Ca^{2+}. This is seen in Fig. 10. The cells recover their volume following hypotonic exposure in Ca^{2+} free medium (upper left), presumably by release of Ca^{2+} from internal stores. The acceleration by addition of external Ca^{2+} (1 mM) suggests an additional role for Ca^{2+} influx. The volume recovery is also accelerated by addition at the time of maximal cell swelling of the Ca^{2+} ionophore A23187 (upper right) and by addition of bradykinin (lower left) and thrombin (lower right). These agonists can be demonstrated to increase cytosolic Ca^{2+}. This is illustrated in Fig. 14 in the case of thrombin.

Separate, conductive pathways for K^+ and Cl^- can also be activated in isotonic cells. Addition of the Ca^{2+} ionophore A23187 in isotonic medium induces fast, transient cell shrinkage both in Ca^{2+}-containing and in Ca^{2+}-free media (see Fig. 11, left and middle). Figure 11 (right) shows that addition of bradykinin also induces transient cell shrinkage which is presumably mediated by an increase in free cytosolic Ca^{2+}. A23187 is demonstrated to induce a substantial increase in both the conductive K^+ and the conductive Cl^- permeability (Hoffmann et al. 1986). The K^+ and Cl^- conductances in A23187-treated Ehrlich cells are estimated at 134 and 30 $\mu S/cm^2$ respectively (Lambert et al. 1989).

Activation of K^+ channels after addition of A23187 is directly demonstrated in Fig. 12 in studies using the patch-clamp technique in the cell-attached mode (Christensen and Hoffmann 1992). In the cell-attached mode

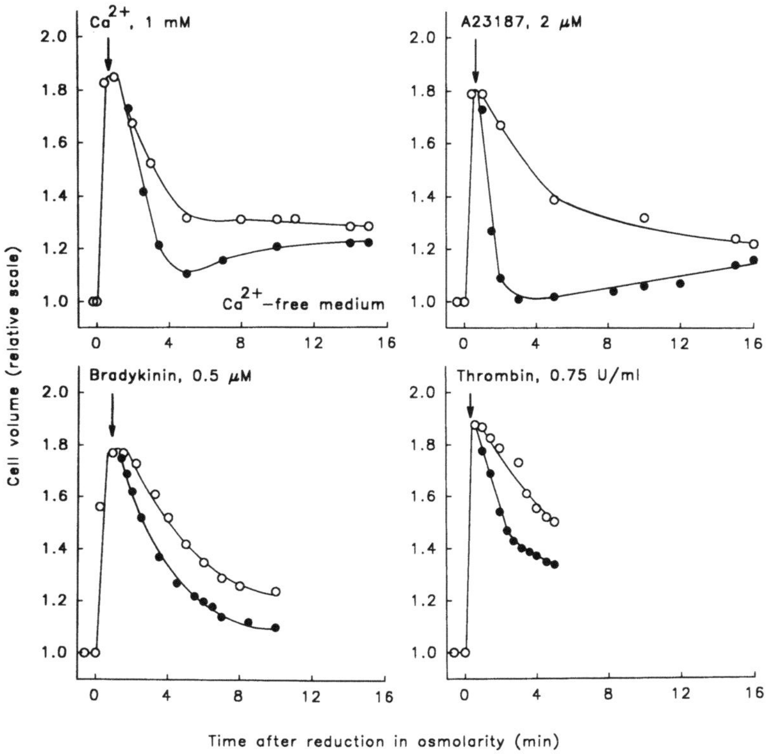

Fig. 10. Acceleration of RVD response in Ehrlich ascites tumour cells exposed to hypotonicity by addition of 1 mM Ca^{2+}, the Ca^{2+} ionophore A23187, and the Ca^{2+}-mobilizing agonists bradykinin and thrombin respectively. The acceleration by addition of external Ca^{2+} is only seen with cells suspended in Ca^{2+}-free medium. The RVD response in control cells exposed to hypotonic 150 mOsmol medium at zero time is shown with *open circles*, and the response in parallel experimental groups with *closed circles*. Data from upper frames are from Hoffmann et al. (1984) and data from lower frames are from Simonsen et al. (1990)

ionophore A23187 plus Ca^{2+}-induced channel activation can also be demonstrated in the case of the small (7-pS) Cl^- channel (Christensen and Hoffmann 1992). In excised inside-out patches the Cl^- channel is not Ca^{2+}-activated (see Sect. 2.1.2). The ionophore A23187-induced activation of Cl^- channels in the cell-attached mode, therefore, suggests an indirect effect of cytoplasmic Ca^{2+}.

The finding that Ca^{2+} is involved in some stage in the activation of the Cl^- transport pathway appears to be at variance with the findings in human lymphocytes where the increase in Cl^- net permeability induced by A23187 plus Ca^{2+} is significant (Grinstein et al. 1982b), but not appreciable compared to that induced by cell swelling (Sarkadi et al. 1984a,b). In human intestinal epithelial cells the Cl^- transport pathway also seems to be essentially Ca^{2+}-independent (Hazama and Okada 1988). On the other hand, a

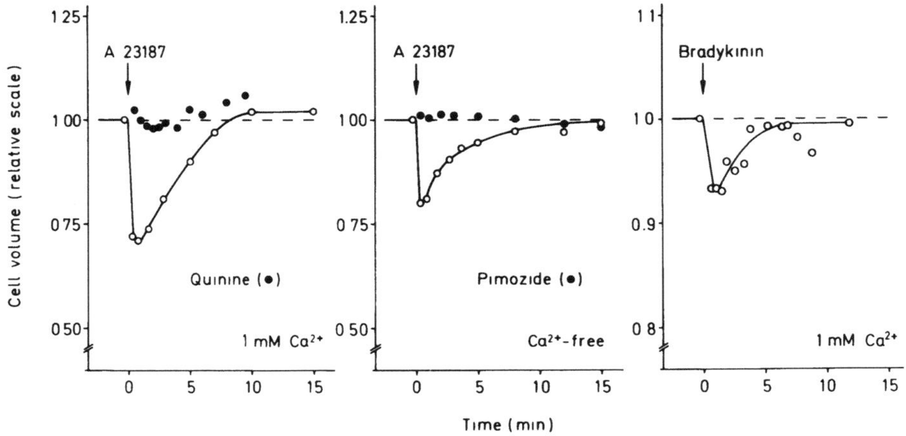

Fig. 11. Cell shrinkage induced in Ehrlich ascites tumour cells suspended in isotonic NaCl medium by addition at zero time of either the Ca^{2+}-ionophore A23187 (2 µM) or the Ca^{2+}-mobilizing agonist bradykinin (0.5 mM). The cell volume (*ordinate*)recorded with a Coulter counter is given relative to the cell volume recorded before ionophore or bradykinin addition. The response in control cells is shown with *open circles*. In parallel experimental groups the media contained, in addition, 1 mM quinine, which would block Ca^{2+}-dependent K^+ channels (*closed symbols, left*), or the anti-calmodulin drug pimozide, 10 µM (*closed symbols, middle*). Data with A23187 are from Hoffmann et al. (1986) and data with bradykinin are from Simonsen et al. (1990)

Ca^{2+}-activated Cl^- transport pathway has recently been reported in several cell types, for example *Xenopus* oocytes (Barish 1983), tracheal epithelial cells (Frizzell et al. 1986), neurons (Owen et al. 1984), and in the luminal membrane of salivary gland epithelial cells (Nauntofte and Poulsen 1984, 1986).

The A23187-induced increase in chloride conductance in Ca^{2+}-free media (which is presumably mediated by release of Ca^{2+} from internal stores) is transient like the activation during the volume response, whereas the activation of P_{Cl} is persistent in Ca^{2+}-containing media (Hoffmann et al. 1986). In the latter case the Ca^{2+} pump is probably swamped by the A23187-induced Ca^{2+} leak across the cell membrane, despite the powerful Ca^{2+} pump in Ehrlich cells (Cittadini et al. 1982; Klaven et al. 1983). In Ca^{2+}-free media, A23187-induced release of Ca^{2+} from internal stores (see Arslan et al. 1985), and a subsequent hysteretic activation of the Ca^{2+} pump (see Scharff et al. 1983), are likely to produce a transient increase in cytosolic Ca^{2+}. These findings suggest that it may be a transient increase in free cytosolic Ca^{2+} which accounts for the transient activation of the Cl^- transport pathway during RVD.

A number of drugs reported to inactivate the Ca^{2+}-binding protein calmodulin have been demonstrated to inhibit the KCl loss induced by cell swelling or by A23187 in lymphocytes (Grinstein et al. 1982b,c) and in Ehrlich cells (Hoffmann et al. 1984, 1986). The inhibition of RVD by

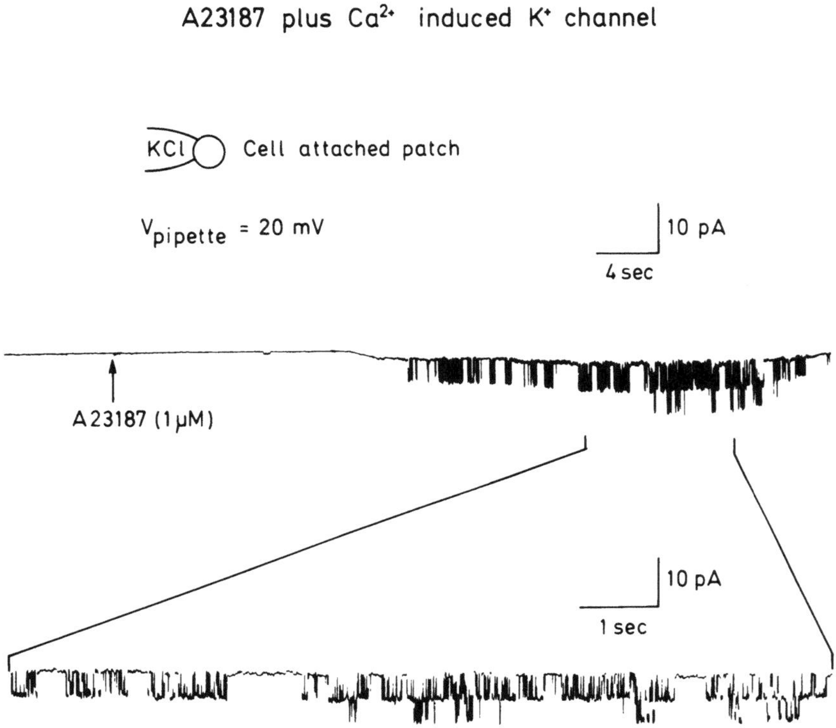

Fig. 12. Patch current from a K^+ channel in a cell-attached patch with KCl in the electrode. Inward K^+ currents are shown as negative (downward) deflections. At the time indicated by the *arrow*, 1 μM of the Ca^{2+} ionophore A23187 is added to the bath solution. The *bottom panel* shows details on an expanded time scale. (Christensen and Hoffmann 1992)

pimozide, a diphenylbutyl piperidine anti-calmodulin drug, is illustrated in Fig. 20 (upper right panel). Inhibition of the ionophore A23187-induced volume response is illustrated in Fig. 11 (middle panel). There is evidence that calmodulin is also involved in the RVD response in *Amphiuma* red cells (Cala et al. 1986) and in *Necturus* gallbladder epithelial cells (Foskett and Spring 1985).

In Ehrlich cells pimozide blocks RVD also in the presence of anions with a high conductive permeability (nitrate or thiocyanate), and also when a high K^+ net permeability is ensured by addition of gramicidin (Hoffmann et al. 1986). These findings demonstrate that the volume-induced activation of both K^+ and Cl^- transport pathways is inhibited by pimozide. A similar inhibition by pimozide was observed in the case of ionophore A23187 plus Ca^{2+}-induced cell shrinkage (see Fig. 11, middle). Moreover, the inhibition by pimozide of the A23187-activated K^+ transport pathway is directly demonstrated by ^{42}K flux experiments (Hoffmann et al. 1986). Although pimozide is not selective as an anti-calmodulin drug, the findings suggest that calmodulin is involved in the activation of both the K^+ and Cl^- transport pathway in Ehrlich cells. Calmodulin has been demonstrated in the cytoplasm of Ehrlich cells, and the cell membranes bind calmodulin in a Ca^{2+}-dependent manner (Aabin and Kristensen 1987). Taken together, the above findings support a role for calmodulin in the Ca^{2+} activation of the channels. Evidence for the involvement of calmodulin in the operation of Ca^{2+}-activated K^+ channels in mouse fibroblasts has been presented (Okada et al. 1987).

Patch-clamp studies have shown, as discussed above (see Sect. 2.1.1), that the same inward rectifier potassium channel is activated in Ehrlich cells after cell swelling and after addition of the Ca^{2+} ionophore A23187. Figure 13 shows the Ca^{2+} dependence of the gating for the inward rectifier K^+ channel from Ehrlich cells in excised inside-out patches bathed in KCl on both sides and at a membrane potential of $-60\,mV$. It can be seen that activation of the K^+ channel takes place after an increase in $[Ca^{2+}]_i$ from 10^{-7} to $10^{-6}\,M$, which is in the physiological range, supporting the notion that the volume-induced K^+ transport is mediated by the Ca^{2+}-activated K^+ channels.

3.4 Agonist-Induced Activation of K^+ and Cl^- Channels in Ehrlich Ascites Tumour Cells

Several hormones, neurotransmitters and other agonists induce Ca^{2+} signalling by receptor-mediated activation of phosphoinositide-specific phospholipase C (phosphoinositidase C) with partial depletion of cell membrane phosphatidyl-inositol 4,5-bisphosphate [PtdIns(1,4,5)P_2] and release of inositol 1,4,5-trisphosphate [Ins(1,4,5)P_3] and diacylglycerol (DAG). Ins(1,4,5)P_3 in turn releases Ca^{2+} from internal stores and produces a transient increase in free cytosolic Ca^{2+} (see e.g. Sekar and Hokin 1986; Berridge 1987; and a series of papers edited by Berridge and Michell 1988).

We have found that bradykinin, thrombin and histamine all induce a transient increase in free cytosolic Ca^{2+} in Ehrlich cells, as measured using the fluorescent Ca^{2+} probe Fura-2 and the "ratio" method based on dual wavelength excitation. The agonist-induced Ca^{2+} signal is illustrated in Fig. 14 in the case of thrombin. Similar Ca^{2+} signals were recorded after stimulation with bradykinin and histamine (not illustrated). We have previously reported a thrombin- and bradykinin-induced release of Ins(1,4,5)P_3 (Simonsen et al. 1990; Svane and Hoffmann 1991). Bradykinin and thrombin accelerate the RVD response in hypotonic medium when added at the time of maximal cell swelling as illustrated in Fig. 10 (lower panels). In isotonic medium the agonists induce a fast, transient cell shrinkage, similar to the volume response seen after addition of the Ca^{2+} ionophore A23187, as illustrated in Fig. 11.

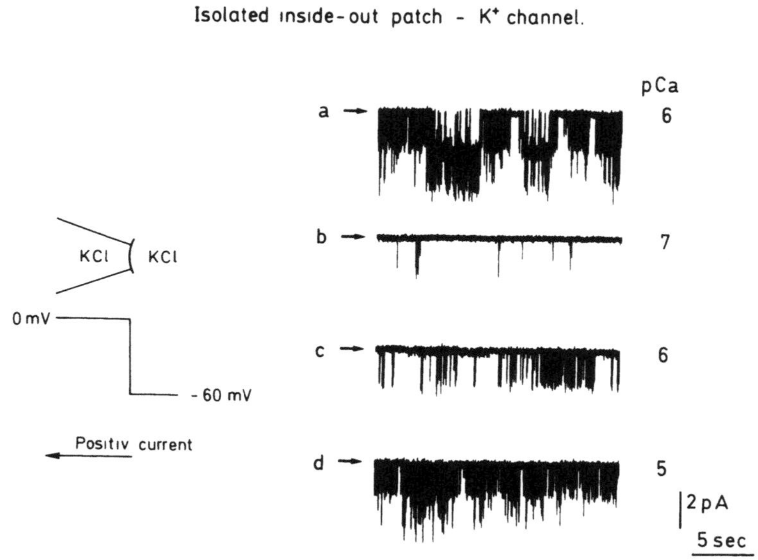

Fig. 13. Calcium dependence of gating for an inside-out excised patch bathed in symmetrical KCl solutions. Membrane potential was −60 mV. The data in *lanes a to d* are sections of a continuous recording displayed in chronological order. Note change in channel activity with time at pCa = 6. Christensen and Hoffmann (1992)

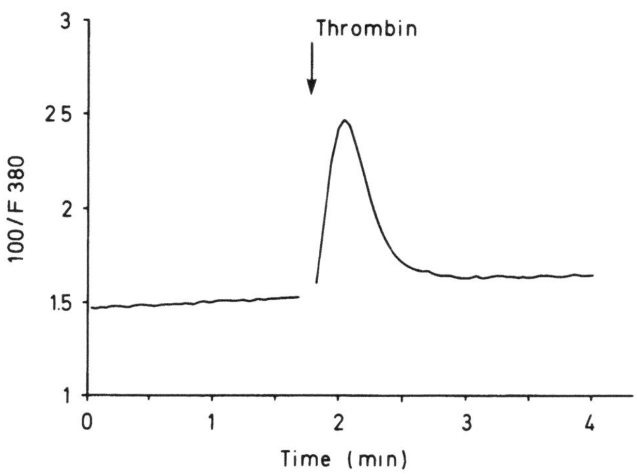

Fig. 14. Thrombin-induced transient increase in cytosolic Ca^{2+} in Ehrlich ascites tumour cells. Thrombin (1 unit/ml) was added at the time indicated by the *arrow* to a suspension of cells loaded with the fluorescent Ca^{2+} probe Fura-2, and the fluorescence recorded with excitation at 380 nm, emission at 510 nm. The *ordinate* shows the reciprocal value of the recorded fluorescence. This value will increase with an increase in the concentration of free cytosolic Ca^{2+}. In other experiments free cytosolic Ca^{2+} was measured by dual wavelength excitation at 340 and 380 nm and the "ratio" method. In the experiment illustrated here the fluorescence was recorded with excitation at 380 nm only, in order to improve the time resolution. Results are from Harbak and Simonsen (1991)

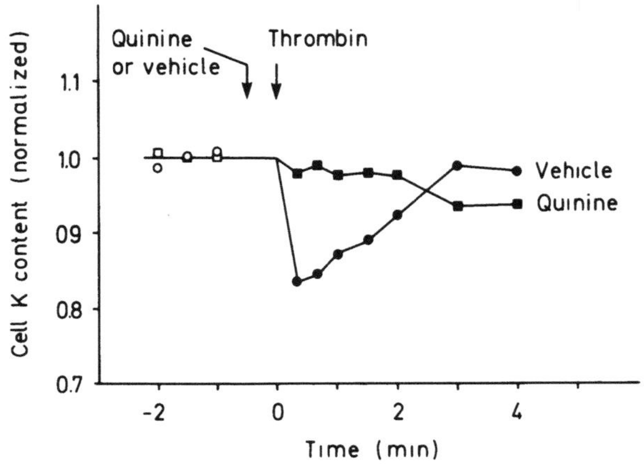

Fig. 15. Inhibition by quinine of the net loss of cell K^+ from Ehrlich ascites tumour cells induced by thrombin. Thrombin (1 unit/ml) was added to the cell suspension at zero time in the presence and absence of quinine (1 mM). Results are from Harbak and Simonsen (1991)

Measurements of the cell K^+ content show that thrombin and histamine induce a rapid initial loss of cell K^+ which is followed by a gradual recovery towards control values within about 5 min (see Figs. 15, 16). The initial K^+ loss amounts to about 15% and is reached within 15 s.

The initial agonist-induced K^+ loss is blocked by quinine which blocks Ca^{2+}-activated K^+ channels (see Fig. 15). A similar inhibition of the agonist-induced K^+ loss is seen in the presence of Ba^{2+} (5 mM) which also blocks K^+ channels (not illustrated).

In cells in which cell chloride had been replaced by nitrate by preincubation in nitrate medium, the initial K^+ loss induced by the Ca^{2+} ionophore ionomycin plus Ca^{2+} (Fig. 16, left) and by histamine (Fig. 16, right) was unaffected. Similar results were obtained with thrombin (not illustrated). Moreover, the thrombin-induced K^+ loss was inhibited by the Cl^- channel blocker NPPB (not illustrated).

These findings are consistent with the loss of KCl via K^+ channels and non-selective Cl^- channels, rather than via Cl^--dependent cotransport systems, similar to the KCl loss induced by the Ca^{2+} ionophore A23187 plus Ca^{2+} (Hoffmann et al. 1986).

The KCl re-uptake following the initial KCl loss is inhibited in nitrate media (see Fig. 16, left and right). The K^+ re-uptake is presumably mediated by activation of the Na^+, K^+, $2Cl^-$ cotransport system (see Sect. 2.2).

The KCl loss induced by Ca^{2+}-mobilizing agonists shows similarities to the RVD response in hypotonic media (see Sect. 2.1). These findings suggest that activation of phospholipase C and the inositol phosphate signalling system could be involved in the activation of the RVD response (Simonsen et al. 1990; Harbak and Simonsen 1991; Svane and Hoffmann 1991; see also Sect. 3.5).

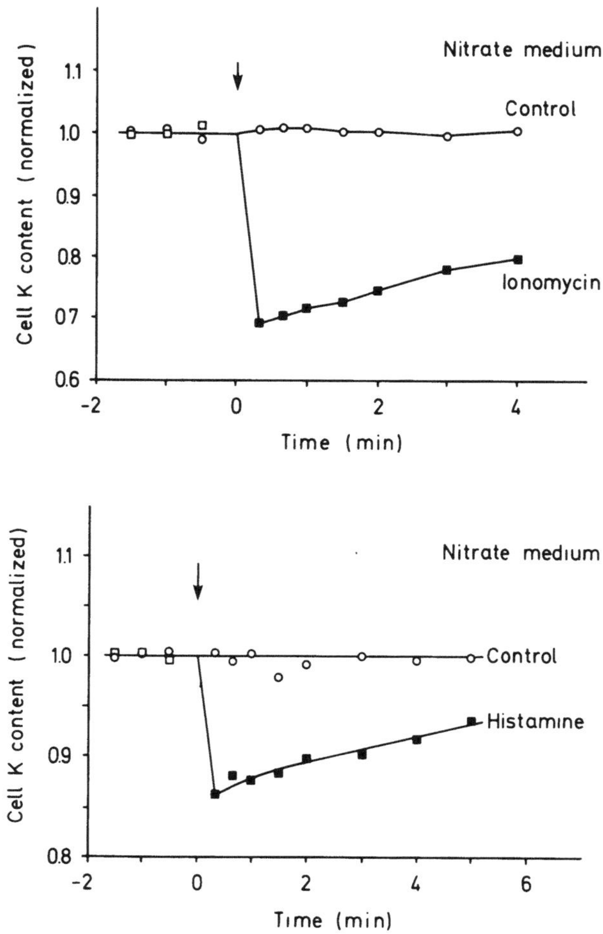

Fig. 16. Net loss of cell K^+ from Ehrlich ascites tumour cells in nitrate medium (with 1 mM Ca^{2+}), induced by ionomycin (1 μM) (*left*) and by histamin (100 μM) (*right*), added at zero time as indicated by *arrow*. Results are from Harbak and Simonsen (1991)

3.5 Inositol Phosphate Cycle

In Ehrlich ascites tumour cells Christensen et al. (1988) have presented evidence suggesting activation of the inositol phosphate cycle during the RVD response in hypotonic medium. They found that the RVD response in Ehrlich cells is associated with a rapid decrease in membrane PtdIns(4,5)P_2 and a concomitant, substantial increase in cytoplasmic Ins(1,4,5)P_3. This is seen in Fig. 17. These results support the notion that Ins(1,4,5)P_3 may play a role as second messenger in Ca^{2+} mobilization during the RVD response in Ehrlich cells. Recently, Baquet et al. (1991) have reported a significant, sustained elevation of Ins(1,4,5)P_3 in hepatocytes after cell swelling induced by hypotonic exposure or by Na^+-dependent amino acid accumulation.

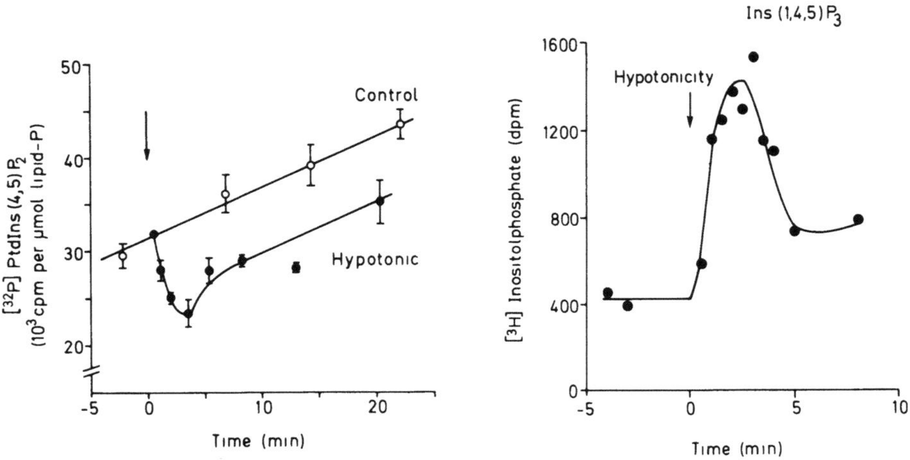

Fig. 17. Time course of cellular [^{32}P]-PtdIns(4,5)P_2 (*left*) and [^{3}H]-Ins(1,4,5)P_3 (*right*) in hypotonically swollen Ehrlich cells. *Left* The cells were preloaded with ^{32}P for 60 min, and at zero time (*arrow*) diluted with 1 vol of P_i-free hypotonic (50 mOsmol) Ringer's solution to give a final osmolarity of 175 mOsmol. The total cellular lipids were extracted with chloroform-methanol-HCl, and their water-soluble deacylation products separated on Dowex anion exchange columns. Total cell lipid P was analyzed as a measure of cell mass. The *error bars* indicate ± SEM for tri- or quadruplicate samples (Christensen et al. 1988). *Right* The cells were preloaded with [^{3}H]-*myo*-inositol for 22 h and after addition of 10 mM LiCl diluted with hypotonic Ringer's. Samples were precipitated with trichloroacetic acid (final conc. 5%), and after ether extraction analyzed by high-performance liquid chromatography on a SAX-10 column, eluted by an ammonium formate gradient. (Hoffmann and Kolb 1991; results are from Svane et al. 1990)

It is unknown whether the diacylglycerol (DAG) also released from PtdIns(4,5)P_2 plays an additional role in the RVD response, e.g. via a protein kinase C-mediated modulation of the Ca^{2+} sensitivity of the transport pathways. A role for protein kinase C as activator of volume-sensitive K^+/H^+ exchange in *Amphiuma* red blood cells has been proposed by Cala (1986) based on a stimulation of K^+/H^+ exchange by phorbol esters. In red cells of the little skate hypotonic cell swelling produces an increase in DAG (Musch and Goldstein 1990), predominantly by stimulation of phosphatidyl choline hydrolysis, and an increase in InsP, whereas the level of $InsP_2$ and $InsP_3$ was unaffected (McConnel and Goldstein 1988). In villus epithelial cells it has been demonstrated that inhibition of protein kinase C inhibited RVD after cell swelling induced by L-alanin uptake, whereas RVD after hypotonic cell swelling was unaffected (MacLeod 1991).

3.6 Eicosanoids

The diacylglycerol (DAG) released by phosphoinositide hydrolysis by phospholipase C is a source of arachidonic acid (Berridge 1984). In addition,

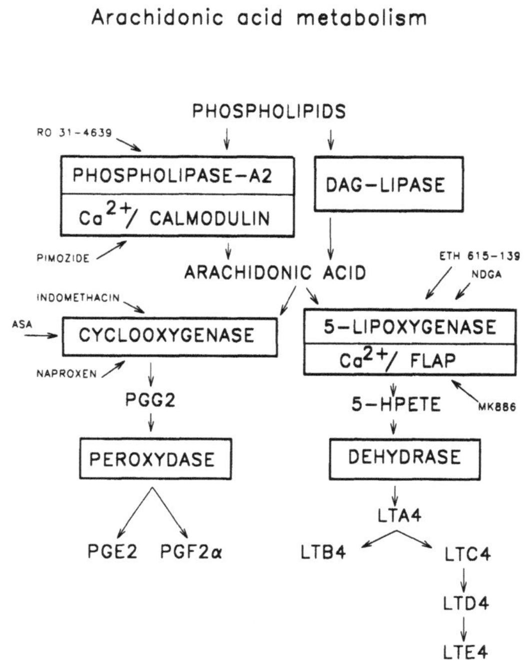

Fig. 18. Pathways of the arachidonic acid cascade which are relevant to the present investigation. The Ca^{2+}-dependent steps are the Ca^{2+}/calmodulin-regulated phospholipase-A_2 and the 5-lipoxygenase which is activated by the Ca^{2+}-dependent binding to the 5-lipoxygenase binding protein (*FLAP*). Inhibitors: the phospholipase-A_2 is inhibited directly by RO 31-4639 and indirectly by pimozide via its binding to the Ca^{2+}, calmodulin system. The cyclooxygenase is inhibited by acetylsalicylic acid (*ASA*), indomethacin and naproxen. The 5-lipoxygenase is inhibited directly by nordihydroguaiaretic acid (*NDGA*) and *ETH 615–139*, and indirectly by *MK 886* via its binding to *FLAP*

calcium and calmodulin play an essential role in the release of arachidonic acid from membrane phospholipids (see Wong and Cheung 1979; van den Bosch 1980) and in the conversion of arachidonic acid into leukotrienes via activation of the 5-lipoxygenase pathway (Feinstein and Sha'afi 1983; see Fig. 18). It has recently been demonstrated (Lambert et al. 1987) that Ehrlich ascites cells possess the ability to synthesize and release the arachidonic acid metabolites, prostaglandins and leukotrienes, and that addition of arachidonic acid stimulates the production of both compounds.

3.6.1 Role in RVD

During RVD, the synthesis of leukotrienes is stimulated while, concomitantly, prostaglandin synthesis is reduced (see Fig. 19). Addition of leukotriene LTD_4 accelerates RVD (which is rate-limited by the K^+ conductance) (see Fig. 20, upper left), while addition of inhibitors of leukotriene synthesis blocks the volume response, even when a high K^+ conductance has been ensured by the presence of gramicidin (not illustrated). These find-

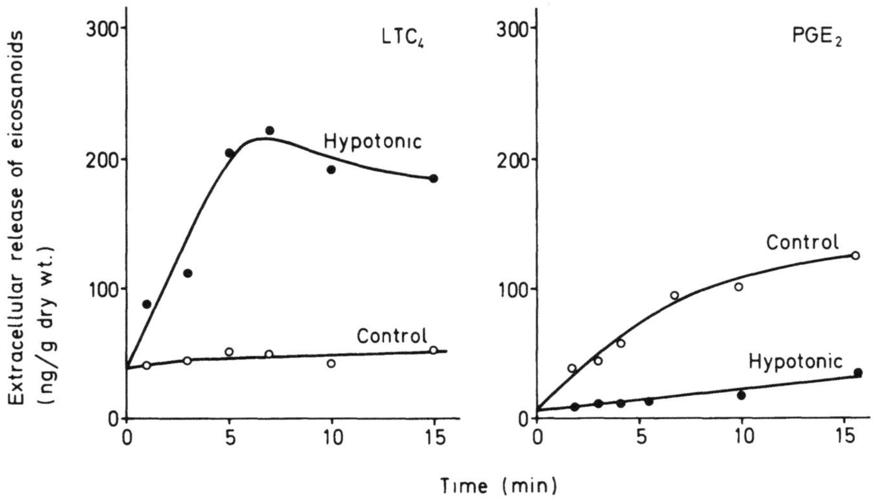

Fig. 19. LTC_4 and PGE_2 synthesis in Ehrlich cells suspended in standard incubation medium and in hypotonic (150 mOsmol) incubation medium. The release of LTC_4 and PGE_2 from the cells was followed with time by serially isolating cell-free medium by centrifugation and measuring the LTC_4 and PGE_2 concentrations in the supernatant by radioimmunoassay. The extracellular eicosanoid release (ng/g cell dry wt.) was calculated from the measured concentration in the medium (ng/ml) and the cell density of the cell suspension (g dry wt./ml). (Lambert et al. 1987)

ings suggest that the activation of both K^+ and Cl^- transport pathways after hypotonic cell swelling involves an increase in leukotriene synthesis (Lambert et al. 1987). Moreover, it has been proposed that the effect of Ca^{2+} and calmodulin on the activation of the Cl^- transport pathway (see Sect. 3.3) is indirect and mediated via stimulation of leukotriene synthesis (Lambert 1987, 1989; Lambert et al. 1987a). This notion was based on the observation that the inhibition of the RVD response seen in the presence of the anti-calmodulin drug pimozide can be lifted by addition of LTD_4, as seen in Fig. 20 (upper right).

Figure 20 shows that the RVD response is inhibited by RO 31-4639 (lower left), which inhibits the phospholipase-A_2 (see Henderson et al. 1989), and by ETH 615-139 (lower right), which inhibits the 5-lipoxygenase (Kirstein et al. 1991). The inhibition by the phospholipase-A_2 inhibitor can be lifted by LTD_4, whereas LTD_4 fails to lift the inhibition by ETH 615-139, presumably because this drug, in addition to the inhibition of 5-lipoxygenase, also acts as an LTD_4-receptor antagonist.

Addition of prostaglandin-E_2 (PGE_2) inhibits the RVD response in hypotonic medium containing Na^+, but not in sodium-free hypotonic media, suggesting that PGE_2 increases the passive Na^+ permeability in Ehrlich cells (Lambert et al. 1987). The reduced prostaglandin synthesis during RVD (see Fig. 19) could thus account for the concomitant reduction in the passive

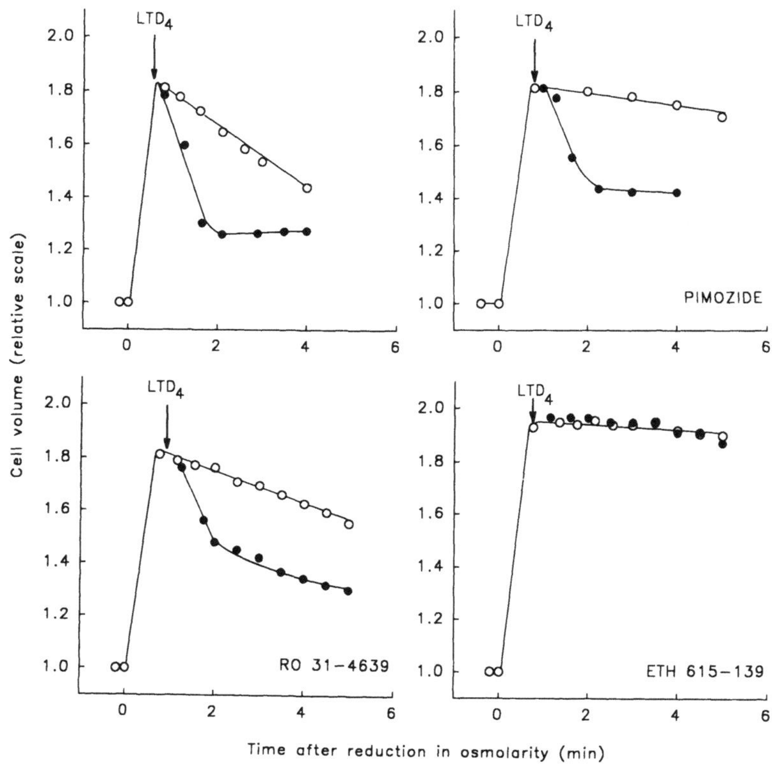

Fig. 20. Effect of *LTD_4* on cell volume in Ehrlich cells swollen in hypotonic medium in the presence and absence of the anti-calmodulin drug pimozide, the phospholipase-A_2 inhibitor RO 31-4639, and the 5-lipoxygenase inhibitor (and presumably LTD_4 receptor antagonist) ETH 615-139, respectively. Cells were prepared and exposed to hypotonicity at zero time, as described in the legend to Fig. 5. Pimozide (10 µM, *upper right*), RO 31-4639 (0.5 µM, *lower left*) and ETH 615-139 (4 µM, *lower right*), respectively, were added within the first seconds after reduction in osmolarity. LTD_4 (100 nM; *closed symbols*) was added at the time indicated by *arrows*. Cell volume was followed with time using a Coulter counter and given relative to the initial cell volume in isotonic media. Data are from Lambert (1989) and Lambert and Hoffmann (1991)

permeability for Na^+ previously reported by Hoffmann (1978). Moreover, the increased synthesis of PGE_2 following addition of arachidonic acid could well explain the stimulating effect of arachidonic acid on the Na^+ conductance, and hence the observed inhibition of RVD by arachidonic acid (Lambert 1987a).

3.6.2 Role in Activation of Taurine Leak Pathway

Taurine efflux from Ehrlich cells suspended in isotonic medium is unaffected by addition of PGE_2 (Lambert and Hoffmann 1993), but strongly stimulated by addition of LTD_4 (Fig. 21), and also stimulated following depolarization of the cell membrane (Lambert and Hoffmann 1990, 1993). The taurine

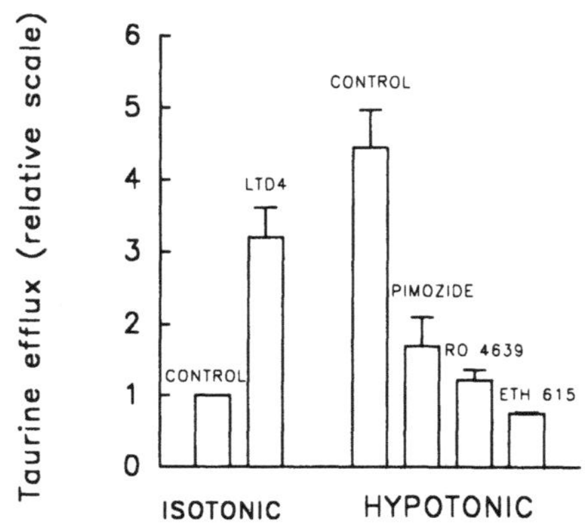

Fig. 21. The effect of LTD_4 under isotonic, and of hypotonic cell swelling with and without *pimozide*, a phospholipase-A_2 inhibitor (*RO 4639*), and a 5-lipoxygenase inhibitor (*ETH 615*), respectively, on taurine release in Ehrlich ascites tumour cells. Ehrlich cells, equilibrated with ^{14}C-taurine in standard NaCl medium were transferred to isotonic (300 mOsmol) or hypotonic (150 mOsmol) solutions. The taurine efflux was followed with time as shown in Fig. 22. Pimozide (12 µM), RO 31-4639 (2.4 µM), ETH 615-139 (10 µM), and LTD_4 (4 µM) were added at time zero. The taurine efflux is given relative to the taurine efflux in isotonic standard medium. Data are from Lambert and Hoffmann (1993)

efflux is activated by cell swelling (see Fig. 21). The volume-induced taurine efflux is dramatically reduced when the phospholipase-A_2 is inhibited directly by addition of RO 31-4639 or indirectly by addition of the anti-calmodulin drug pimozide (see Fig. 21). The inhibition of the volume-induced taurine efflux is in both cases lifted by addition of LTD_4 (see Fig. 22). The swelling-induced taurine efflux is also inhibited by addition of ETH 615-139 which inhibits 5-lipoxygenase and presumably in Ehrlich cells also acts as an LTD_4 receptor antagoinst (see Fig. 21). It is therefore concluded that the swelling-induced activation of the taurine leak pathway involves activation of phospholipase-A_2 with release of arachidonic acid from membrane phospholipids, and an increased oxidation of arachidonic acid into leukotrienes via the 5-lipoxygenase pathway (see Fig. 18). LTD_4 seems to act as a second messenger for the activation of the taurine leak pathway either directly or perhaps indirectly via the depolarization of the cell membrane potential by LTD_4-induced activation of the Cl^- channels (Lambert and Hoffmann 1993).

In skate red cells the swelling-induced release of taurine can be mimicked by the Ca^{2+} ionophore A23187 (Leite and Goldstein 1987). Calmodulin is probably not involved in the response to hypotonicity, since anti-calmodulin drugs failed to inhibit the taurine efflux (Leite and Goldstein 1987). As mentioned in Section 3.5, hypotonic cell swelling in red cells of the little skate produces an increase in DAG, predominantly by stimulation of phosphatidylcholine hydrolysis (Musch and Goldstein 1990). DAG is a precursor for arachidonic acid (see Fig. 18) and stimulates protein kinase C

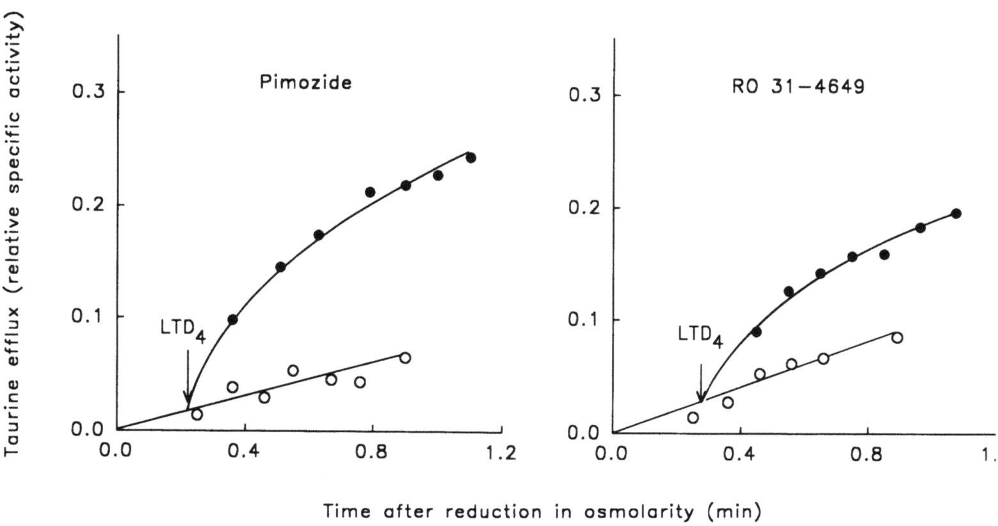

Fig. 22. Effect of LTD_4 on the taurine efflux in hypotonic medium in the presence of *pimozide* (12 µM, *left*) or *RO 31-4639* (2.4 µM, *right*). Ehrlich cells, equilibrated with ^{14}C-taurine for 60 min in isotonic NaCl medium were transferred to hypotonic (Ca^{2+}-free) NaCl medium with half-osmolarity at time zero. *Taurine efflux* is shown as the extracellular relative specific activity, $(a_t^m - a_o^m)/a_o^c$ plotted versus the time, where a_t^m and a_o^m are the ^{14}C-taurine activity in the efflux medium at time *t* and time zero, respectively, and a_o^c is the cellular ^{14}C-taurine activity at time zero. LTD_4 (4 µM, *closed symbols*) was added at the time indicated by the *arrow*. Data are from Lambert and Hoffmann (1993)

(see Fig. 25). The relative role of these putative signalling systems in controlling taurine efflux in skate red cells is at present unclear.

3.6.3 Effect of Polyunsaturated Fatty Acids

The Ehrlich mouse ascites tumour cell (which is grown in the abdominal cavity of a host mouse) has been used as a model in the study of possible effects of dietary fish oil supplement on regulation of cellular salt and water content. Dietary MaxEPA supplement (7.5% wt/wt) for 2 weeks significantly increased the relative content of eicosapentaenoic acid (EPA, C20:5, n-3) and significantly decreased the relative content of arachidonic acid (AA, C20:4, n-6) in cellular phosphatidylcholine as well as in phosphatidylethanolamine, compared with an olive oil-control diet (see Table 7). Cells grown in the abdominal cavity of fish oil-fed mice showed an accelerated RVD response (see Fig. 23, left) and net Cl^- loss following hypotonic exposure. Moreover, in Na^+-free isotonic medium these cells show a more pronounced ionophore A23187 plus Ca^{2+}-induced cell shrinkage (see Fig. 23, right). The fish oil-dietary supplement was found to result in an increased Cl^- conductance during RVD, demonstrated as an increased initial rate of volume recovery after gramicidin addition to the hypotonically swollen cells in order to ensure a high K^+ conductance (not illustrated). As

Table 7. Effect of fish oil dietary supplement on the percentage distribution of fatty acids (arachidonic acid, n-6; AA) and eicosapentaenoic acid (n-3; EPA) in phosphatidylcholine and in phosphatidylethanolamine fractions of cell membrane phospholipids[a]

				Dietary supplement		
				Olive oil	Fish oil	
Membrane phospholipid	Polyunsaturated fatty acids			Fatty acid distribution (Peak percentage)[b]		P
Phosphatidyl-choline	AA	C20:4	n-6	6.4 ± 0.8	3.7 ± 0.5	<0.01[c]
	EPA	C20:5	n-3	0.2 ± 0.04	2.8 ± 0.5	<0.001[d]
Phosphatidyl-ethanolamine	AA	C20:4	n-6	12.0 ± 1.4	5.9 ± 0.5	<0.01[d]
	EPA	C20:5	n-3	0.6 ± 0.3	5.1 ± 0.5	<0.001[c]

[a] Values, from Lauritzen et al. (1993).
Mean ± SEM for 8 and 10 expts. for phosphatidyl cholin and phosphatidyl ethanolamine, respectively.
[b] Peak percentage is the area of the specific peak in percentage of the area of all peaks in the gas chromatogram.
[c] Student's t-test.
[d] D-test, because the F-test showed that the conditions for a Student's t-test were not fulfilled.

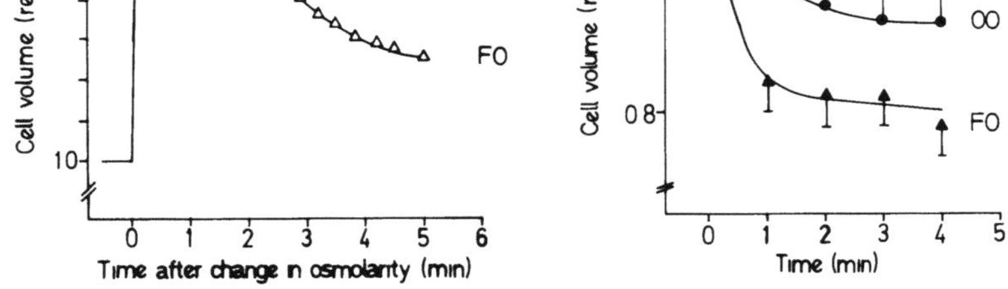

Fig. 23. Effect of dietary fish oil (*FO*) and olive oil (*OO*) supplement on the RVD response in Ehrlich ascites cells in hypotonic medium, and on ionophore A23187-induced cell shrinkage in isotonic medium. In order to modify the fatty acid composition of the phospholipids, host mice were fed either a fish oil or an olive oil supplement diet for 2 weeks. *Left* The cells were handled and the cell volume measured as described in Fig. 5. *Right* The cells were pre-incubated in isotonic NaCl medium, and the cell suspension diluted about 300-fold in isotonic Na^+-free, choline medium. A23187 (2 µM) was added at the time indicated by the *arrow*, and the cell volume followed with time in a Coulter counter. Cell volumes are given relative to the cell volume of unperturbed cells in isotonic NaCl medium. (Lauritzen et al. 1993)

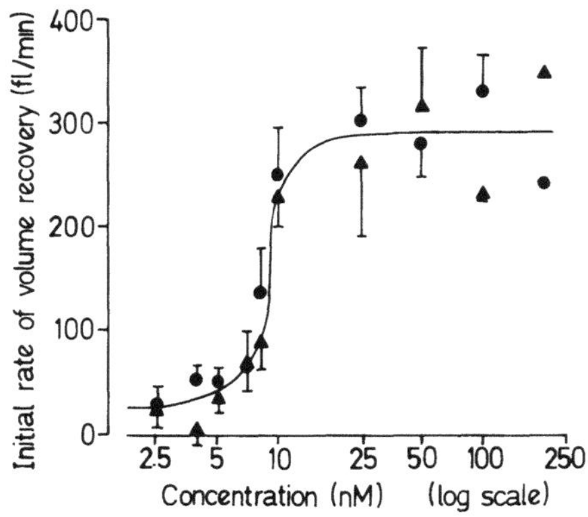

Fig. 24. Saturation curves for the acceleration by LTD_4 (●) and LTD_5 (▲) of the volume recovery in Ehrlich ascites tumour cells after swelling in hypotonic NaCl medium (150 mOsmol) in the presence of pimozide (10 μM). Experimental protocol for preparation of cells as in Figs. 5 and 20. The volume recovery (fl/min) was estimated from curves similar to the one shown in Fig. 20 (upper right), using cell volumes measured within the first minute following addition of LTD_4/LTD_5. Lauritzen et al. (1993)

discussed above (see Sect. 3.6), the RVD response is inhibited by the anti-calmodulin drug pimozide, but the inhibition can be lifted by addition of leukotrienes (see Fig. 20, upper right). Figure 24 shows that LTD_4 (synthesized from AA) and LTD_5 (synthesized from EPA, see below) are equipotent in lifting the inhibition by pimozide. We conclude that dietary (n-3) fatty acid supplement results in a more effective cell volume recovery during hypotonic stress due to an increase in the volume-induced Cl^- conductance. Since LTD_4 and LTD_5 are found to be equipotent in activating the volume response, it is proposed as a simple interpretation that the total leukotriene synthesis (LTD_4 plus LTD_5) in cells enriched with EPA is larger than the total leukotriene synthesis (predominantly LTD_4) in control cells.

EPA is both a substrate for the 5-lipoxygenase and the dehydrase (Needleman et al. 1986), and pentaene leukotrienes are produced by intact cells (Budowsky 1988). EPA and arachidonic acid have been reported to be equivalent as substrates for leukotriene synthesis (Murphy et al. 1981). Recent evidence indicates that EPA may even be a better substrate than arachidonic acid for leukotriene synthesis (Careaga-Houck and Sprecher 1990). These findings are in accord with the proposal that the improved cell volume regulation seen in Ehrlich cells from mice fed on an EPA-enriched diet could result from synthesis of LTD_5 and perhaps increased total leukotriene synthesis. In addition, EPA is a poor substrate for the cyclooxygenase pathway (see Fig. 18) and a competitive inhibitor of arachidonic acid metabolism via this pathway (Needleman et al. 1979). This means that the volume-induced shift during RVD from prostaglandin to leukotriene synthesis (see Fig. 19) could conceivably be augmented in cells from mice fed on an EPA-enriched diet. Prostaglandins inhibit the RVD

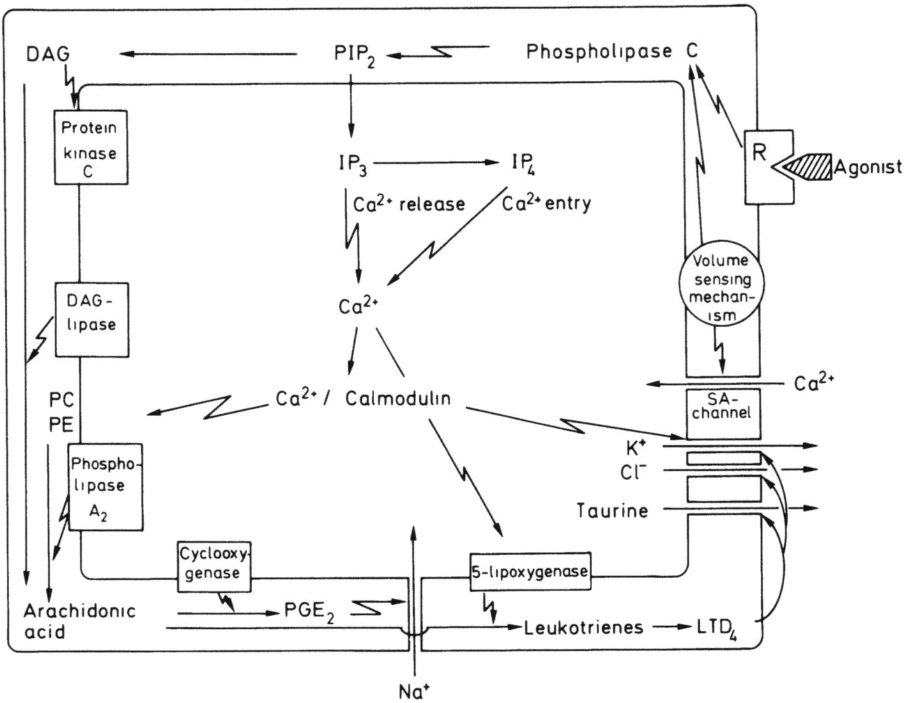

Fig. 25. Activation mechanisms during RVD in Ehrlich ascites tumour cells. *PE* Phosphatidylethanolamine; *PC* phosphatidylcholine; *DAG* diacylglycerol; *PIP₂* PtdIns (4,5) P_2; *IP₃* Ins(1,4,5)P_3; *IP₄* Ins(1,3,4,5)P_4; *R* agonist receptor. (Adapted and extended from Hoffmann et al. 1988)

response (see Sect. 3.6.1), and hence an augmented shift from prostaglandin to leukotriene synthesis will accelerate volume recovery. Such a mechanism could well play a major role in the observed effects of an EPA-enriched diet on cell volume regulation.

3.7 Model for Intracellular Transmission

In conclusion, the following model for activation of the RVD response in Ehrlich cells is proposed (see Fig. 25). Cell swelling activates unselective cation channels producing Ca^{2+} entry and an increase in free cytosolic Ca^{2+} which is proposed to be part of the volume-activated signal system. Moreover, the activation of unselective cation channels will tend to depolarize the cell membrane which in some cells will activate voltage-gated K^+ channels. Moreover, cell swelling activates (by an unknown mechanism) phospholipase C with resulting hydrolysis of cell membrane PtdIns(4,5)P_2 and release of Ins(1,4,5)P_3 and DAG. The increase in cytoplasmic Ins(1,4,5)P_3 stimulates the release of Ca^{2+} from intracellular stores and the formation of the Ca^{2+}-calmodulin complex. The rise in free cytosolic Ca^{2+} activates the small inwardly rectifying Ca^{2+}-gated K^+ channels and, moreover, activates phospholipase-A_2 and 5-lipoxygenase, converting arachidonic acid [released from phosphatidylcholine (PC) and phosphatidylethanolamine (PE)] to LTD_4. The DAG released from PtdIns(4,5)P_2 (PIP_2) hydrolysis also produces arachidonic acid which is converted to LTD_4. LTD_4 then activates both Cl^- and K^+ conductances as well as the taurine leak pathway. In several cell types the Cl^- conductance increases more than the K^+ conductance, resulting in a depolarization of the cell membrane which, as discussed above, may play a role in the activation of voltage-gated K^+ channels. Whether Ca^{2+}/calmodulin is involved in the decreased production of prostaglandins following the down-regulation of cyclooxygenase is unknown.

The end result of this sequence of events is a loss of KCl and taurine, and also an inhibition of the Na^+ uptake mechanism(s). Thus, one can state that the Ehrlich ascites cells during the RVD response have provided optimal conditions for net loss of cell osmolytes by the concomitant stimulation of osmolyte loss and inhibition of osmolyte gain.

4 Activation Mechanism for RVI

The actual mechanism of activation of the RVI response in osmotically shrunken cells is unknown. In frog skin epithelial cells Ussing (1982b) has proposed that reduced internal Cl^- may play a permissly role and modulate the response to cell shrinkage. A similar role may be played by low internal Cl^- in the activation of the RVI response in Ehrlich cells (Hoffmann et al. 1983) and lymphocytes (Grinstein et al. 1983). In squid giant axons, cell shrinkage at high external osmolarity has recently been reported to stimulate influx via the Na^+, K^+, Cl^- cotransporter by shifting the relationship between internal Cl^- concentration and influx via the cotransport system, hence reducing inhibition of the cotransporter by internal Cl^- (Breitweiser et al. 1990).

In avian red cells, cell shrinkage may influence the operation of the Na^+, K^+, $2Cl^-$ cotransport system by inducing a conformational change similar to that induced by cAMP (see Palfrey and Rao 1983; Siebens 1985). The activation of the cotransport system by catecholamines is mediated via cAMP and involves phosphorylation of a 230-kDa protein (Alper et al. 1980a). Cell shrinkage, however, does not increase cAMP levels in avian red cells (Kregenow et al. 1976), nor does it increase phosphorylation in these cells of the 230-kDa protein, which is phosphorylated during catecholamine-induced activation of the cotransport system (Alper et al. 1980b).

The cAMP level does not seem to play any role in the volume-sensitive cotransport in Ehrlich cells (Geck and Pfeiffer 1985). Theophylline and other alkylxanthines have been reported to inhibit the furosemide-sensitive cotransport in Ehrlich cells, presumably by reducing the critical cell volume above which the cotransport system becomes quiescent. The cotransport inhibition by these drugs is not mediated by cAMP or cGMP (Geck and Pfeiffer 1985).

The role of cAMP in the regulation of the Na^+, K^+, $2Cl^-$ cotransport system is ambiguous. Apart from avian red cells (see above) cAMP stimulates the cotransport system in cultured myocardial cells from chicken (Liu et al. 1989), whereas cAMP inhibits the cotransport system in vascular smooth muscle cells (Smith and Smith 1987), vascular endothelial cells (Brock et al. 1986), fibroblasts (Owen and Prastein 1985), human erythrocytes (Garey 1982) and ferret red cells (Mercer and Hoffman 1985). As discussed previously (see Sect. 2.2), there is evidence that cAMP may convert K^+-independent Na^+, Cl^- cotransport into K^+-dependent Na^+, K^+, $2Cl^-$ cotransport in mouse medullary thick ascending limb cells (Sun et al. 1991).

cGMP has also been implicated in the regulation of the Na^+, K^+, Cl^- cotransporter. The cotransporter is stimulated by cGMP in vascular smooth muscle cells (O'Donnell and Owen 1986), and is inhibited in endothelial cells (O'Donnell 1989) and in HeLa cells (Kort and Koch 1990).

Ca^{2+}/calmodulin and protein kinase C have also been assigned a role in the activation of the Na^+, K^+, $2Cl^-$ cotransporter. In accord with this the cotransporter can be activated by several agonists which stimulate phospholipase C, resulting in release of internal Ca^{2+} and activation of protein kinase C. This is illustrated in Table 8 in the case of bradykinin stimulation of Ehrlich cells. Similar results have been reported in several cell types after stimulation by bradykinin, thrombin, vasopressin and angiotensin II (see Brock et al. 1986; Paris and Pouysségur 1986; Smith and Smith 1987; O'Donnell 1989; Owen and Ridge 1989; Homma et al. 1990; Klein and O'Neill 1990). The stimulation by the above agonists could conceivably be mediated by an increase in cytosolic Ca^{2+} and/or activation of protein kinase C.

Paris and Pouysségur (1986) have suggested that the stimulation by α-thrombin of the Na^+, K^+, $2Cl^-$ cotransport system in hamster fibroblasts is mediated via proteinkinase C, since the cotransporter can also be activated by the phorbol ester TPA. TPA also stimulates the Na^+, K^+, $2Cl^-$ cotransport system in Ehrlich cells (see Table 8). Protein kinase C is also reported to activate the Na^+, K^+, $2Cl^-$ cotransporter in human fibroblasts (Panet and Atlan 1990), whereas the cotransporter is inhibited in vascular smooth muscle (Owen 1985). There is some evidence that protein kinase C is also involved in a negative feedback control of the cotransporter (Paris and Pouysségur 1986).

The stimulation of the Na^+, K^+, $2Cl^-$ cotransporter by thrombin and angiotensin II in rat glomerular mesangial cells has also been reported to involve both protein kinase C and Ca^{2+}/calmodulin (Homma et al. 1990). Ca^{2+} has been reported to be involved in the activation of the cotransporter

Table 8. Activation mechanisms for the Na^+, K^+, $2CL^-$ cotransport system in Ehrlich ascites tumour cells[a,b]

	Bumetanide-sensitive unidirectional K^+ influx[c] ($\mu mol \cdot g\ dry\ wt^{-1} \cdot min^{-1}$)	
	Physiological steady state	RVI
Control	0.5 ± 0.3 (3)	19 ± 3 (4)
Pimozide (120 μM)[d]	–	5 ± 2 (3)
Phorbol ester (TPA, 100 nM)[d]	4.5 ± 0.4 (3)	
Bradykinin (1 μM)[d,e]	17.1 ± 0.5 (3)	

[a] Data from Jensen and Hoffmann (1993). The values are given as mean ± SEM with the number of experiments in parentheses.
[b] The RVI response was initiated at zero time by dilution of the cell suspension with 1/3 volume hypertonic NaCl medium to give a final osmolarity at 450 mOsmal. In parallel groups at physiological steady state the cell suspension was diluted with 1/3 volume isotonic NaCl medium.
[c] The unidirectional K^+ influx was measured as the initial ^{86}Rb uptake, with ^{86}Rb being added with the diluting medium at time zero. The bumetanide-sensitive component was determined as the difference between the uptake in the absence and in the presence of bumetanide (30 μM).
[d] The drugs were added with the diluting medium to give the final concentration indicated.
[e] In the experiments with bradykinin, Ba^{2+} (2 mM) was present (and the medium sulphate-free) in order to block K^+ channels and inhibit bradykinin-induced cell shrinkage (cf. Fig. 11).

in vascular smooth muscle cells (Smith and Smith 1987), in endothelial cells (O'Donnell 1989) and in rabbit tracheal epithelial cells (Liedtke 1990). In vascular smooth muscle the stimulation of the Na^+, K^+, $2Cl^-$ cotransport system by angiotensin II has been reported to be mediated by an increase in free cytosolic Ca^{2+}, involving both Ca^{2+} entry via receptor-operated Ca^{2+} channels and Ca^{2+} release from internal stores (Owen and Ridge 1989).

The activation of the Na^+, K^+, $2Cl^-$ cotransporter during RVI is inhibited by pimozide in Ehrlich cells (see Table 8), suggesting the involvement of Ca^{2+}/calmodulin in the activation of the cotransporter in these cells. As discussed above, the stimulation of the cotransporter by TPA (see Table 8) indicates that protein kinase C may also be involved in the activation.

For discussion of the mechanisms of stimulation of the Na^+/H^+ exchange after cell shrinkage, see Rotin et al. (1991).

5 Induction of Organic Osmolyte Transport Systems

The immediate response (within seconds to minutes) to cell shrinkage is mainly accounted for by cellular electrolyte uptake (short-term RVI). On a longer time scale (hours to days) the volume maintenance mainly depends upon accumulation within the cells of non-perturbing, "compatible" organic osmolytes (long-term RVI). This osmolyte accumulation can result from increased synthesis and/or reduced degradation of cellular osmolytes, or from induction of increased membrane transport activity (see Garcia-Perez and Burg 1991).

In renal medulla the degradation of glycerophosphocholine (GPC) by GCP: choline phosphodiesterase is inhibited by hypertonicity (see Garcia-Perez and Burg 1991). Moreover, expression of the genes encoding for aldose reductase rises in response to extracellular hypertonicity, resulting in increased synthesis of sorbitol from glucose, and cellular osmolyte accumulation (Cowley et al. 1990). Expression of the genes encoding for the Na^+-dependent transporters for betaine and *myo*-inositol also rises within several hours in response to hypertonicity, resulting in cellular accumulation of these osmolytes (Burg 1991; Handler 1991). When the hypertonic cells were returned to isotonic medium, the mRNA levels for betaine and *myo*-inositol transporters returned to isotonic levels within 24 h (Handler 1991).

Cultured human fibroblasts shrink following exposure to hypertonic media, and then recover their initial volume (RVI) when the medium contains a substrate for the highly concentrative, Na^+-dependent amino acid transport system A. Substrates of the barely concentrative system L are unable to sustain the RVI response. The activity of system A is low in fibroblasts that are incubated in an isotonic amino acid-supplemented medium, but increases substantially upon incubation in hypertonic medium. It is suggested that de novo protein synthesis is involved in the activation, because no activation occurs in the presence of the protein-synthesis inhibitor cycloheximide. A slow time course for the activation also indicates involvement of the genome (Gazzola et al. 1991).

Similar regulatory mechanisms also play a role under isotonic conditions. In human fibroblasts replacement of the complete culture medium with an amino acid-free salt solution results in a loss of intracellular organic osmolytes accompanied by cell shrinkage. In this phase, transport system A enhances its capacity through the synthesis of new transporters. Reinstatement of the complete medium then causes a massive influx of amino acids through the activated system A. The steep rise in the concentration of these organic osmolytes leads to cellular swelling that is promptly counteracted by the release of intracellular solutes (inorganic ions, such as K^+ and Cl^-, and organic osmolytes). On a longer time scale, the activity of amino acid transport system A returns progressively to basal values.

For reviews on the volume-regulatory role of non-electrolytes, see Fugelli and Rohrs (1980), Gilles (1979, 1983, 1987), Goldstein and Kleinzeller (1987), Chamberlin and Strange (1989), Law and Burg (1991), Law (1985, 1991) and Garcia-Perez and Burg (1991).

6 Biological Role of Volume Regulation

Cell volume regulation conceivably plays an essential role in cell growth and proliferation which include a balanced increase in the cell content of organic osmolytes and ions, in cell volume, and in the cell membrane area. Moreover, cell volume regulatory processes appear to be an integrated part of the function of secreting and absorbing epithelia. Finally, it has been demonstrated that many hormones and growth factors also affect the ion transport mechanisms involved in volume regulation, and via the same "second messengers".

6.1 Secreting Epithelia

A simplified model for epithelial secretion in salivary and pancreatic acinar cells has recently been discussed by Hoffmann and Ussing (1992) and is shown in Fig. 26. In the model for NaCl secretion in the shark rectal gland (see Greger et al. 1988), the second messenger involved is cAMP, otherwise the same ion transport systems are operating. In the resting state, the apical membrane is virtually tight to all ions, but possesses dormant Cl^- channels (see Greger et al. 1988; Petersen and Gallacher 1988). The basal membrane in the gland cells possesses the Na^+, K^+ pump, a Na^+, K^+, $2Cl^-$ cotransport system (see Greger et al. 1988; Kinne 1988; Petersen) and K^+ channels which, however, are mostly closed in exocrine acinar cells (see Petersen and Gallacher 1988). The Na^+ taken up by the Na^+, K^+, $2Cl^-$ cotransporter is replaced by K^+ via the Na^+, K^+ pump. In the resting state, the cells will attain a steady state with a high concentration of chloride and a large cell volume.

Now, when the glands are stimulated and secretion is initiated (by agonist-induced activation of the inositol phosphate cycle and Ca^{2+} signalling), there is a simultaneous opening of the basal K^+ channels and the apical Cl^- channels (see Petersen and Gallacher 1988; Osipchuck et al. 1990). Thus, KCl is rapidly lost and the cells shrink. The increase in free cytoplasmic Ca^{2+} as well as the resulting cell shrinkage are demonstrated in Fig. 27. A similar cell shrinkage has been demonstrated in perfused glands (Nakahari et al. 1990). This cell shrinkage leads to activation of the Na^+, K^+, $2Cl^-$ cotransport system, but even though there is a rapid uptake of Na^+, K^+, and Cl^-, the cells stay shrunk as long as the stimulation is maintained (see Fig. 27B). A steady state is re-established, but with a lower cellular Cl^- and K^+ level, a slightly increased Na^+ level, and reduced cell volume. In the new steady state the net process is a secondary active chloride secretion at the expense of a strongly increased pumping rate of the Na^+, K^+ pump, and an increased metabolic rate. This secondarily active

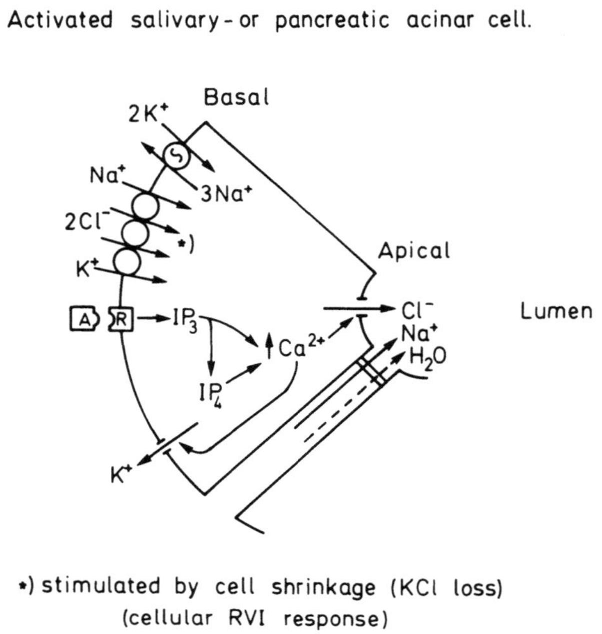

Fig. 26. Activation of a gland cell. When the agonist (*A*) reacts with the receptor (*R*), inositoltrisphosphate (IP_3) is released. This, in turn, releases Ca^{2+} from an intracellular pool, and Ca^{2+} then activates apical chloride and basolateral potassium channels. In the resting cell, the chloride and potassium concentrations are so high that the Na^+, K^+, $2Cl^-$ cotransporter of the basolateral membrane is virtually standing still (i.e. the forward and backward rates are about equal). The opening of the Cl^- and K^+ channels leads to rapid loss of KCl (and concomitant cell shrinkage), and the Na^+, K^+, $2Cl^-$ cotransporter is activated. The apical exit of Cl^- and the basolateral exit of K^+ creates an electric potential drop between blood and lumen which drives Na^+ (and water) through the cation-selective junction between the gland cells. The Na^+/K^+ pump maintains the ionic gradients, so that the secretion of Na^+ and Cl^- can continue as long as the stimulation lasts. The model thus accounts for the electrolyte secretion. However, the formation of an isotonic secretion requires recycling of Na^+ via apical channels. (Hoffmann and Ussing 1992)

chloride secretion in turn creates the potential difference between blood and lumen, which drives the counterions to Cl^- (mostly Na^+) along the paracellular route between the gland cells.

The model discussed above implies that the level at which the cell volume is maintained by volume regulatory processes in glandular cells depends upon the functional state of the cell, and that the transport systems involved in cell volume regulation are also involved in epithelial secretion. For a detailed discussion of the role of cell volume regulatory processes in absorbing epithelia and in isotonic secretion, see Hoffmann and Ussing (1992).

A similar sequence of events as discussed above for secretory epithelia can be demonstrated in other cell types, e.g. the Ehrlich ascites tumour cell.

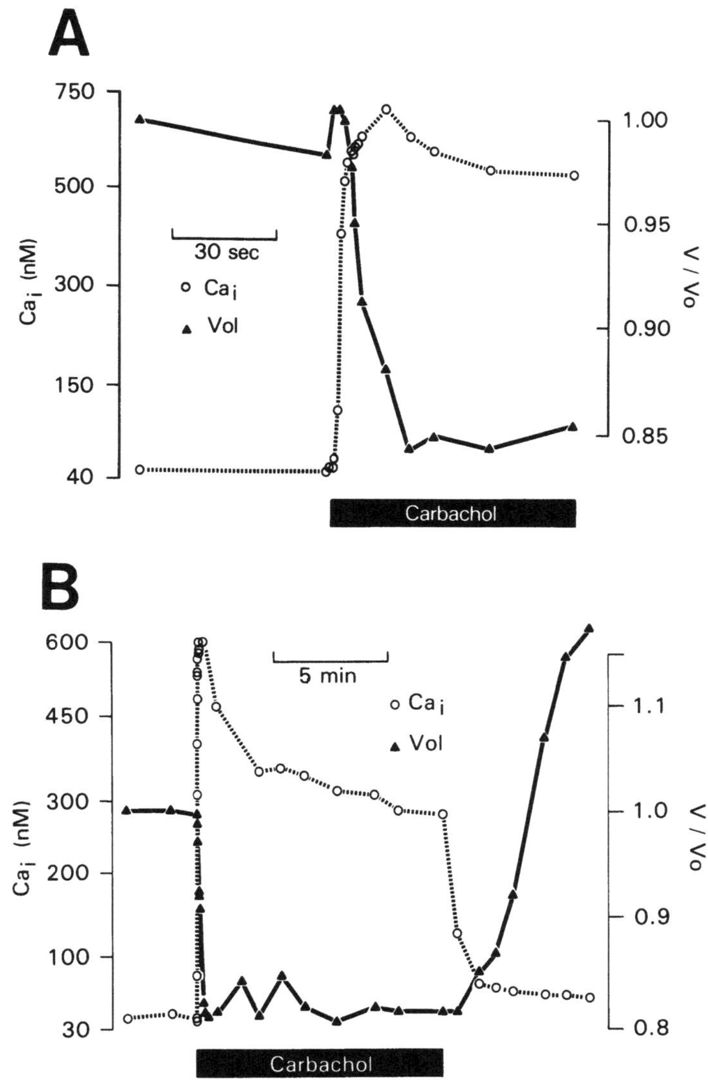

Fig. 27. Time course of cytosolic Ca^{2+} and cell volume (normalized) in a single rat parotid ascinar cell following stimulation by carbachol (10 μM). Recordings with high temporal resolution are shown in *A*, and the response to prolonged exposure to carbachol in *B*. $[Ca^{2+}]_i$ (o) was determined by recording fluorescence ratios from Fura-2 loaded cells by fluorescence microscopy. The cell volume (▲) was measured by differential interference contrast microscopy. (Slightly modified from Foskett and Melvin 1989)

The agonists thrombin, histamine and bradykinin induce inositol phosphate-mediated Ca^{2+} signalling and an initial KCl loss via K^+ channels and non-selective Cl^- channels (see Sect. 3.4 and Fig. 28). Bradykinin-induced inositol trisphosphate release, Ca^{2+} signalling and activation of K^+ channels have also recently been demonstrated in MDCK cells (Lang et al. 1991).

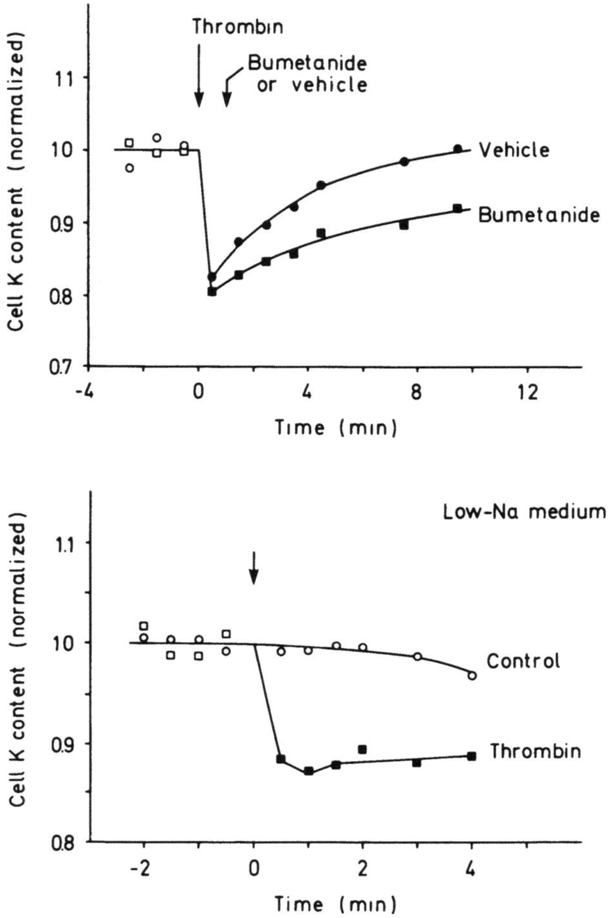

Fig. 28. Time course of the thrombin-induced net K^+ loss in Ehrlich ascites tumor cells showing inhibition of the re-uptake of KCl in the presence of bumetanide (*left*) and in low-Na^+ medium (*right*). *Left* Thrombin (1 unit/ml) was added at zero time as indicated by the *arrow* and the cell K^+ content followed with time in the presence or absence of bumetanide (25 μM). *Right* The cells were suspended in a low-Na^+ medium with the Na^+ concentration reduced to about 30 mM, with the Na^+ being partially replaced by N-methyl-D-glucammonium ($NMDG^+$). Thrombin (1 unit/ml) or vehicle was added at zero time (*arrow*) and the cell K^+ content followed with time. Results are from Harbak and Simonsen (1991)

In the Ehrlich cells the agonist-induced KCl loss is transient, and the subsequent KCl re-uptake is presumably mediated by activation of a Na^+-coupled, bumetanide-sensitive cotransport system, with Na^+ being finally replaced by K^+ via the Na^+, K^+ pump. This is illustrated in Fig. 28, which shows that the KCl re-uptake is inhibited by bumetanide (left panel) and in low Na^+ medium (right panel). The KCl re-uptake is also inhibited in nitrate media (see Fig. 16, right panel).

In conclusion, volume regulatory processes are an integrating part of a variety of cellular functions. Studying the molecular structure and function and the biochemical regulation of the passive membrane transport systems involved in cell volume regulation can thus provide useful knowledge with important implications.

References

Aabin B, Hoffmann EK (1986) Inhibition of the volume- and Ca^{2+}-activated Cl^- transport pathway in Ehrlich ascites tumor cells. Acta Physiol Scand 128: 42A

Aabin B, Kristensen BI (1987) Indentification of calmodulin in Ehrlich ascites tumor cells and its Ca^+-dependent binding to the cell membrane. Acta Physiol Scand 129: 18A

Adorante JS, Miller SS (1990) Potassium-dependent volume regulation in retinal pigment epithelium is mediated by Na, K, Cl cotransport. J Gen Physiol 96: 1153–1175

Alper SL, Beam KG, Greengard P (1980a) Hormonal control of Na, K-cotransport in turkey erythrocytes: multiple site phosphorylation of globin, a high molecular weight protein of the plasma membrane. J Biol Chem 225: 4864–4871

Alper SL, Palfrey HC, Deriemer SA, Greengard P (1980b) Hormonal control of protein phosphorylation in turkey erythrocytes: phosphorylation by cAMP-dependent and Ca^{2+}-dependent protein kinases of distinct sites in globin, a high molecular weight protein of the plasma membrane. J Biol Chem 255: 11029–11039

Arslan P, Virgilio FD, Beltrame M, Tsien RY, Pozzan T (1985) Cytosolic Ca^{2+} homeostasis in Ehrlich and Yoshida carcinomas. A new, membrane-permeant chelator of heavy metals reveals that these ascites tumor cell lines have normal cytosolic free Ca^{2+}. J Biol Chem 260: 2719–2727

Bakker-Grunwald T, Ogden P, Lamb JF (1982) Effects of ouabain and osmolarity on bumetanide-sensitive potassium transport in simian virus-transformed 3T3 cells. Biochim Biophys Acta 687: 333–336

Bankir L, de Rouffinac C (1985) Urinary concentrating ability: insights from comparative anatomy. Am J Physiol 249: R643–R666

Baquet A, Meijer AJ, Hue L (1991) Hepatocyte swelling increases inositol 1,4,5-trisphosphate, calcium and cyclic AMP concentration but antagonizes phosphorylase activation by Ca^{2+}-dependent hormones. FEBS 278(1): 103–106

Barish ME (1983) A transient calcium-dependent chloride current in the immature *Xenopus* oocyte. J Physiol (Lond) 342: 309–325

Berridge MJ (1984) Inositol trisphosphate and diacylglycerol as second messengers. Biochem J 220: 345–360

Berridge MJ (1987) Inositol trisphosphate and diacylglycerol: two interacting second messengers. Annu Rev Biochem 56: 159–193

Berridge MJ, Michell RH (eds) (1988) Inositol lipids and transmembrane signalling. Philos Trans R Soc Lond B 320: 235–436

Blatz AL, Magleby KL (1983) Single voltage-dependent chloride-selective channels of large conductance in cultured rat muscle. Biophys J 43: 237–241

Borgese F, Garcia-Romeu F, Motais R (1987) Control of cell volume and ion transport by β-adrenergic catecholamines in erythrocytes of rainbow trout, *Salmo gairdneri*. J Physiol (Lond) 382: 123–144

Bosma MM (1989) Anion channels with multiple conductance levels in a mouse B lymphocyte cell line. J Physiol 410: 67–90

Breitweiser GE, Altamirano AA, Russell JM (1990) Osmotic stimulation of Na^+-K^+-Cl^- cotransport in squid giant axon is $[Cl^-]_i$ dependent. Am J Physiol 258: C749–C753

Brock TA, Brugnara C, Canessa M, Gimbrone MA (1986) Bradykinin and vasopressin stimulate Na, K, Cl cotransport in cultured endothelial cells. Am J Physiol 250: C888–C895

Budowski P (1988) ω3-Fatty acids in health and disease. World Rev Nutr Diet 57: 214–274

Burg MB (1991) Renal medullary organic osmolytes. Fogarty International Center Conference on Cell volume regulation from bacteria to mammalian tissue and tumors. National Institutes of Health, Bethesda, Maryland

Cala PM (1980) Volume regulation by *Amphiuma* red blood cells. The membrane potential and its implications regarding the nature of the ion-flux pathways. J Gen Physiol 76: 683–708

Cala PM (1983) Volume regulation by red blood cells: mechanisms of ion transport. Mol Physiol 4: 33–52

Cala PM (1986) Volume-sensitive alkali metal-H transport in *Amphiuma* red blood cells. In: Aronson PS, Boron WF (eds) Na^+-H^+ exchange, intracellular pH, and cell function. Curr Top Membr Transp 26: 79–99

Cala PM, Mandel LJ, Murphy E (1986) Volume regulation by *Amphiuma* red blood cells: cytosolic free Ca and alkali metal-H exchange. Am J Physiol 250: C423–C429

Careaga-Houck M, Sprecher H (1990) Effects of a fish oil diet on the metabolism of endogenous (n-6) and (n-3) fatty acids in rat neutrophils. Biochim Biophys Acta 1947: 29–34

Chamberlin ME, Strange K (1989) Anisosmotic Cell volume regulation: a comparative view. Am J Physiol 257: C159–C173

Cheung RK, Grinstein S, dosch H-M, Gelfand EW (1982) Volume regulation by human lymphocytes: characterization of the ionic basis for regulatory volume decrease. J Cell Physiol 112:189–196

Chipperfield AR (1986) The (Na^+-K^+-Cl^-) co-transport system. Clin Sci 71: 465–476

Christensen O (1987) Mediation of cell volume regulation by Ca^{2+} influx through stretch-activated channels. Nature 330: 66–68

Christensen O, Hoffmann EK (1992) Cell swelling activates K^+- and Cl^--channels as well as nonselective stretch-activated cation channels in Ehrlich ascites tumor cells. J Membr Biol 129: 13–36

Christensen S, Hoffmann EK, Saermark T, Simonsen LO (1988) Inositol trisphosphate may be a second messenger in regulatory volume decrease in Ehrlich mouse ascites-tumour cells. J Physiol (Lond) 403: 109P

Christensen O, Simon M, Randlev T (1989) Anion channels in a leaky epithelium. A patch clamp study of choroid plexus. Pflügers Arch 415: 37–46

Christophersen P (1991) The Ca^{2+}-activated K^+ channel from human erythrocyte membranes: single channel rectification and selectivity. J Membr Biol 119: 75–83

Cittadini A, Dani AM, Wolf F, Bossi D, Calviello G (1982) Calcium permeability of Ehrlich ascites tumour cell plasma membrane in vivo. Biochim Biophys Acta 686: 27–35

Cornet M, Delpire E, Gilles R (1987) Study of microfilaments network during volume regulation process of cultured PC12 cells. Pflügers Arch 410: 223–225

Cornet M, Lambert IH, Hoffmann EK (1993) Relation between cytoskeleton, hypo-osmotic treatment and volume regulation in Ehrlich ascites tumor cells. Biol 131: 55–66, 1993

Cowley BD, Ferraris JD, Carper D, Burg MB (1990) In vivo osmotic regulation of aldose reductase mRNA, protein, and sorbitol content in rat renal medulla. Am J Physiol 258: F154–F161

Davis CW, Finn AL (1982) Sodium transport inhibition by amiloride reduces basolateral membrane potassium conductance in tight epithelia. Sciences (NY) 216: 525–527

Davis CW, Finn AL (1985) Cell volume regulation in frog urinary bladder. Fed Proc 44: 2520–2525

Delpire E, Gilles R, Duchéne C, Goessens G (1985) Effects of osmotic shocks on the ion content and ultrastructure of rat pheochromocytoma cells of line PC 12. Mol Physiol 8: 293–307

DiStefano A, Wittner M, Schlatter E, Lang HJ, Englert H, Greger R (1985) Diphenylamine-2-carboxylate, a blocker of the Cl^--conductive pathway in Cl^--transporting epithelia. Pflügers Arch 405 (Suppl 1): S95–S100

Doppler W, Maly K, Grunicke H (1986) Role of the Na^+/H^+ antiport in the regulation of the internal pH of Ehrlich ascites tumor cells in culture. J Membr Biol 91: 147–155
Drewnowska K, Baumgarten CM (1991) Regulation of cellular volume in rabbit ventricular myocytes: bumetanide, chlorothiazide, and ouabain. Am J Physiol 260: C122–C131
Duhm J, Göbel BO (1984) Na^+-K^+ transport and volume of rat erythrocytes under dietary K^+ deficiency. Am J Physiol 246: C20–C29
Dürr JE, Hviid Larsen E, (1986) Indacrinone (MK-196) – a specific inhibitor of the voltage-dependent Cl^- permeability in toad skin. Acta Physiol Scand 127: 145–153
Eveloff JL, Warnock DG (1987) Activation of ion transport systems during cell volume regulation. Am J Physiol 252: F1–F10
Falke LC, Misler S (1989) Activity of ion channels during volume regulation by clonal N1E115 neuroblastoma cells. Proc Natl Acad Sci USA 86: 3919–3923
Feinstein MB, Sha'afi RI (1983) Role of calcium in arachidonic acid metabolism and in the actions of arachidonic acid-derived metabolites. In: Chung WY (ed) Calcium and cell function, vol IV. Academic Press, New York, pp 337–376
Filipovic D, Sackin H (1991) A calcium-permeable stretch-activated cation channel in renal proximal tubule. Am J Physiol 260: F119–F129
Filipovic D, Sackin H (1992) Stretch and volume activated channels in isolated proximal tubule cells. Am J Physiol 262: F857–870
Fincham DA, Wolowyk MW, Young JD (1987) Volume-sensitive taurine transport in fish erythrocytes. J Membr Biol 96: 45–56
Foskett JK, Melvin JE (1989) Activation of salivary secretion: coupling of cell volume and $[Ca^{2+}]_i$ in single cells. Science 244: 1582–1585
Foskett JK, Spring KR (1985) Involvement of calcium and cytoskeleton in gallbladder epithelial cell volume regulation. Am J Physiol 248: C27–C36
Freeman CJ, Bookchin RM, Ortiz OE, Lew VL (1987) K-permeabilized human red cells lose an alkaline, hypertonic fluid containing excess K^+ over diffusible anions. J Membr Biol 96: 235–241
Frelin C, Chassande O, Lazdunski M (1986) Biochemical characterization of the $Na^+/K^+/Cl^-$ cotransport in chick cardiac cells. Biochem Biophys Res Commun 134: 326–331
Friedrich F, Paulmilch M, Kolb HA, Lang F (1988) Inward rectifier K channel in renal epithelial cells (MDCK) activated by serotonin. J Membr Biol 106: 149–155
Frizzell RA, Field M, Schultz SG (1979) Sodium-coupled chloride transport by epithelial tissues. Am J Physiol 236: F1–F8
Frizzell RA, Rechkemmer G, Shoemaker RL (1986) Altered regulation of airway epithelial cell chloride channels in cystic fibrosis. Sciences (NY) 233: 558–560
Fugelli K, Rohrs H (1980) The effect of Na^+ and osmolality on the influx and steady state distribution of taurine and gamma-aminobutyric acid in flounder (*Platichthys flesus*) erythrocytes. Comp Biochem Physiol 67A: 545–551
Fugelli K, Thoroed SM (1986) Taurine transport associated with cell volume regulation in flounder erythrocytes under anisosmotic conditions. J Physiol (Lond) 374: 245–261
Garay RP (1982) Inhibition of Na^+/K^+ cotransport system by cyclic AMP and intracellular Ca^{2+} in human red cells. Biochim Biophys Acta 688:786–792
Garcia-Perez A, Burg MB (1991) Renal medullary organic osmolytes. Physiol Rev 71: 1081–1115
Gargus JJ, Slayman CW (1980) Mechanism and role of furosemide-sensitive K^+ transport in L cells: a genetic approach. J Membr Biol 52: 245–256
Gazzola GC, Dall'Asta V, Nucci FA, Rossi PA, Bussolati O, Hoffmann EK, Guidotti GG (1991) Role of amino acid transport system A in the control of cell volume in cultured human fibroblasts. Cell Physiol Biochem 1: 131–142
Geck P (1990) Volume regulation in Ehrlich cells. In: Beyenbach KW (ed) Cell volume regulation. Karger, Basel pp 26–59
Geck P, Pfeiffer B (1985) $Na^+ + K^+ + 2Cl^-$ cotransport in animal cells – its role in volume regulation. In: Semenza G, Kinne R (eds) Membrane transport driven by ion gradients. Ann NY Acad Sci 456: 166–182

Geck P, Pietrzyk C, Burckhardt B-C, Pfeiffer B, Heinz E (1980) Electrically silent cotransport of Na^+, K^+ and Cl^- in Ehrlich cells. Biochim Biophys Acta 600: 432–447
Gerencser GA, Lee S-H (1983) Cl^--stimulated adenosine triphosphatase: existence, location and function. In: Treherhe JE (ed) Epithelial and cellular mechanisms in osmoregulation. J Exp Biol 106: 143–161
Gilles R (1979) Intracellular organic osmotic effectors. In: Gilles R (ed) Mechanisms of osmoregulation in animal maintenance of cell volume. Wiley, New York, pp 111–154
Gilles R (1983) Volume maintenance and regulation in animal cells: some features and trends. Mol Physiol 4: 3–16
Gilles R (1987) Volume regulation in cells of eurohaline invertebrates. In: Gilles R, Kleinzeller A, Bolis L (eds) Cell volume control: fundamental and comparative aspects in animal cells. Curr Top Membr Transp 30: 205–247
Gilles R, Hoffmann EK, Bolis L (1991) Volume and osmolality control in animal cells. Advances in comparative and environmental physiology 9. Springer Berlin Heidelberg New York
Goldstein L, Kleinzeller A (1987) Cell volume regulation in lower vertebrates. In: Gilles R, Kleinzeller A, Bolis L (eds) Cell volume control: fundamental and comparative aspects in animal cells vol 30. Academic Press, Orlando, pp 181–204
Greger R (1985) Ion transport mechanisms in the thick ascending limb of Henle's loop of mammalian nephron. Physiol Rev 65: 760–797
Greger R, Gögelein H, Schlatter E (1988) Stimulation of NaCl secretion in the rectal gland of the dogfish *Squalus acanthias*. Comp Biochem Physiol 90A: 733–739
Grinstein S, Smith JD (1989) Calcium induces charybdotoxin-sensitive membrane potential changes in rat lymphocytes. Am J Physiol 22: C197–C206
Grinstein S, Smith JD (1990) Calcium-independent cell volume regulation in human lymphocytes. Inhibition by charybdotoxin. J Gen Physiol 95: 97–120
Grinstein S, Clarke CA, DuPre A, Rothstein A (1982a) Volume-induced increase of anion permeability in human lymphocytes. J Gen Physiol 80: 801–823
Grinstein S, Clarke CA, Rothstein A (1982b) Increased anion permeability during volume regulation in human lymphocytes. Philos Trans R Soc Lond B 299: 509–518
Grinstein S, DuPre A, Rothstein A (1982c) Volume regulation by human lymphocytes. Role of calcium. J Gen Physiol 79: 849–868
Grinstein S, Cohen S, Sarkadi B, Rothstein A (1983) Induction of ^{86}Rb fluxes by Ca^{2+} and volume changes in thymocytes and their isolated membranes. J Cell Physiol 116: 352–362
Grinstein S, Rothstein A, Sarkadi B, Gelfand EW (1984) Responses of lymphocytes to anisotonic media: volume-regulating behavior. Am J Physiol 246: C204–C215
Grygorczyk R, Schwarz W, Passow H (1984) Ca^{2+}-activated K^+ channels in human red cells. Comparison of single-channel currents with ion fluxes. Biophys J 45: 693–698
Guharay F, Sachs F (1984) Stretch-activated single ion channel currents in tissue-cultured embryonic chick skeletal muscle. J Physiol (Lond) 352: 685–701
Haas M (1989) Properties and diversity of (Na-K-Cl) cotransporters. Annu Rev Physiol 51: 443–457
Hamill OP (1983) Potassium and chloride channels in red blood cells. In: Sakmann B, Neher E (eds) single-channel recording. Plenum Press, New York, pp 451–471
Handler JS (1991) Cell and molecular biology of sodium-coupled transporters of organic osmolytes. Fogarty International Center Conference on Cell volume regulation from bacteria to mammalian tissue and tumors. National Institutes of Health, Bethesda, Maryland
Harbak H, Simonsen LO (1991) Thrombin and histamin induce Ca^{2+}-signalling and KCl loss in Ehrlich mouse ascites tumor cells. Acta Physiol Scand 143: 40A
Hazama A, Okada Y (1988) Ca^{2+} sensitivity of volume-regulatory K^+ and Cl^- channels in cultured human epithelial cells. J Physiol (Lond) 402: 687–702
Hazama A, Okada Y (1990) Biphasic rises in cytosolic free Ca^{2+} in association with activation of K^+ and Cl^- conductance during the regulatory volume decrease in cultured human epithelial cells. Pflügers Arch 416: 710–714

Henderson LM, Chappel JB, Jones OTG (1989) Superoxide generation is inhibited by phospholipase A2 inhibitors. Role for phospholipase A2 in the activation of the NADPH oxidase. Biochem J 264: 249–255

Hendil DB, Hoffmann EK (1974) Cell volume regulation in Ehrlich ascites tumor cells. J Cell Physiol 84: 115–125

Hladky SB, Rink TJ (1977) pH equilibrium across the red cell membrane. In: Ellory JC, Lev VL (eds) Membrane transport in red cells. Academic Press London, pp 115–135

Hoffmann EK (1978) Regulation of cell volume by selective changes in the leak permeabilities of Ehrlich ascites tumor cells. In: Jørgensen CB, Skadhauge E (eds) Alfred Benzon Symposium XI. Osmotic and volume regulation. Munksgaard, Copenhagen, pp 397–417

Hoffmann EK (1983) Volume regulation by animal cells. In: Cossins AR, Shetterline P (eds) Cellular acclimatisation to environmental change. Soc Exp Biol Semin Ser 17 Cambridge University Press, Cambridge, pp 55–80

Hoffmann EK (1985) Role of separate K^+ and Cl^- channels and of Na^+/Cl^- cotransport in volume regulation in Ehrlich cells. Fed Proc 44: 2513–2519

Hoffmann EK, Hendil KB (1976) The role of amino acids and taurine in isosmotic intracellular regulation in Ehrlich ascites mouse tumour cells. J Comp Physiol 108: 279–286

Hoffmann EK, Kolb A (1991) The mechanisms of activation of regulatory volume responses after cell swelling. In: Gilles R, Hoffmann EK, Bolis L (eds) Volume and osmolality control in animal cells. Advances in comparative physiology 9. Springer, Berlin Heidelberg New York, pp 140–185

Hoffmann EK, Lambert IH (1983) Amino acid transport and cell volume regulation in Ehrlich ascites tumour cells. J Physiol (Lond) 338: 613–625

Hoffmann EK, Simonsen LO (1989) Membrane mechanisms in volume and pH regulation in vertebrate cells. Physiol Rev 69: 315–382

Hoffmann EK, Ussing HH (1992) Membrane mechanisms in volume regulation in vertebrate cells and epithelia. In: Giebisch GH, Schafer JA, Ussing HH, Kristensen P (eds) Membrane transport in biology, vol 5. Springer Berlin Heidelberg New York 317–399

Hoffmann EK, Sjøholm C, Simonsen LO (1983) Na^+, Cl^- co-transport in Ehrlich ascites tumor cells activated during volume regulation (regulatory volume increase). J Membr Biol 76: 269–280

Hoffmann EK, Simonsen LO, Lambert IH (1984) Volume-induced increase of K^+ and Cl^- permeabilities in Ehrlich ascites tumor cells. Role of internal Ca^{2+}. J Membr Biol 78: 211–222

Hoffmann EK, Lambert IH, Simonsen LO (1986) Separate, Ca^{2+}-activated K^+ and Cl^- transport pathways in Ehrlich ascites tumor cells. J Membr Biol 91: 227–244

Hoffmann EK, Lambert IH, Simonsen LO (1988) Mechanisms in volume regulation in Ehrlich ascites tumor cells. Renal Physiol Biochem 11: 221–247

Homma T, Burns KD, Harris RC (1990) Agonist stimulation of $Na^+/K^+/Cl^-$ cotransport in rat glumerular mesangial cells. Evidence for protein kinase C-dependent and Ca^{2+}/calmodulin-dependent pathways. J Biol Chem 265: 17613–17620

Hudson RL, Schultz SG (1988) Sodium-coupled glycine uptake by Ehrlich ascites tumor cells results in an increase in cell volume and plasma membrane channel activities. Proc Natl Acad Sci USA 85: 279–283

Jacobs MH, Stewart DR (1942) The role of carbonic anhydrase in certain ionic exchanges involving the erythrocyte. J Gen Physiol 25: 539–552

Jensen BS, Hoffmann EK (1990) On the K-dependence of the cation-anion cotransport system in Ehrlich ascites tumor cells. Acta Physiol Scand 140: 34A

Jensen BS, Hoffmann EK (1991) Na^+, Cl^- or Na^+, K^+, $2Cl^-$ cotransport during regulatory volume increase in Ehrlich cells. Acta Physiol Scand 143: 32A

Jensen BS, Jessen F, Hoffmann EK (1993) Na^+, K^+, Cl^- cotransport and its regulation in Ehrlich ascites tumor cells. Ca^{2+}/calmodium and protein kinase (dependent pathways. J Memb Biol 131 (submitted)

Jessen F, Sjøholm C, Hoffmann EK (1986) Identification of the anion exchange protein of Ehrlich cells: a kinetic analysis of the inhibitory effects of 4,4'-diisothiocyano-2,2'-stilbene-disulfonic acid (DIDS) and labelling of membrane proteins with ^{3}H-DIDS. J Membr Biol 92: 195–205

Jessen F, Hoffmann EK, Aabin B (1989a) Functional reconstitution of an isolated K^{+} channel from Ehrlich ascites tumor cells. Acta Physiol Scand 136: 10A

Jessen F, Cherksey BD, Zeuthen T, Hoffmann EK (1989b) Isolation and reconstitution of furosemide-binding proteins from Ehrlich ascites tumor cells. J Membr Biol 108: 139–151

Kimelberg HK (1991) Swelling and volume control in brain astroglial cells. In: Gilles R, Hoffmann EK, Bolis L (eds) Volume and osmolality control in animal cells. Advances in comparative and environmental physiology 9. Springer, Berlin Heidelberg New York, pp 81–117

Kimelberg HK, Frangakis MV (1986) Volume regulation in primary astrocyte cultures. Adv Biosci 61: 177–186

Kimelberg HK, O'Connor ER (1988) Swelling-induced depolarization of astrocyte potentials. Glia 1: 219–224

King PA, Goldstein L (1983) Organic osmolytes and cell volume regulation in fish. Mol Physiol 4: 53–66

Kinne R (1988) Sodium co-transport systems in epithelial secretion. Comp Biochem Physiol 90A: 721–727

Kirk KL, DiBona DR, Schafer JA (1987a) Regulatory volume decrease in perfused proximal nephron: evidence for a dumping of cell K^{+}. Am J Physiol 252: F933–F942

Kirk KL, Schafer JA, DiBona DR (1987b) Cell volume regulation in rabbit proximal straight tubule perfused in vitro. Am J Physiol 252: F922–932

Kirstein D, Thomsen MK, Ahnfelt-Rønne I (1991) Inhibition of leukotriene biosynthesis and polymorphonuclear leukocyte functions by orally active quinolylmethoxyphenyl-amines. Pharmacol Toxicol 68: 125–130

Klaven NB, Pershadsingh HA, Henius GV, Laris PC, Long JW Jr, McDonald JM (1983) A high-affinity, calmodulin-sensitive (Ca^{2+} + Mg^{2+})-ATPase and associated calcium-transport pump in the Ehrlich ascites tumor cell plasma membrane. Arch Biochem Biophys 226: 618–628

Klein JD, O'Neill WC (1990) Effect of bradykinin on Na-K-2Cl cotransport and bumetanide binding in aortic endothelial cells. J Biol Chem 265: 22238–22242

Knapp, HR, Salem N Jr (1989) Formation of PGI3 in the rat during dietary fish oil supplementation. Prostaglandins 38(5): 509–521

Knepper MA (1982) Measurement of osmolality in kidney slices using vapor pressure osmometry. Kidney Int 21: 653–655

Kolb HA, Ubl J (1987) Activation of anion channels by zymosan particles in membranes of peritoneal macrophages. Biochim Biophys Acta 899: 239–246

Kort JJ, Koch G (1990) The Na^{+}, K^{+}, $2Cl^{-}$-cotransport system in HeLa cells: aspect of its physiolopgical regulation. J Cell Physiol 145: 253–261

Kramhøft B, Hoffmann EK, Feit PW (1984) The effect of 'loop-diuretics' on a Na,Cl cotransport system activated during regulatory volume increase in Ehrlich ascites tumor cells. Acta Physiol Scand 121: P2

Kramhøft B, Lambert IH, Hoffmann EK, Jørgensen F (1986) Activation of Cl-dependent K transport in Ehrlich ascites tumor cells. Am J Physiol 251: C369–C379

Kramhøft B, Lambert IH, Hoffmann EK (1988) Na^{+}/H^{+} exchange in Ehrlich ascites tumor cells. Activation by cytoplasmatic acidification and by treatment with cupric sulphate. J Membr Biol 102: 35–48

Kramhøft B, Lambert IH, Pedersen R (1989) Mercuric chloride activates latent, anion-dependent cation transport systems in the plasma membrane of Ehrlich ascites tumour cells. Pharmacol Toxicol 64: 421–425

Kregenow FM (1973) The response of duck erythrocytes to norepinephrine and an elevated extracellular potassium: volume regulation in isotonic media. J Gen Physiol 61: 509–527

Kregenow FM (1981) Osmoregulatory salt transporting mechanisms: control of cell volume in anisotonic media. Annu Rev Physiol 43: 493–505
Kregenow FM, Robbie DE, Orloff J (1976) Effect of norepinephrine and hypotertonicity on K influx and cyclic AMP in duck erythrocytes. Am J Physiol 231: 306–312
Kristensen LØ (1986) Associations between transports of alanine and cations across cell membrane in rat hepatocytes. Am J Physiol 251: G575–G584
Kristensen LØ, Folke M (1984) Volume-regulatory K^+ efflux during concentrative uptake of alanine in isolated rat hepatocytes. Biochem J 221: 265–268
Kristensen P, Ussing HH (1985) Epithelial organization. In: Seldin DW, Giebisch G (eds) The kidney: physiology and pathophysiology. Raven Press, New York, pp 173–188
Lambert IH (1984) Na^+-dependent taurine uptake in Ehrlich ascites tumour cells. Mol Physiol 6: 233–246
Lambert IH (1985) Taurine transport in Ehrlich ascites tumour cells. Specificity and chloride dependence. Mol Physiol 7: 323–332
Lambert IH (1987a) Effect of arachidonic acid, fatty acids, prostaglandins, and leukotrienes on volume regulation in Ehrlich ascites tumor cells. J Membr Biol 98: 207–221
Lambert IH (1987b) Activation of the volume- or ionophore A23187 plus Ca^{2+}-induced Cl^- transport pathway is Ehrlich mouse ascites tumour cells is mediated via leucotriene-D4. J Physiol (Lond). 394: 85P
Lambert IH (1989) Leukotriene-D4 induced cell shrinkage in Ehrlich ascites tumor cells. J Membr Biol 108: 165–176
Lambert IH, Hoffmann EK (1990) Taurine transport and cell volume regulation in a mammalian cell. In: Posantes-Morales H, Martin DL, Shain W, Martin del Rio R, (eds) Taurine: functional neurochemistry, physiology and cardiology. Wiley-Liss, New York, pp 267–276
Lambert IH, Hoffmann EK (1991) The role of phospholipase-A2 and 5-lipoxygenase in the activation of K and Cl channels and the taurine leak pathway in Ehrlich ascites tumor cells. Acta Physiol Scand 143: 33A
Lambert IH, Hoffmann EK (1993) Regulation of taurine transport in Ehrlich ascites tumor cells. J Membr Biol 131: 67–79
Lambert IH, Simonsen LO, Hoffmann EK (1984) Volume regulation in Ehrlich ascites tumor cells: pH sensitivity of the regulatory volume decrease, and role of the Ca^{2+}-dependent K^+ channel. Acta Physiol Scand 120: 46A
Lambert IH, Hoffmann EK, Christensen P (1987) Role of prostaglandins and leukotrienes in volume regulation by Ehrlich ascites tumor cells. J Membr Biol 98: 247–256
Lambert IH, Hoffmann EK, Jørgensen F (1989) Membrane potential, anion and cation conductances in Ehrlich ascites tumor cells. J Membr Biol 111: 113–132
Lang F, Paulmichl M, Voelkl H, Gstrein E, Friedrich F (1987) Electrophysiology of cell volume regulation. In: Kovacevic Z, Guder Wa (eds) Molecular nephrology. Biochemical aspects of kidney function. de Gruyter, Berlin, pp 133–139
Lang F, Völke H, Häussinger D (1990) General principles in cell volume regulation. In: Beyenbach KW (ed) Cell volume regulation. Karger, Basel, pp 1–26
Lang F, Paulmichl M, Pfeilschifter J, Friedrich F, Wöll E, Waldegger S, Ritter M, Tschernko E (1991) Cellular mechanisms of bradykinin-induced hyperpolarization in renal epitheloid MDCK-cells. Biochim Biophys Acta 1073: 600–608
Lansman JB, Hallam TJ, Rink TJ (1987) Single stretch-activated ion channels in vascular endothelial cells as mechanotransducers? Nature 325: 811–813
Larson M, Spring KR (1987) Volume regulation in epithelia. In: Gilles R, Kleinceller A, Bolis L (eds) Cell volume control: fundamental and comparative aspects in animal cells. Curr Top Membr Transp 30: 105–123
Lau KR, Hudson RL, Schultz SG (1984) Cell swelling increases a barium-inhibitable potassium conductance in the basolateral membrane of *Necturus* small intestine. Proc Natl Acad Sci USA 81: 3591–3594
Lauf PK (1985a) On the relationship between volume- and thiol-stimulated K^+ Cl^- fluxes in red cell membranes. Mol Physiol 8: 215–234

Lauf PK (1985b) K^+: Cl^- cotransport: Sulphydryls, divalent cations, and the mechanism of volume activation in a red cell. J Membr Biol 88: 1–13
Lauf PK, McManus TJ, Haas M, Forbush B III, Duhm J, Flatman PW, Saier MH, Russel JM (1987) Physiology and biophysics of chloride and cation cotransport across cell membranes. Fed Proc 46: 2377–2394
Lauritzen L, Hoffmann EK, Hansen HS, Jensen B (1993) Dietary N-3 and NV-6 fatty acids are equipotent in stimulating volume regulation in Ehrlich ascites tumor cells. Am J Physiol 264: C109–C117
Law RO (1985) Volume regulation by mammalian renal cells exposed to anisosmotic media. Mol Physiol 8: 143–160
Law RO (1991) Amino acids as volume-regulatory osmolytes in mammalian cells. Comp Biochem Physiol 99A: 263–277
Law RO, Burg MB (1991) The role of organic osmolytes in the regulation of mammalian cell volume. In: Gilles R, Hoffmann EK, Bolis L (eds) Volume and osmolality control in animal cells. Advances in comparative and environmental physiology 9. Springer, Berlin Heidelberg New York, pp 189–226
Leite MV, Goldstein L (1987) Ca^{2+} ionophore and phorbol ester stimulate taurine efflux from skate erythrocytes. J Exp Zool 242: 95–97
Levinson C (1987) Volume regulatory activity of the Ehrlich ascites tumor cell and its relationship to ion transport. J Membr Biol 100: 183–191
Levinson C (1990) Regulatory volume increase in Ehrlich ascites tumor cells. Biochem Biophys Acta 1021: 1–8
Levinson C (1991) Regulatory volume increase in Ehrlich ascites tumor cells is mediated by the 1Na: 1K: 2Cl cotransport system. Faseb J 4: A564
Lew VL, Bookchin RM (1986) Volume, pH, and ion-content regulation in human red cells: analysis of transient behavior with an integrated model. J Membr Biol 92: 57–74
Lewis SA, de Moura JLC (1982) Incorporation of cytoplasmic vesicles into apical membrane of mammalian urinary bladder epithelium. Nature 297: 685–688
Lewis SA, Donaldson P (1990) Ion channels and cell volume regulation: chaos in an organized system. News Physiol Sci 5: 112–119
Li M, McCann JD, Anderson MP, Clancy JP, Liedtke CM, Nairn AC, Greengard P, Welsh MJ (1989) Regulation of chloride channel by protein kinase C in normal and cystic fibrosis airway epithelia. Science 244: 1353–1356
Liedtke CM (1990) Calcium and α-adrenergic regulation of Na-Cl(K) cotransport in rabbit tracheal epithelial cells. Am J Physiol 259: L66–L72
Liu S, Jacob R, Piwnica-Worms D, Liberman M (1989) Interaction of Na, K, 2Cl contransport and the Na/K pump in cultured chick cardial myocytes. Mol Cell Biochem 89: 147–150
Livne A, Hoffmann EK (1990) Cytoplasmic acidification and activation of Na^+/H^+ exchange during regulatory volume decrease in Ehrlich ascites tumour cells. J Membr Biol 114: 153–157
Livne A, Grinstein S, Rothstein A (1987) Volume-regulating behavior of human platelets. J Cell Physiol 131: 354–363
Lohr JW, Grantham JJ (1986) Isovolumetric regulation of isolated S2 proximal tubules in anisotonic media. J Clin Invest 78: 1165–1172
Macknight ADC (1983) Volume regulation in mammalian kidney cells. Mol Physiol 4: 17–31
Macknight ADC, Leaf A (1985) Cellular responses to extracellular osmolality. In: Seldin DW, Giebisch G (eds) The kidney: physiology and pathophysiology. Raven Press, New York, pp 117–132
MacLeod RJ (1991) Regulatory differences in villus and crypt epithelial cells. Fogarty International Center Conference on Cell volume regulation from bacteria to mammalian tissues and tumors. National Institute of Health, Bethesda, Maryland
MacLeod RJ, Hamilton JR (1990) Regulatory volume increase in isolated mammalian jejunal villus is due to bumetanid sensitive NaKCl cotransport. Am J Physiol 258: G665–G674

MacLeod RJ, Hamilton JR (1991) Separate K and Cl transport pathways are activated for regulatory volume decrease in jejunal villus cells. Am J Physiol 260: G405–G415
MacRobbie EAC, Ussing HH (1961) Osmotic behaviour of the epithelial cells of frog skin. Acta Physiol Scand 53: 348–365
Marty A, Tan YP, Trautmann A (1984) Three types of calcium-dependent channels in rat lacrimal glands. J Physiol (Lond) 257: 293–325
NcCarty NA, O'Neil RG (1990) Dihydropyridine-sensitive cell volume regulation in proximal tubule: the calcium window. Am J Physiol 259 (Renal Fluid Electrolyte Physiol 28): F950–F960
McConnell FM, Goldstein L (1988) Intracellular signals and volume regulatory response in skate erythrocytes. Am J Physiol 255 (Regulatory Integrative Comp Physiol 24): R982–R987
Mercer RW, Hoffmann JF (1985) Bumetanide-sensitive Na/K cotransport in ferret red blood cells. Biophys J 47: 157a
Mills JW (1987) The cell cytoskeleton: possible role in volume control. Curr Top Membr Transp 30: 75–101
Mills JW, Lubin M (1986) Effect of adenosine 3',5'-cyclic monophosphate on volume and cytoskeleton of MDCK cells. Am J Physiol 250: C319–C324
Mills JW, Skiest DJ (1985) Role of cyclic AMP and the cytoskeleton in volume control in MDCK cells. Mol Physiol 8: 247–262
Montrose-Rafizadeh C, Guggino WB (1991) Role of intracellular calcium in volume regulation by rabbit medullary thick asccending limb cells. Am J Physiol 260 (Renal Fluid Electrolyte Physiol 29) F402–409
Morris CE (1990) Mechanosensitive ion channels. J Membr Biol 113: 93–107
Murphy RC, Pickett WC, Culp BR, Lands WEM (1981) Tetraene and pentaene leukotrienes: selective production from murine mastocytoma cells after dietary manipulation. Prostaglandins 22: 612–622
Musch MW, Field M (1989) K-independent Na-Cl cotransport in bovine tracheal epithelial cells. Am J Physiol 256: C658–C665
Musch MW, Goldstein L (1990) Hypotonicity stimulates phosphatidylcholine hydrolysis and generates diacylglycerol in erythrocytes. J Biol Chem 265: 13055–13059
Nakahari T, Murakami M, Yoshida H, Miyamoto M, Sohma Y, Imai Y (1990) Decrease in rat submandibular acinar cell volume during ACh stimulation. Am J Physiol 258: G878–G886
Nauntofte B, Poulsen JH (1984) Chloride transport in rat parotid acini: furosemide-sensitive uptake and calcium-dependent release. J Physiol (Lond) 357: 61P
Nauntofte B, Poulsen JH (1986) Effects of Ca^{2+} and furosemide on Cl^- transport and O_2 uptake in rat parotid acini. Am J Physiol 251: C175–C185
Needleman GJ, Raz A, Minkes MS, Ferrendelli JA, Sprecher H (1979) Triene prostaglandins: prostacyclin and thromboxane biosynthesis and unique biological properties. Proc Natl Acad Sci USA 76(2): 944–948
Needleman GJ, Turk PJ, Bakschik BA, Morrison AR, Lefkowith JB (1986) Arachidonic acid metabolism. Annu Rev Biochem 55: 69–102
Nellans HN, Frizzell RA, Schultz SG (1973) Coupled sodium-cloride influx across the brush border of rabbit ileum. Am J Physiol 225: 467–475
O'Donnell ME (1989) Regulation of Na-K-Cl cotransport in endothelial cells by atrial natriuretic factor. Am J Physiol 257: C36–C44
O'Donnell ME, Owen NE (1986) Role of cyclic GMP in atrial natriuretic factor stimulation of Na^+,K^+,Cl^- cotransport in vascular smooth cells. J Biol Chem 261: 15461–15466
O'Grady SM, Palfrey HC, Field M (1987) Characteristics and functions of Na-K-Cl cotransport in epithelial tissues. Am J Physiol 253: C177–C192
Okada Y, Hazama A (1989) Volume-regulatory ion channels in epithelial cells. NIPS 4: 238–242
Okada Y, Yada T, Ohno-Shosaku T, Oiki S (1987) Evidence for the involvement of calmodulin in the operation of Ca-activated K channels in mouse fibroblasts. J Membr Biol 96: 121–128

Okada Y, Hazama A, Yuan W (1990) Stretch-induced activation of Ca^{2+}-permeable ion channels is involved in the volume regulation of hypotonically swollen epithelial cells. Neurosci Res 12: S5–S13
Olea RS, Pasantes-Morales H, Lázaro A, Cereijido M (1991) Osmolarity-sensitive release of free amino acids from cultured kidney cell (MDCK). J Membr Biol 121: 1–9
Osipchuck YV, Wakui M, Yule DI, Gallacher DV, Petersen OH (1990) Cytoplasmic Ca^{2+} oscillations evoked by receptor stimulation, G-protein activation, internal application of inositol trisphosphate or Ca^{2+}: simultaneous microfluorimetry and Ca^{2+} dependent Cl^- current recoriding in single pancreatic acinar cells. EMBO 9(3): 697–704
Owen NE (1985) Effect or TPA on ion fluxes and DNA synthesis in vascular smooth muscle cells. J Cell Biol 101: 454–459
Owen NE, Prastein ML (1985) Na/K/Cl cotransport in cultured human fibroblasts. J Biol Chem 260: 1445–1451
Owen NE, Ridge KM (1989) Mechanism of angiotensin II stimulation of Na-KCl cotransport of vascular smooth muscle cells. Am J Physiol 257: C629–C636
Owen DG, Segal M, Barker JL (1984) A Ca-dependent Cl-conductance in cultured mouse spinal neurones. Nature 311: 567–570
Palfrey HC, Rao MC (1983) Na/K/Cl cotransport and its regulation. J Exp Biol 106: 43–54
Panet R, Atlan H (1990) Bumetanide-sensitive $Na^+/K^+/Cl^-$-transporter is stimulated by phorbol ester and different mitogens in quiscent human skin fibroblasts. J Cell Physiol 145: 30–38
Parent L, Cardinal J, Sauvé R (1988) Single-channel analysis of a K channel at basolateral membrane of rabbit proximal convoluted tubule. Am J Physiol 254: F1–F9
Paris S, Pouysségur J (1986) Growth factors activate the bumetanide-sensitive $Na^+/K^+/Cl^-$ cotransport in hamster fibroblasts. J Biol Chem 261(14): 6177–6183
Pasantes-Morales H, Schousboe A (1988) Volume regulation in astrocytes: a role for taurine as osmoeffector. J Neurosci Res 20: 505–509
Petersen OH, Gallacher DV (1988) Electrophysiology of pancreatic and salivary acinar cells. Annu Rev Physiol 50: 65–80
Pierce SK, Politis AD (1990) Ca^{2+}-activated cell volume recovery mechanisms. Annu Rev Physiol 52: 27–42
Rink TJ, Sanchez A, Grinstein S, Rothstein A (1983) Volume restoration in osmotically swollen lymphocytes does not involve changes in free Ca^{2+} concentration. Biochim Biophys Acta 762: 593–596
Rossi JPFC, Schatzmann HJ (1982) Is the red cell calcium pump electrogenic? J Physiol (Lond) 327: 1–15
Rotin D, Mason MJ, Grinstein S (1991) Channels, antiports, and regulation of cell volume in lymphoid cells. In: Gilles R, Hoffmann EK, Bolis K (eds) Volume and osmolality in animal cells. Advances in comparative and environmental physiology 9. Springer, Berlin Heidelberg New York, pp 118–139
Rugolo M, Mastrocola T, Flamigni A, Lenaz G (1989) Chloride transport in human fibroblasts is activated by hypotonic shock. Biochem Biophys Res Commun 160(3): 1330–1338
Sachs F (1987) Baroreceptor mechanisms at the cellular level. Fed Proc 46: 12–16
Sachs F (1988) Mechanical transduction in biological systems. CRC Crit Rev Biomed Eng 16: 141–169
Sackin H (1987) Stretch-activated potassium channels in renal proximal tubule. Am J Physiol 253: F1253–F1262
Sackin H (1989) A stretch-activated K^+ channel sensitive to cell volume. Proc Natl Acad Sci USA 86: 1731–1735
Saltin B, Sj¢gaard G, Strange S, Juel C (1987) Redistribution of K^+ in the human body during muscular exercise; its role to maintain whole body homeostasis. In: Shiraki K, Yousef MK (eds) Man in stressful environments. Thermal and work physiology. Charles C Thomas, Springfield, pp 247–267

Sarkadi B, Mack E, Rothstein A (1984a) Ionic events during the volume response of human peripheral blood lymphocytes to hypotonic media. I. Distinctions between volume-activated Cl^- and K^+ conductance pathways. J Gen Physiol 83: 497–512

Sarkadi B, Mack E, Rothstein A (1984b) Ionic events during the volume response of human peripheral blood lymphocytes to hypotonic media. II. Volume- and time-dependent activation and inactivation of ion transport pathways. J Gen Physiol 83: 513–527

Sarkadi B, Cheung R, Mack E, Grinstein S, Gelfand EW, Rothstein A (1985) Cation and anion transport pathways in volume regulatory response of human lymphocytes to hyposmotic media. Am J Physiol 248: C480–C487

Sauvé R, Simoneau C, Monette R, Roy G (1986) Single-channel analysis of the potassium permeability in HeLa cancer cells: evidence for a calcium-activated potassium channel of small unitary conductance. J Membr Biol 92: 269–282

Sauvé R, Parent L, Simoneau C, Roy G (1988) External ATP triggers a biphasic activation process of a calcium-dependent K^+ channel in cultured bovine aortic endothelial cells. Pflügers Arch 412: 469–481

Scharff O, Foder B, Skibsted U (1983) Hysteretic activation of the Ca^{2+} pump revealed by calcium transients in human red cells. Biochim Biophys Acta 730: 295–305

Schultz SG (1989a) Intracellular sodium activities and basolateral membrane potassium conductances of sodium-absorbing epithelial cells. In: Schultz SG (ed) Cellular and molecular biology of sodium transport. Curr Top Membr Transp 3: 21–44

Schultz SG (1989b) Volume preservation. Then and now. News Physiol Sci 4: 169–172

Schultz SG, Hodson RL, Lapointe J-Y (1985) Electrophysiological studies of sodium cotransport in epithelia: towards a cellular model. In: Semenza G, Kinne R (eds) Membrane transport driven by ion gradients. Ann NY Acad Sci 456: 127–135

Schwarze W, Kolb HA (1984) Voltage-dependent kinetics of an ion channel of large conductance in macrophages and myotube membranes. Pflügers Arch 402: 281–291

Seker MC, Hokin LE (1986) The role of phosphoinositides in signal transduction. J Membrane Biol 89: 193–210

Siebens AW (1985) Cellular volume control. In: Seldin DW, Giebisch G (eds) The kidney: physiology and pathophysiology. Raven Press, New York, pp 91–115

Simmons NL (1984) Epithelial cell volume regulation in hypotonic fluids: studies using a model tissue culture renal epithelial cell system. Q J Exp Physiol 69: 83–95

Simonsen LO, Brown AM, Christensen S, Harbak H, Svane PC, Hoffmann EK (1990) Thrombin and bradykinin mimic the volume response induced by cell swelling in Ehrlich mouse ascites tumour cells. Renal Physiol Biochem 13: 176

Smith JB, Smith L (1987) $Na^+/K^+/Cl^-$ cotransport in cultured vascular smooth muscle cells: stimulation by angiotensin II and calcium ionophores, inhibition by cyclic AMP and calmodulin antagonists. J Membr Biol 99: 51–63

Solis JM, Herranz AS, Herreras O, Menedez N, Martin del Rio R (1990) Weak organic acids induce taurine release through an osmotic-sensitive process in in vivo rat hippocampus. J Neurosci Res 26: 159–167

Spring KR, Hoffmann EK (1992) Cell volume control. In: Seldin DW, Giebisch G (eds) The kidney: physiology and pathophysiology 2nd edn. Raven Press, New York, pp 147–169

Stockbridge LL, French AS (1988) Stretch-activated cation channels in human fibroblasts. Biophys J 54: 187–190

Sun A, Grossman EB, Lombard M, Hebert SC (1991) Vasopressin alters the mechanism of apical Cl^- entry from Na^+, Cl^- to Na^+, K^+, $2Cl^-$ cotransport in mouse medullary thick ascending limb. J Membr Biol 120: 83–94

Svane PC, Hoffmann EK (1991) Formation of inositol (1,4,5) trisphosphate following hypotonic shock or agonist stimulation of the Ehrlich mouse ascites tumor cell, demonstrated by HPLC. Acta Physiol Scand 143: 8A

Svane P, Saermark TS, Stuer-Lauridsen F, Hoffmann EK (1990) Transient increase in Ins (1,4,5) P3 and Ins (1,3,4,5) induced by cell swelling in Ehrlich mouse ascites tumor cells demonstrated by high pressure liquid chromatography. Renal Physiol Biochem 13: 177

Tanglietti V, Teselli M (1988) A study of stretch activated channels in the membrane of frog oocyte: interaction with Ca^{2+} ions. J Physiol (Lond) 407: 311–328

Thurston JH, Hauhart RE, Nacarato EF (1981) Taurine: possible role in osmotic regulation of mammalian heart. Sciences (NY) 214: 1373–1374

Ubl J, Murer H, Kolb H-A (1988) Ion channels activated by osmotic and mechanical stress in membranes of opossum kidney cells. J Membr Biol 104: 223–232

Ussing HH (1982) Volume regulation of frog skin epthelium. Acta Physiol Scand 114: 363–369

Ussing HH (1982b) Pathways for transport in epithelia. In: Corradino RA (ed) Functional regulation at the cellular and molecular levels. Elsevier North Holland, Amsterdam, pp 285–297

Ussing HH (1985) Volume regulation and basolateral co-transport of sodium, potassium, and chloride ions in frog skin epithelium. Pflügers Arch 405 (Suppl 1): S2–S7

Ussing HH (1986) Epithelial cell volume regulation illustrated by experiments in frog skin. Renal Physiol 9: 38–46

van den Bosch H (1980) Intracellular phospholipases A. Biochim Biophys Acta 604: 191–246

van Rossum GDV, Russo MA (1981) Ouabain-resistant mechanism of volume control and the ultrastructural organization of liver slices recovering from swelling in vitro. J Membr Biol 59: 191–209

Vestergaard-Bogind B, Stampe P, Christophersen P (1985) Single-file diffusion through the Ca^{2+}-activated K^+ channel of human red cells. J Membr Biol 88: 67–75

Vislie T (1983) Cell volume regulation in fish heart ventricles with special reference to taurine. Comp Biochem Physiol 76A: 507–514

Walz W (1989) Role of glial cells in the regulation of the brain ion microenvironment. Prog Neurobiol 33: 309–333

Warnock DG, Greger R, Dunham PB, Benjamin MA, Frizzell RA, Field M, Spring KR, Ives HE, Aronson PS, Seifter J (1984) Ion transport processes in apical membranes of epithelia. Fed Proc 43: 2473–2487

Welling PA, O'Neil RG (1990) Cell swelling activates basolateral membrane Cl and K conductance in rabbit proximal tubule. Am J Physiol 258: F951–F962

Swelling PA, Linshaw MA, Sullivan LP (1985) Effect of barium on cell volume regulation in rabbit proximal straight tubules. Am J Physiol 249: F20–F27

Welsh MJ (1986) An apical-membrane chloride channel in human tracheal epithelium. Science 232: 1648–1650

Wiener E, Dubyak G, Scarpa A (1986) Na^+/H^+ exchange in Ehrlich ascites tumor cells. Regulation by extracellular ATP and 12-O-tetradecanoylphorbol 13 acetate. J Biol Chem 261: 4429–4534

Wong SME, Chase HS (1986) Role of intracellular calcium in cellular volume regulation. Am J Physiol 250: C841–C852

Wong, PY-K, Cheung WY (1979) Calmodulin stimulates human platelet phospholipase A2. Biochem Biophys Res Commun 90: 473–480

Yamauchi A, Uchida S, Kwon HM, Preston AS, Robey RB, Garcia-Perez A, Burg MB, Handler JS (1992) Cloning of a Na- and Cl-dependent betaine transporter that is regulated by hypertonicity. J Biol Chem 267: 649–652

Yang X-C, Sachs F (1989) Block of stretch activated ion channels in *Xenopus* oocytes by gadolinium and calcium ions. Science 243: 1068–1071

Chapter 8

Regulation of Volume-Modulating Ion Transport Systems by Growth Promoters

L. Bianchini[1] and *S. Grinstein*[1,2]

Contents

1 Introduction

Exposure of quiescent cells to a variety of growth factors and pharmacological mitogens leads to a cascade of biochemical events and ionic changes that occur within minutes and are believed to play crucial roles in the initiation of cell proliferation. The biochemical events induced by growth promoters include increased turnover of inositol phospholipids, protein phosphorylation and increased transcription of several protooncogenes. Modification of monovalent ion (e.g. Na^+ and K^+) fluxes across the plasma membrane is one of the earliest changes observed in response to growth

[1] Division of Cell Biology and Department of Biochemistry, Hospital for Sick Children, 555 University Ave., Toronto, M5G 1X8, Canada
[2] Correspondence to: Sergio Grinstein, Division of Cell Biology, Hospital for Sick Children, 555 University Ave., Toronto, M5G 1X8, Canada

Advances in Comparative and Environmental Physiology, Vol. 14

promoters. These ionic fluxes traverse the membrane via several carrier-mediated pathways. One of these pathways is the amiloride-sensitive Na^+/H^+ antiport. Activation of the Na^+/H^+ antiport by various mitogens has been well characterized in a number of cell types (Table 1). This activation results, on the one hand, in cytoplasmic alkalinization and, on the other, in an increased influx of Na^+. The resulting increase in intracellular Na^+ content leads to stimulation of the Na^+/K^+ ATPase, which in turn increases the intracellular K^+ content (Rozengurt 1981). In addition to the flux catalyzed by the Na^+/K^+ ATPase, increased K^+ influx in proliferating cells has also been attributed to a ouabain-resistant, bumetanide-sensitive K^+ transport system identified as $Na^+/K^+/Cl^-$ cotransport (Panet et al. 1982; Amsler et al. 1985; Panet 1085). Activation of the antiports, pumps or cotransporters can lead to a net uptake of inorganic solutes, accompanied by osmotically obliged water. The associated increase in cell volume can be considered one of the early responses of quiescent cells to growth promoters.

The purpose of this review is to examine the effects of various growth promoters on Na^+/H^+ exchange and on $Na^+/K^+/Cl^-$ cotransport. The basic properties of these systems are reviewed, followed by a discussion of the nature and mechanism of activation by mitogens. Finally, we consider the consequences of activation of these transport pathways on cell volume and the potential relevance of these volume modifications to the mitogenic process.

A variety of agents shown to be involved in growth and mitogenesis are considered in the present review. These include growth factors, mitogenic hormones and tumour promoters. Growth factors are generally polypeptides that induce replicative DNA synthesis and cell division in their target cells. They initiate their action by binding to specific cell surface receptors that frequently possess intrinsic protein tyrosine kinase activity, which is essential for signal transduction. Tumor promoters include phorbol esters, non-peptide mitogens that are known to bind and activate protein kinase C by substituting for endogenous diacylglycerol. In addition, this review focusses on the effects of a recently discovered tumour promoter, okadaic acid, on Na^+/H^+ antiport and $Na^+/K^+/Cl^-$ cotransport. Okadaic acid is a polyether fatty acid that has proven to be a very potent and selective inhibitor of protein phosphatases 2A and 1, two of the four major protein phosphatases in the cytosol of eukaryotic cells, that dephosphorylate phosphoserine and phosphothreonine residues. The discovery of okadaic acid provides an important tool for investigation of the role of protein dephosphorylation in the regulation of ion transport systems.

2 Effect of Growth Promoters on Na^+/H^+ Exchange

2.1 Characteristics of the Na^+/H^+ Antiport

One of the primary mechanisms involved in the extrusion of H^+ from vertebrate cells is the Na^+/H^+ antiport, originally described by Murer et al. (1976) in epithelial membranes. The Na^+/H^+ antiport, recently identified as a phosphoglycoprotein of molecular weight ≈110 kDa (Sardet et al. 1990) has been detected in the plasma membrane of virtually all mammalian cell types studied to date. It has been implicated in multiple cellular functions including the regulation of intracellular pH, the control of cell volume and the initiation of growth and proliferation.

This system catalyzes the exchange of Na^+ for H^+ and can be blocked by the potassium-sparing diuretic amiloride. The stoichiometry of the exchange reaction is one for one (Cala 1980; Boron and Boulpaep 1983; Grinstein et al. 1984). Consistent with this, the system is electroneutral and insensitive to maneuvers that alter the membrane potential (Kinsella and Aronson 1980). Under physiological conditions, the antiport catalyzes the net uptake of Na^+, which is coupled to the efflux of cellular H^+. Because the Na^+ gradient is the primary driving force for H^+ extrusion, the operation of the exchanger is therefore indirectly dependent on the Na^+/K^+ ATPase, which ensures the maintenance of the Na^+ gradient. However, the direction of the exchange can be reversed by inverting the ionic gradients, e.g. by removal of extracellular Na^+ (Moolenaar et al. 1983; Cassel et al. 1984). In addition to Na^+ and H^+, the antiport can transport Li^+ in most systems (Aronson 1985; Mahnnensmith and Aronson 1985) and in some cases, NH_4^+ as well (Kinsella and Aronson 1981). In contrast, K^+, Rb^+, Cs^+ and organic cations such as choline$^+$, tetramethylammonium$^+$ and N-methyl-D-glucammonium$^+$ are transported very poorly, if at all.

The properties of the Na^+/H^+ antiporter and its physiological functions have been explored using the diuretic compound amiloride and its derivatives, particularly those disubstituted at position 5 of the pyrazine ring. With few exceptions (Ives et al. 1983; Moran et al. 1986), amiloride has been found to be a purely competitive inhibitor with respect to external Na^+, suggesting that the blocker acts at or near the Na^+-binding site (see Benos 1988 for review). However, the Na^+/H^+ antiporter has been shown to exhibit varying sensitivity to amiloride in different cell types (Gurich and Warnock 1986; Pouyssegur et al. 1987; Haggerty et al. 1988), suggesting the existence of multiple isoforms of the exchanger or binding of the inhibitor to different sites. The structure-activity relationship of the inhibitory effect of amiloride has been deduced in various cell types from studies comparing the potency of a variety of its structural analogues: some of the 5-disubstituted derivatives have been shown to be up to 100 times more potent than amiloride itself in blocking Na^+/H^+ exchange.

The dependence of the rate of transport on the extracellular Na^+ concentration follows Michaelis-Menten kinetics, consistent with one single binding site with an apparent K_m value usually in the millimolar range. In contrast, the dependence of forward Na^+_o/H^+_i transport on $[H^+]_i$ is considerably steeper. Such behaviour is not predicted for a simple, saturable, one-for-one exchanger and suggests that additional effects must exist. This apparent inconsistency has been attributed to the existence of a second cytoplasmic H^+-binding site which allosterically controls the activity of the antiport. This hypothesis, initially introduced by Aronson and colleagues (1982) is supported by observations showing that, under certain conditions, increasing $[H^+]_i$ can stimulate Na^+_i efflux (Aronson et al. 1982), an effect opposite to the expected competition for the internally oriented transport site. Stimulation of transport by protonation of an allosteric "modifier" site can, however, account for these observations.

The existence of a modifier site is essential for understanding how the exchanger exerts its physiological function. In quiescent cells, despite the large inward Na^+ gradient, transport through the exchanger is barely measurable at normal intracellular pH (pH_i) (Moolenaar et al. 1983, 1984; Paris and Pouyssegur 1983; Grinstein et al. 1984). This peculiar behaviour has been attributed to the kinetic control exerted by the modifier site on the rate of countertransport, which confers to the exchanger an activity threshold or "set point". Below this pH_i threshold, protonation of the modifier site greatly enhances the rate of transport. On the other hand, by curtailing the activity of the exchanger above the set point, the modifier protects the cell from approaching the very alkaline pH_i at which the system is predicted to attain thermodynamic equilibrium (approximately $pH_i = 8.3$ for extracellular pH (pH_o) = 7.3 and a $[Na^+]_o/[Na^+]_i$ ratio of 10).

Estimates of the density of Na^+/H^+ antiport sites in the membrane have only been reported in a few studies. In all cases, the number of sites was estimated by using equilibrium binding of potent disubstituted amiloride analogues. Vigne et al. (1984) determined the binding of 5-(N-ethyl-N-[^{3}H]propyl)amiloride to detergent-solubilized renal membranes. The maximal number of binding sites determined was 2 pmol/mg protein and the turnover, estimated using rates of transport measured in intact vesicles, was ≈2500 per second. Dixon et al. (1987) used 5-(N-methyl-N-[^{3}H]isobutyl) amiloride to quantify the antiport density in rat thymic lymphocytes. Scatchard analysis of the binding data indicated a maximum of 8000 antiports per cell. Based on estimates of transport obtained under comparable conditions, the maximally activated lymphocyte antiport was calculated to cycle at least 2000 times per second. Comparable results were recently obtained in human platelets, using a similar approach (Rosskopf et al. 1991). The number of antiports has also been studied under conditions where transport is stimulated. Activation of Na^+/H^+ antiport in lymphocytes treated with phorbol esters was not accompanied by a change in the number of antiports (estimated as high affinity amiloride analogue binding sites; Dixon et al.

1987), suggesting that with this stimulus, activation of transport is due to an increased rate of ion translocation by a fixed number of exchangers, as opposed to an increase in the number of antiporters. However, binding of the amiloride analogues to inactive transporters cannot be ruled out. Stimulation might then result from activation of exchangers which are functionally quiescent, yet capable of binding the inhibitors.

2.2 Molecular Identification of the Mammalian Na^+/H^+ Antiport

One isoform of the human Na^+/H^+ antiport (called NHE-1) has been fully identified by Pouyssegur and his colleagues. These authors recently cloned a human cDNA encoding the amiloride-sensitive, growth factor-activatable Na^+/H^+ antiporter by complementation of a Na^+/H^+ antiport-deficient cell line (Sardet et al. 1989). The first step towards the molecular cloning of the transporter consisted of isolation of mouse fibroblast mutants lacking antiport activity, which was accomplished by the proton suicide method, based on the reversibility of the antiport (see Pouyssegur et al. 1984 for details). These antiport-deficient mutants were then transfected with human genomic DNA. Transformants re-expressing high levels of the antiport could then be selected. This selection was achieved by incubating the transfected cells under conditions where Na^+/H^+ exchange is essential for survival, i.e. acute acid-loading in HCO_3^--free media. Preparations of genomic DNA from independent primary or secondary mouse transfectants were used successively for second- and third-cycle transformation to eliminate human DNA sequences not required for the maintenance of the Na^+/H^+ exchange activity. This complementation of the mouse fibroblast antiport-deficient mutant with human genomic DNA allowed isolation of a 0.8 kb genomic probe from a third-cycle mouse transformant. This probe was found to detect gene amplification in Na^+/H^+ antiport overexpressers and a single class of mRNA of ca. 516 kb in human, mouse and hamster cells. The probe allowed isolation of a cDNA from a library constructed from a mouse transformant in which the transfected human gene was amplified. This cDNA includes a non-coding leader sequence of 407 bp, a 2682-bp open reading frame, and a 3′ non-coding sequence containing a mouse B1-repeated element. The amino acid sequence predicts a protein of 815 amino acids which contains two distinct domains: an N-terminal domain that is amphipathic and contains $\approx$10 transmembrane-spanning segments plus 2 potential glycosylation sites, and a second (C-terminus) domain which is highly hydrophilic and presumably represents a large cytoplasmic region of the antiport. The latter has a net positive charge and several potential phosphorylation sites.

Stable expression of the transfected cDNA in Na^+/H^+ antiporter-deficient cells was shown to fully restore the biochemical and physiological features of the transporter: H^+-activated Na^+ influx, pH_i sensitivity and inhibition by amiloride. To confirm the model extrapolated from the nucleotide sequence of the cDNA, a fusion protein of *Escherichia coli* β-galactosidase and the last 157 amino acids of the hydrophilic domain of the antiporter was constructed. Antibodies raised against this fusion protein detected a single band of 105 to 110 kDa only in cells expressing Na^+/H^+ exchange activity (Sardet et al. 1990).

2.3 Modulation of Na^+/H^+ Antiport by Growth Promoters

2.3.1 Activation of Na^+/H^+ Antiport by Growth Promoters

Most agents displaying mitogenic potential, either alone or in combination with other growth promoters, induce activation of Na^+/H^+ exchange. Modulation of Na^+/H^+ antiport by growth promoters has been observed in a variety of cells including those of invertebrates, fish, amphibia and various mammalian species. The effect has been noted in different tissues, as well as in cultures of primary cells or established lines.

Table 1 presents a detailed list of those mitogens which have been demonstrated to activate Na^+/H^+ exchange as judged by: (1) an increased amiloride- or amiloride analogue-sensitive influx of Na^+; (2) an amiloride- or amiloride analogue-sensitive cytoplasmic alkalinization; (3) a Na^+-dependent cytoplasmic alkalinization, or by a combination of these criteria. Mitogens that stimulate the antiport include growth factors, G-protein activators such as fluoroaluminate and vanadate, tumour promoters like phorbol esters and the recently discovered, non-phorbol tumour promoter okadaic acid. Modulators of the Na^+/H^+ antiport differ in the time course of their action: growth promoters are short-term effectors that activate

Table 1. Activation of Na^+/H^+ antiport by growth promoters

Agent	Cell type	References
Serum	Mouse neuroblastoma	Moolenaar et al. (1981)
	Human fibroblasts	Moolenaar et al. (1983)
	Hamster fibroblasts	L'Allemain et al. (1984)
	Swiss 3T3 fibroblasts	Burns and Rozengurt (1983)
	3T3 variant, NR6	Cassel et al. (1983)
	C-6 glioma	Sapirstein and Benos (1984)
	Rat smooth muscle	Owen (1984)
	Rat myoblasts	Vigne et al. (1985)
	Human epidermal carcinoma	Cassel et al. (1985)
Epidermal growth factor	Human epidermal carcinoma	Cassel et al. (1985)
	Human fibroblasts	Muldoon et al. (1985)
	C-6 glioma	Benos and Sapirstein (1983)
	Neuroblastoma	Benos and Sapirstein (1983)
	Rat pheochromocytoma	Boonstra et al. (1983)
	Cultured porcine thyroid cells	Takasu et al. (1989)
	Rat hepatocytes	Moule and McGivan (1990)

Table 1. Continued

Agent	Cell type	References
Platelet-derived growth factor	Swiss 3T3 fibroblasts	Daniel and Ives (1987)
	3T3 variant, NR6	Cassel et al. (1983)
	Rat smooth muscle	Owen (1984)
Interleukin 2	Rat lymphocytes	Mills et al. (1985)
	Rat NB2 lymphoma	Too et al. (1987)
Colony stimulating factor (CSF-1)	Macrophages	Vairo and Hamilton (1988)
Nerve growth factor	Rat pheochromocytoma	Boonstra et al. (1983)
Insulin	Frog skeletal muscle	Putnam (1985)
	Cultured myocytes	Klip et al. (1986)
	Hepatocytes	Fehlman and Freychet (1981)
	Swiss 3T3 fibroblasts	Schuldiner and Rozengurt (1982)
	Hamster lung fibroblasts	L'Allemain et al. (1984)
	Human fibroblasts	Muldoon et al. (1985)
Bombesin	Swiss 3T3 fibroblasts	Mendoze et al. (1986)
Phorbol esters and diacylglycerol	Vascular smooth muscle	Owen (1985)
	Rat myoblasts	Vigne et al. (1985)
	Sea urchin eggs	Swann and Whitaker (1985)
	Human epidermal carcinoma	Whiteley et al. (1984)
	Human leukemic cells (HL-60)	Besterman and Cuatrecasas (1984)
	Human leukemic T-cell (HPB-All)	Rosoff and Cantley (1985)
	Pre-B-lymphoblasts (70Z/3)	Rosoff et al. (1984)
	Rat NB2 lymphoma	Too et al. (1987)
	C-6 glioma	Sapirstein and Benos (1984)
	Swiss 3T3 fibroblasts	Mendoza et al. (1986)
	Human fibroblasts	Moolenaar et al. (1984)
	HeLa	Moolenaar et al. (1984)
	Neuroblastoma	Moolenaar et al. (1984)
	Porcine neutrophils	Grinstein et al. (1985c)
	Rat thymocytes	Grinstein et al. (1985b)
	Rat hepatocytes	Stewart (1988)
Thrombin	Human platelets	Greenberg and Simons (1984)
	Hamster fibroblasts	L'Allemain et al. (1984)
	Human fibroblasts (WS-1)	Hendey et al. (1989)
Lipopolysaccharides	Pre-B-lymphoid line	Rosoff and Cantley (1985)
Lectins	Murine T lymphocytes	Hesketh et al. (1985)
	Porcine blood lymphocytes	Prasad et al. (1987)
	Bovine lymphocytes	Kakinuma et al. (1987)
Vanadate	Human epidermal carcinoma	Whiteley et al. (1984)
	Human fibroblasts	Muldoon et al. (1987)
	Swiss 3T3 fibroblasts	Daniel and Ives (1987)
	Hamster fibroblasts	Paris et al. (1987)
Fluoroaluminate	Hamster fibroblasts	Paris et al. (1987)
Okadaic acid	Rat thymocytes	Bianchini et al. (1991)
	Hamster fibroblasts	Sardet et al. (1991)
	Human platelets	Livne et al. (1991)

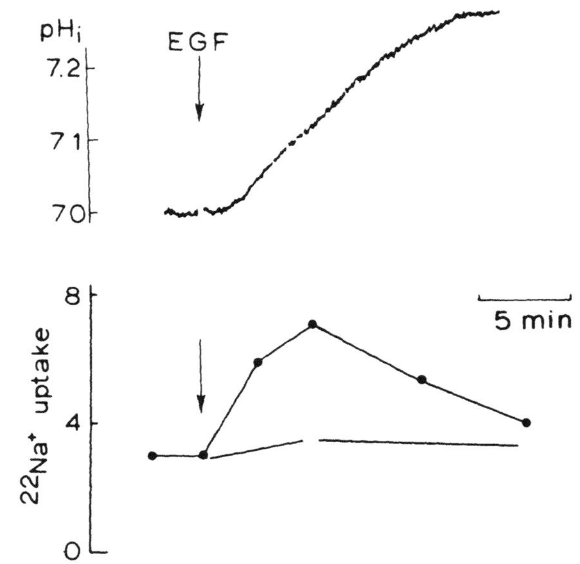

Fig. 1. Activation of Na^+/H^+ exchange by *EGF* in human fibroblasts. *Above* Shift in pHi induced by EGF (10 ng/ml) as measured by BCECF fluorescence. *Below* Time course of EGF-induced $^{22}Na^+$ uptake in the presence (*lower trace*) and absence (*upper trace*) of 1 mM amiloride. (Moolenaar et al. 1988)

the antiport within seconds to minutes. The alkalinization produced usually ranges from 0.15 to 0.30 pH units. Figure 1 (above) shows a typical example of the cytosolic alkalinization induced by addition of epidermal growth factor (EGF) to quiescent human fibroblasts loaded with the pH-sensitive fluorescent dye biscarboxyethyl-carboxyfluorescein (BCECF). The alkalinization is accompanied by a transient increase in the amiloride-sensitive component of $^{22}Na^+$ uptake (below).

Stimulation of the antiport by mitogens and phorbol esters is thought to be associated with activation of protein kinase activity. Due to the lack of specific inhibitors, the role of phosphatases in the modulation of Na^+/H^+ exchange has been largely ignored. Recently, however, okadaic acid was introduced as a potent and selective inhibitor of phosphoprotein phosphatases 1 and 2A (see Sect. 1). Unlike other phosphatase inhibitors such as vanadate which are rather unspecific, okadaic acid specifically inhibits phosphatases 2A and 1, without affecting other enzymes including protein kinase C (PKC), cAMP-dependent protein kinase, Ca^{2+}/calmodulin-dependent kinases and inositol 1,4,5-triphosphatase (Bialojan and Takai 1988; Haystead et al. 1989). We therefore decided to study the effects of this phosphatase antagonist on intracellular pH using rat thymic lymphocytes. As shown in Fig. 2, treatment of lymphoid cells with 1 µM okadaic acid resulted in a rapid and sustained cytoplasmic alkalinization ($\Delta pH = 0.15 \pm 0.015$, $n = 18$). This pH_i increase was shown to be Na^+-dependent and was

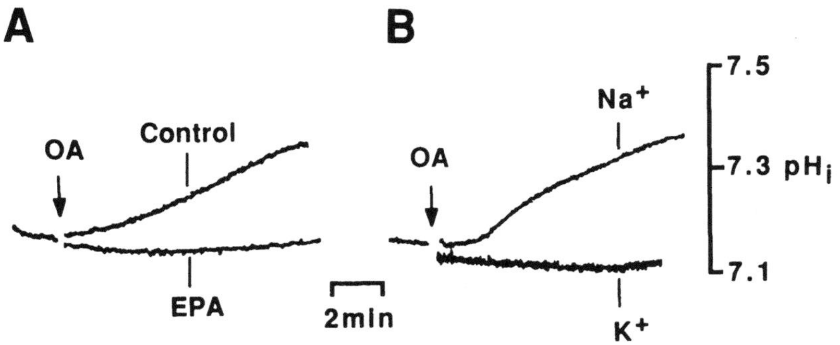

Fig. 2. A Effect of 5-(N-ethyl-N-propyl)amiloride (*EPA*) on okadaic acid-induced cytoplasmic alkalinization. BCECF-loaded thymocytes were suspended in Na^+ solution with or without 5 μM EPA. Where indicated, 1 μM okadaic acid was added to both samples and recording was resumed. *B* Na^+ dependence of okadaic acid-induced alkalinization. BCECF-loaded thymocytes were suspended in either normal Na^+ solution or in K^+ solution (Na^+ free). Where indicated, 1 μM okadaic acid was added to both samples and recording resumed. (Bianchini et al. 1991)

inhibited by 5,N-disubstituted analogues of amiloride, indicating mediation by the Na^+/H^+ antiport. Activation of the Na^+/H^+ exchanger by okadaic acid has also been reported in human platelets (Livne et al. 1991) and in fibroblasts (Sardet et al. 1991), where the effect of the phosphatase inhibitor was found to be additive with that of growth factors.

The finding that the antiport is activated by treatment with okadaic acid alone, a purportedly specific phosphatase inhibitor, implies that one or more kinases are constitutively active in otherwise unstimulated cells. The identity of this stimulatory kinase remains to be identified: experiments using kinase inhibitors suggested that protein kinase C, Ca^{2+}/calmodulin-dependent kinase and cyclic nucleotide-dependent kinases are not responsible for the effects of okadaic acid (Bianchini et al. 1991).

The mechanism of activation of the antiport has been studied in various systems (Grinstein and Rothstein 1986; Moolenaar 1986). Many studies have reported that mitogenic agents activate the antiport by shifting the pH dependence of the modifier site, adjusting the set point upward by 0.15–0.30 pH units, which is the magnitude of the alkalosis recorded in most cases (Villereal 1981; Schuldiner and Rozengurt 1982; Paris and Pouyssegur 1984; Grinstein et al. 1989). As a result, the exchanger is activated, but only temporarily, returning to near quiescence when pH_i attains the new set point value.

What is the molecular mechanism underlying the activation of the antiport by mitogens? It had been speculated repeatedly that the alteration in the set point was mediated by phosphorylation of the antiport itself or of an ancillary protein. This speculation was based on the following observations: first, the biological effects of most, if not all, of the mitogens listed in Table 1 are believed to be mediated by protein kinases; second, in the cases where the metabolic dependence of the response was studied (Grinstein et al. 1985a; Bianchini et al. 1991), the stimulation of the antiport was shown to depend on the availability of cellular ATP, the putative substrate of the kinases; finally, in the cases of agents thought to act via stimulation of protein kinase C, depletion of this enzyme prior to addition of the stimulus precluded the activation of the antiport (Grinstein et al. 1986; Daniel and Ives 1987).

Direct evidence confirming the above speculations was recently reported by Sardet et al. (1990). Performing immunoprecipitation experiments using specific polyclonal antibodies directed against the Na^+/H^+ antiport, they were able to study the effect of various mitogens on the phosphorylation of the antiport. Sardet and colleagues demonstrated that mitogenic activation of hamster fibroblasts and A431 human epidermoid cells with growth factors

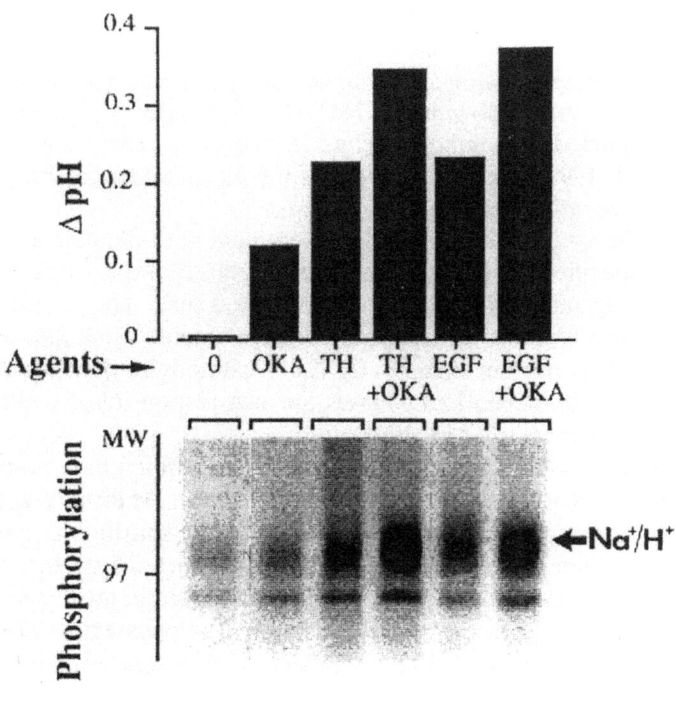

Fig. 3. Okadaic acid and growth factors induce intracellular alkalinization concomitant with in vivo NHE1 phosphorylation in ER22 cells. For the pH determination, G0-arrested cells were treated with 1 μM okadaic acid (*OKA*) for 15 min. Intracellular pH (*above*) and in vivo phosphorylation (*below*) were measured on parallel G0-arrested ER22 cell cultures and under the same conditions of stimulation by 1 μM OKA and/or 10 nM α-thrombin (*TH*), 100 nM *EGF*, for 15 min. (Sardet et al. 1991)

or phorbol esters is accompanied by phosphorylation of the Na^+/H^+ antiport, with a time course similar to the rise in intracellular pH (see Fig. 3). Similar results have been obtained for phorbol ester or thrombin-stimulated human platelets. Increased phosphorylation of the antiport was also found to occur upon stimulation with okadaic acid in human bladder carcinoma cells (Bianchini et al. 1991), hamster fibroblasts (Sardet et al. 1991) and human platelets (Livne et al. 1991).

EGF and thrombin activate distinct transmembrane signalling pathways (receptor tyrosine kinase in the case of EGF; G-protein-coupled activation of phospholipase C, followed by stimulation of protein kinase C for thrombin); yet, both mitogens have been shown to stimulate phosphorylation of the antiport exclusively on serine residues (Sardet et al. 1990). Moreover, tryptic phosphopeptide maps of the Na^+/H^+ antiport immunoprecipitated from cells treated with EGF or thrombin showed a common pattern of phosphorylation (Sardet et al. 1991).

Taken together, these results are consistent with the notion that direct phosphorylation of the exchanger is the triggering mechanism that controls its set point. They also suggest that the final step in Na^+/H^+ activation is mediated by an as yet unidentified serine protein kinase that can integrate signals from receptor tyrosine kinases and G-protein-coupled receptors.

2.3.2 Mitogen-Induced Activation of Na^+/H^+ Exchange and pH_i Changes

Although observed in most cell types (Table 1), mitogen-induced cytoplasmic alkalinization is not a universal phenomenon, even in media devoid of bicarbonate. Some cases where these events have been dissociated are listed next. Addition of serum to quiescent Ehrlich ascites tumour cells failed to increase amiloride-sensitive $^{22}Na^+$ uptake and did not produce cytoplasmic alkalinization (Doppler et al. 1986). Vigne et al. (1985) reported that the Na^+/H^+ antiport is activated by serum and phorbol esters in proliferating myoblasts but not in differentiated myotubes. In human T-lymphocytes, mitogenic stimulation with lectin resulted in a small ($\approx$0.05 pH units) cytoplasmic *acidification* (Gelfand et al. 1988). A sharp and rapid acidification (0.25–0.30 pH units), followed by a slow recovery to a level below the initial resting pH_i, was also observed upon addition of serum to LLC-PK1 cells (Haggerty et al. 1987). Finally, Bierman et al. (1987) reported that the pH of embryonal carcinoma P19 cells failed to respond to serum, growth factors or phorbol esters. It should be emphasized, however, that in autonomously growing cells Na^+/H^+ exchange may be in a permanently activated state, resulting in a much higher resting pH_i value, possibly precluding further stimulation of the antiport by mitogens.

Most studies showing mitogen-induced activation of Na^+/H^+ exchange and concomitant cytoplasmic alkalinization (Table 1) have been performed in the nominal absence of bicarbonate. In the presence of bicarbonate, at least two additional mechanisms (cation-independent HCO_3^-/Cl^- exchange and Na^+-dependent HCO_3^-/Cl^- exchange) can become operative and participate in pH_i homeostasis in concert with Na^+/H^+ exchange. Thus, different effects of mitogens on pH in the presence and absence of bicarbonate would not be unexpected. Only a few studies have investigated the effects of growth promoters on Na^+/H^+ exchange and pH_i in more physiological, bicarbonate-containing media. Comparisons made at a comparable pH_o indicate that, with few exceptions (L'Allemain et al. 1985), pH_i in resting cells is significantly higher (from 0.10 to 0.25 units) in the presence of bicarbonate than in its absence. This effect of bicarbonate on resting pH_i has been observed in various cell types such as A431 (Cassel et al. 1985), 3T3 fibroblasts (Bierman and Moolenaar 1987), NIH 3T3 (Szwergold et al. 1988), rat mesangial cells (Ganz et al. 1989), MES-1 fibroblasts (Bierman et al. 1988) and BALB 3T3 cells (Gillies and Martinez-Zaguilan 1991). All of these studies reveal that transition to media containing bicarbonate (generally ≈25 mM) not only raises the resting pH_i, but in addition abolishes the alkalinizing effect of mitogens and phorbol esters (see Fig. 4). In fact, some authors report growth promoter-induced cytoplasmic acidification in the presence of bicarbonate (Ganz et al. 1989; Gillies and Martinez-Zaguilan 1991).

It has been hypothesized that the failure of the growth promoters to induce cytoplasmic alkalinization in the presence of bicarbonate is directly correlated to the higher basal pH_i observed under these conditions (Grinstein et al. 1989). As discussed in preceding sections, an upward shift in the set point of the Na^+/H^+ antiport underlies its stimulation by growth promoters. If the presence of bicarbonate raises resting pH_i to a level comparable or higher than the stimulated set point, no further activation of the antiport would be expected upon addition of the stimuli. In support of this hypothesis, Bierman et al. (1988) demonstrated that, in bicarbonate-containing solu-

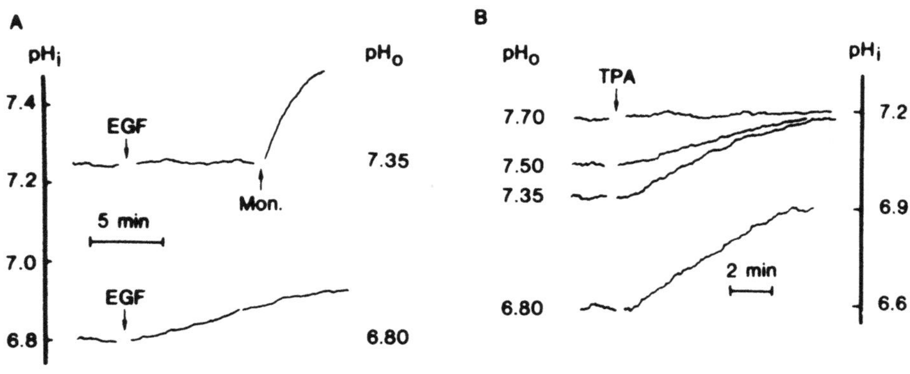

Fig. 4. Mitogen-induced alkalinization in bicarbonate- and Hepes-buffered media as a function of pH_i. *A* Serum-deprived MES-1 cells, loaded with BCECF, were stimulated with EGF (100 ng/ml) at pH_o 7.35 and 6.80 respectively. Monensin (*Mon.* 5 μM) was added at the time indicated. *B* Serum-deprived MES-1 cells were stimulated in Hepes-buffered solution with TPA (100 ng/ml) at different pH_o values as indicated. (Bierman et al. 1988)

tions, artificially lowering steady state pH_i by 0.4 units restored the pH_i response to growth promoters through activation of Na^+/H^+ exchange. Conversely, in bicarbonate-free media, when steady state pH_i was elevated to the value normally observed in the presence of bicarbonate, no mitogen-induced rise in pH_i was detectable. Similarly, Gillies and Martinez-Zaguilan (1991) reported that, in BALB/c3T3 cells, the presence of bicarbonate and/or high extracellular pH values can raise intracellular pH to a value that is high enough to preclude further alkalinization upon stimulation by mitogens. These authors found that the pH_o value at which there is no effect of serum is lower in the presence of bicarbonate (7.3) than in its absence (7.5). Because a pH_o of 7.3 in the presence of bicarbonate is typical for cell culture conditions, it was concluded that the pH response to mitogens cannot per se be a stimulus for proliferation under these conditions. However, many of the cells commonly used for ex vivo culture were obtained from tissues where a lower pH_o (in the range of 6.9–7.2) prevails (Wike-Hooley et al. 1985). Under such conditions of comparatively low extracellular pH, mitogenic stimulation is expected to produce a cytoplasmic alkalinization which might therefore influence proliferation.

2.4 Na^+/H^+ Exchange and Growth Promoter-Induced Changes in Cell Volume

A frequently overlooked consequence of activation of Na^+/H^+ exchange, which is particularly relevant in bicarbonate-containing media, is the associated increase in cell volume. This phenomenon has been explored in some detail in the context of cell volume regulation. It is well established that, when cells shrink, the antiport is activated in a manner that closely resembles that noted in mitogenically stimulated cells. Using previously published data, Grinstein and Foskett (1990) compared the pH_i dependence of Cl^-/HCO_3^- exchange and that of the Na^+/H^+ antiport in isotonic and hypertonic conditions. They noted that, whereas at physiological tonicity and pH_i ($\approx$7.1) both Na^+/H^+ exchange and Cl^-/HCO_3^- exchange are relatively inactive, in hypertonically treated cells, the set point of the antiport is shifted upward and the exchanger remains active at more alkaline levels. As a result, a pH_i range exists where both Na^+/H^+ and Cl^-/HCO_3^- exchange operate in parallel at substantial rates. This joint operation produces a net gain of NaCl which, together with the uptake of osmotically obliged water, leads to an increase in cell volume. Some degree of swelling can be expected even in the absence of bicarbonate since, though Na^+/H^+ exchange is in principle osmotically ineffectual, the extruded protons are readily replaced from the intracellular buffers, which causes a net osmotic gain.

Because growth promoters, like osmotic shrinking, activate Na^+/H^+ antiport by shifting its set point upward, they are expected to induce a volume gain, especially in bicarbonate-containing media. This has indeed been demonstrated in the case of lymphocytes stimulated with phorbol esters (Grinstein et al. 1985b) and in A431 cells treated with serum and EGF (Cassel et al. 1985). It is tempting to speculate regarding the relevance of such changes in cell volume to the mitogenic process. An increase in the salt and water content of the cell could be considered as a necessary prelude to the synthesis of proteins and other macromolecules that occurs following mitogenic stimulation. Indeed, mitogen-induced swelling can be regarded as providing the intracellular space required to accomodate newly synthesized organelles and macromolecules. Cell swelling might therefore represent an essential step in the cascade of events leading to cellular proliferation.

3 Effect of Growth Promoters on Na/K/Cl Cotransport

3.1 Properties of the Na/K/Cl Cotransport System

Na/K/Cl cotransport is a specific transport system that mediates the coupled, electrically silent transport of Na^+, K^+ and Cl^- ions across the plasma membrane. It has been identified in a wide variety of cell types including epithelial as well as non-epithelial cells (reviewed in O'Grady et al. 1987a; Haas 1989). Na/K/Cl cotransport appears to be involved in a number of physiological functions including transport of salt and water across absorptive and secretory epithelia (Hannafin et al. 1983; Greger 1985; O'Grady et al. 1986), cell volume regulation (Kregenow 1981; Geck and Pfeiffer 1985) and extrarenal regulation of plasma K^+ (Haas and McManus 1985); it is also believed to be involved in the pathogenesis of essential hypertension (O'Donnell and Owen 1988).

A stoichiometry of 1Na:1K:2Cl has been demonstrated in all tissues studied to date (Geck et al. 1980; Haas et al. 1982; McRoberts et al. 1982; Brown and Murer 1985; O'Grady et al. 1986). The only two exceptions are ferret red cells (Hall and Ellory 1985) and squid axons (Russell 1983), where a stoichiometry of 2Na:1K:3Cl has been reported. Regardless of the stoichiometry, however, the cotransport process has been found to be electroneutral in all cases.

The following properties are the hallmarks of Na/K/Cl cotransport: (1) interdependency of the transported ions. K^+ influx through the cotransport pathway is abolished in the absence of either Na^+ or Cl^- and vice versa; (2) strict substrate selectivity. Only Rb^+ can quantitatively substitute for K^+ (O'Grady et al. 1986), Li^+ poorly replaces Na^+ (Palfrey and Greengard 1981) and the anions that can participate are limited to Br^- or Cl^- (Kinne et al. 1985); and (3) specific inhibition by a family of sulphamoylbenzoic acid derivatives known as "loop diuretics". These include furosemide, bumetanide, piretanide and benzmetanide (Palfrey et al. 1980b, 1984; Schlatter et al. 1983). By contrast, cotransport is unaffected by agents such as ouabain, amiloride, stilbene disulphonates and thiazide diuretics (Kregenow 1977; Palfrey et al. 1980a; Palfrey and Greengard 1981; Stokes 1984). Whereas furosemide has been reported to inhibit some other transport systems (Palfrey et al. 1980b; Lauf 1986), bumetanide appears as a more potent and specific inhibitor of Na/K/Cl cotransport. In almost all cases, bumetanide has proven to be 10–100 times more potent than furosemide and does not seem to affect other transport processes at concentrations that block cotransport (the IC_{50} values for inhibition of cotransport range from 0.05 to 0.5 μM in most tissues).

3.1.1 Molecular Identification of the Na/K/Cl Cotransporter

The stucture of the Na/K/Cl cotransporter has not been fully characterized yet. A number of studies have attempted to identify the protein composition of the cotransporter in various tissues, mainly by affinity labelling with photoactivatable analogues of bumetanide (for review, see Haas 1989). Haas and coworkers have identified ≈150 kDa polypeptides that are specifically photolabelled by a tritiated analogue of bumetanide (4-[^{3}H]-benzoyl-5-sulphamoyl-3-(3-thenyloxy)benzoate or [^{3}H]-BSTBA) in three different cell types: dog kidney (Haas and Forbush 1987), duck erythrocytes (Haas and Forbush 1988) and mouse kidney (Haas et al. 1991). These photolabelled proteins were found to bind the probe with an affinity and ion dependence resembling those reported for the cotransporter and are therefore likely to represent at least a component of the Na/K/Cl cotransport system. The apparent molecular mass determined by photolabelling is compatible with recent estimates of the size of the mRNA coding for the rat kidney Na/K/Cl cotransporter, based on expression in the *Xenopus* oocyte system, as reported by Burnham et al. (1990), and with analysis of the sedimentation behaviour of bumetanide-binding proteins in sucrose gradients (Turner and George 1990). Another approach has involved the use of affinity chromatography: Feit et al. (1988) have used a bumetanide-Sepharose affinity column to purify two proteins of molecular masses of ≈39 and ≈82 kDa from solubilized membranes of Ehrlich ascites cells. Antiserum raised against these purified proteins was shown to strongly inhibit Na/K/Cl cotransport, providing evidence that the purified proteins contain constituents of the Na/K/Cl cotransporter (Dunham et al. 1990). This antiserum has recently been found to cross-react with a 150-kDa mouse kidney protein, most likely corresponding to the [^{3}H]-BSTBA photolabelled polypeptide mentioned above (Haas et al. 1991).

3.2 Regulation of Na/K/Cl Cotransport by Growth Promoters

3.2.1 Activation of Na/K/Cl Cotransport by Growth Promoters

One of the earliest responses of quiescent cells to most mitogens is the stimulation of monovalent ion fluxes (primarily Na^+ and K^+) through the plasma membrane. As discussed in the first part of this chapter, mitogens activate the amiloride-sensitive Na^+/H^+ antiport in a number of cell types. This activation results in an increased Na^+ influx which in turn leads to stimulation of the Na^+/K^+ ATPase and subsequent increase in intracellular K^+ content (Rozengurt 1981). It has been proposed that these steps are critical in the initiation of DNA synthesis (Rozengurt 1980, 1981). In particular, it has been suggested that a low intracellular K^+ concentration is one of the key limiting factors preventing quiescent cells from progressing through the cell cycle, and that the rise in $[K^+]_i$ produced through stimulation of the Na^+/K^+ ATPase would release this putative block. K^+ fluxes through the plasma membrane can be mediated by pathways other than the Na^+ pump, yet little attention was given to this component until recent

Table 2. Modulation of Na/K/Cl cotransport by growth promoters

Agents stimulating Na/K/Cl cotransport		
Agent	*Cell type*	*References*
Serum	NIH 3T3 fibroblasts	Panet (1985)
	Swiss 3T3 fibroblasts	Amsler et al. (1985)
	Rat smooth muscle	Owen (1984)
Epidermal growth factor	Human fibroblasts	Owen and Prastein (1985)
	Hamster fibroblasts	Paris and Pouyssegur (1986)
	Rat glomerular mesangial cells	Homma et al. (1990)
	Rat smooth muscle	Owen (1984)
Fibroblast growth factor	BALB/c 3T3 fibroblasts	Panet et al. (1986)
	Human fibroblasts (HSWP)	Owen and Prastein (1985)
Thrombin	Hamster fibroblasts	Paris and Pouyssegur (1986)
	Rat glomerular mesangial cells	Homma et al. (1990)
Insulin	Swiss 3T3 fibroblasts	Amsler et al. (1985)
Phorbol esters	Hamster fibroblasts	Paris and Pouyssegur (1986)
	Rat glomerular mesangial cells	Homma et al. (1990)
	Human colonic adenocarcinoma (short-term effect)	Franklin et al. (1989)
Okadaic acid	Avian erythrocytes	Pewitt et al. (1990)
	Mouse thick ascending limb (short-term effect)	Kaji (1991)
Agents inhibiting Na/K/Cl cotransport		
Phorbol esters	BALB 3T3 preadipocytes	O'Brien and Krzeminski (1983)
	Cultured endothelial cells	O'Donnell (1989a)
	Rat smooth muscle	Owen (1985)
	Swiss 3T3 fibroblasts	O'Brien et al. (1988b)
	Mouse thick ascending limb (long-term effect)	Kaji (1991)
	Human colonic adenocarcinoma (long-term effect)	Franklin et al. (1989)
Okadaic acid	Mouse thick ascending limb (long-term effect)	Kaji (1991)

years despite the fact that, as early as 1975, Rozengurt and Heppel reported mitogen-stimulated K^+ uptake via a ouabain-insensitive pathway. Identification of this mitogen-stimulated pathway as bumetanide-sensitive Na/K/Cl cotransport was initially made in studies using Swiss 3T3 fibroblasts (Panet et al. 1982; Amsler et al. 1985; Panet 1985). As shown in Table 2, Na/K/Cl cotransport, usually assessed as (ouabain-resistant) bumetanide-sensitive ^{86}Rb influx, has since been reported to be activated by a variety of growth promoters in a number of cell types.

Figure 5 shows an example of the stimulation of bumetanide-sensitive ^{86}Rb uptake induced by EGF, 12-O-tetradecanoylphorbol 13, acetate (TPA) or thrombin in lung hamster fibroblasts. EGF- and TPA-induced activation are nearly additive, suggesting mediation by independent pathways. Thrombin was found to be potent enough to give maximal stimulation by itself. One interesting aspect of this study was the use of a mutant cell line lacking Na^+/H^+ exchange activity, which enabled the authors to rule out any indirect contribution of the Na^+/H^+ antiport to the stimulated K^+ uptake.

As reported for Na^+/H^+ antiport, phosphatase inhibitors such as fluoride and okadaic acid have been found capable of activating Na/K/Cl cotransport

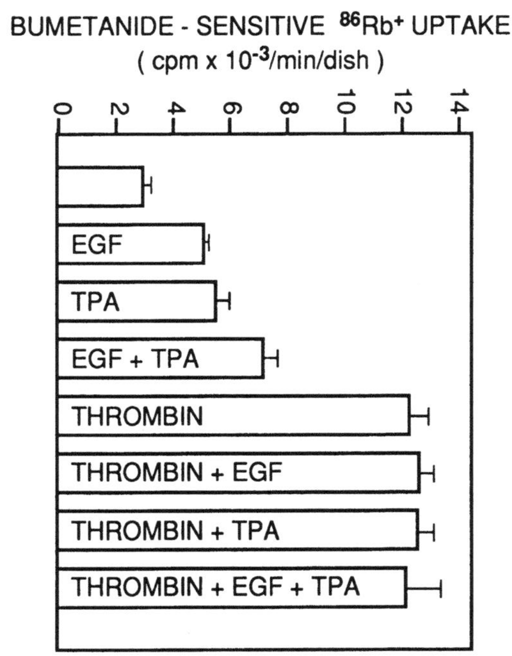

Fig. 5. Comparative effects of *EGF*, *TPA* and *thrombin* on bumetanide-sensitive $^{86}Rb^+$ uptake. Initial rates of bumetanide-sensitive $^{86}Rb^+$ uptake were measured 2 min after addition of the indicated combination of growth factors. EGF, TPA and thrombin were added at 100 ng/ml, 100 ng/ml and 0.1 unit/ml respectively. The first *bar on the left* represents the control. (Paris and Pouyssegur 1986)

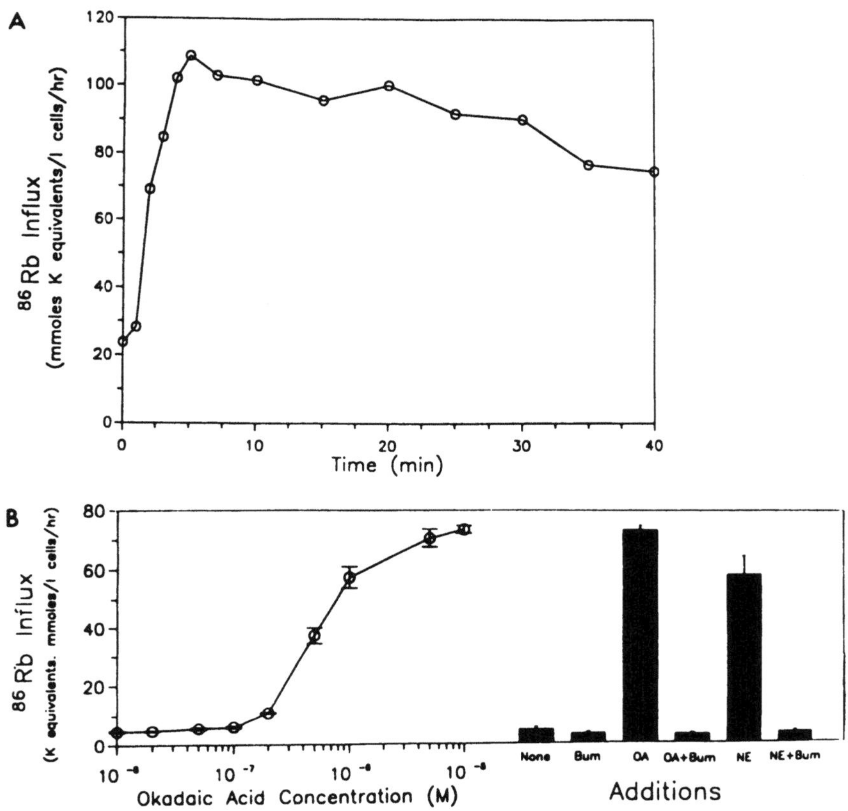

Fig. 6. Kinetics of okadaic acid (OA) effects on Na/K/2Cl cotransport. *A* Time course of OA stimulation of Na/K/2Cl cotransport. Erythrocytes (5% hematocrit) were stimulated with OA (1 μM) at zero time and aliquots were removed for ^{86}Rb influx (30-s uptake) determinations. *B* Dose response and bumetanide sensitivity of okadaic acid stimulation. *Left* ^{86}Rb influx was determined as in *A* after a 7-min pre-incubation with OA (10^{-8}–10^{-5} M). *Right* Bumetanide (10 μM) sensitivity of the ^{86}Rb influx was tested in unstimulated, norepinephrine (NE; 10 μM)- and OA (1 μM)-stimulated red cells. (Pewitt et al. 1990)

by themselves. Figure 6 illustrates the effect of okadaic acid on bumetanide-sensitive ^{86}Rb influx in avian erythrocytes. The time course of the stimulation shows that, after an initial lag of 1 min, the maximum rate of transport was reached 4–5 min after addition of okadaic acid; stimulation of ^{86}Rb influx was still detectable after 40 min of incubation with the phosphatase inhibitor.

Little is known about the molecular mechanism(s) underlying the activation of Na/K/Cl cotransport. Na/K/Cl transport activity has been shown to require intracellular ATP in many cell types (Mc Roberts et al. 1982; Palfrey and Rao 1983; Dagher et al. 1985; Hall and Ellory 1985; Ueberschar and Bakker-Grunwald 1985). This ATP requirement suggests that the symporter may be regulated by phosphorylation/dephosphorylation processes. This hypothesis is supported by the observation that in squid giant axon, intracellular dialysis with ATP-depleting fluids causes bumetanide-sensitive ^{36}Cl influx to decrease (Altamirano et al. 1988). More importantly, these authors demonstrated that fluoride and vanadate can significantly decrease the rate of inactivation of Na/K/Cl cotransport caused by ATP depletion. In addition, treatment with a combination of vanadate and okadaic acid was found to induce a significant increase in the bumetanide-sensitive ^{36}Cl influx (Altamirano et al. 1991). These results led the authors to postulate that phosphorylation of the Na/K/Cl cotransporter, or of a regulatory protein, activates the transport process, whereas dephosphorylation induces inactivation. The same conclusions were reached by Pewitt et al. (1990). Using avian erythrocytes, these investigators also observed stimulation of Na/K/Cl cotransport upon treatment with okadaic acid.

Concerning the effects of growth promoters and phorbol esters, it is noteworthy that these agents are all putative activators of phosphorylation reactions. A limited number of studies have probed the molecular mechanism of activation of the cotransporter. Homma et al. (1990) examined the intracellular mediators of stimulation of Na/K/Cl cotransport by thrombin and EGF in rat glomerular mesangial cells. They demonstrated that stimulation of Na/K/Cl cotransport by thrombin occurs through two separate pathways, one mediated by PKC and the other involving a Ca^{2+}/calmodulin-dependent process, most likely a protein kinase. In similar studies, EGF-induced stimulation was found to be mediated by an as yet unidentified third pathway independent of PKC.

In view of the putative role of PKC in the regulation of Na/K/Cl cotransport, it is interesting to note that phorbol esters have been reported to modulate Na/K/Cl cotransport activity in a differential manner depending on the cell type examined. Moreover, the time of exposure to the stimulus appears to be critical: in some instances the activation of cotransport is only transient and is followed by a progressive inhibition over a period of 2 h (Franklin et al. 1989; Kaji 1991). Short-term inhibitory effects have nevertheless been reported in vascular smooth muscle cells (Owen 1985), BALB/c 3T3 preadipocytes (O'Brien and Krzeminski 1983) and endothelial cells (O'Donnell 1989a). In addition to these direct inhibitory effects, TPA has been shown to inhibit both thrombin- and EGF-induced stimulation of cotransport in hamster fibroblasts (Paris and Pouyssegur 1986), as well as EGF-induced stimulation in vascular smooth muscle cells (Owen 1985). These observations suggest that PKC might exert a feedback control on the stimulation of the cotransporter, this negative control has been tentatively attributed to PKC-mediated phosphorylation of the EGF receptor (Owen 1985). It should be emphasized that the effects of phorbol esters on Na/K/Cl cotransport, whether stimulatory or inhibitory, have been clearly demonstrated to be mediated by activation of PKC (O'Brien et al. 1988a; Owen 1985; Kaji 1991; O'Donnell 1989a).

Binding measurements using radiolabelled loop diuretics have recently provided a new approach for understanding the mechanisms underlying the modulation of Na/K/Cl cotransport activity by various effectors. Studies by Forbush and Palfrey (1983) first characterized the binding of [^{3}H]-bumetanide and [^{3}H]-benzmetanide to microsomal membranes from canine kidney outer medulla. Since then, a number of studies of binding of tritiated loop diuretics to intact cells and membrane preparations from tissues exhibiting Na/K/Cl cotransport have been performed (reviewed in Haas 1989). Binding of radiolabelled diuretics to intact cells (Haas and Forbush 1986), as well as to cell membranes (Forbush and Palfrey 1983; Giesen-Crouse et al. 1985; Giesen-Crouse and McRoberts 1987; O'Grady et al. 1987b), was shown to closely parallel Na/K/Cl cotransport with respect to inhibition by loop diuretics and dependence on Na^+, K^+ and Cl^-, providing convincing evidence that the sites detected by binding studies are the inhibitory moiety of the cotransporter. A few studies have focussed on the correlation between the effects of phorbol esters on Na/K/Cl cotransport and on [^{3}H]-bumetanide binding. O'Donnell (1989b) reported that, in vascular endothelial cells, TPA reduces the saturable binding of [^{3}H]-bumetanide to the same extent that it inhibits Na/K/Cl cotransport. In agreement with these results, Franklin et al. (1989), using human colonic adenocarcinoma cells, demonstrated that phorbol esters produce time- and dose-dependent decreases in specific [^{3}H]-bumetanide binding similar to the observed decreases in cotransport. They concluded that, in these cells, phorbol esters induce a decrease in the number of cotransporters resulting in a reduction in Na/K/Cl cotransport activity. The same correlation between decreased number of specific [^{3}H]-bumetanide binding sites and inhibition of cotransport activity upon treatment with either TPA or okadaic acid has been reported for medullary thick ascending limb cells (Kaji 1991). These results led the authors to postulate that phosphorylation of the cotransporter by PKC would result in inactivation of the cotransporter, whereas dephosphorylation (prevented by the phosphatase inhibitor okadaic acid) would lead to its activation. The inactivation process could involve either a reduction in the synthesis or insertion of new cotransporters into the membrane, or an increased degradation or internalization of the existing cotransporters.

3.2.2 Role of Na/K/Cl Cotransport in Mitogenic Signal Transduction

The dramatic stimulation of Na/K/Cl cotransport observed upon treatment with growth factors in various cell types (Table 2) raises the question of whether cotransport plays a role in mitogenic signal transduction. As early as 1982, Lopez-Rivas and collaborators reported that, in fibroblasts, DNA synthesis is reduced when the intracellular K^+ content of hormonally stimulated cells is artificially kept at (or near) the quiescent level, suggesting that intracellular K^+ may play a role in the proliferative response. In agreement with these results, Panet (1985) found that, by mediating an uphill uptake of K^+, Na/K/Cl cotransport might increase $[K^+]_i$ and thus facilitate protein and DNA synthesis. Meyer et al. (1991) demonstrated that furosemide is capable of inhibiting the mitogenic response following expression of the transforming oncogene Ha-ras or addition of bombesin, indicating involvement of the furosemide-sensitive K^+ uptake in mitogenic signal transduction. An interesting and complementary approach to this question was introduced by O'Brien and colleagues, who tested the ability of phorbol esters to stimulate the growth of quiescent BALB/c 3T3 mutants defective in Na/K/Cl cotransport (Sussman and O'Brien 1985; O'Brien and Prettyman 1987). They reported that the TPA-induced inhibitory effects on Na/K/Cl cotransport and the associated volume decrease (≈20%) observed in the parental cell line were absent in the mutant cells. They also showed that the proliferative response of the Na/K/Cl cotransport-defective cells to TPA is considerably reduced compared to the parental cell line, suggesting again that the signals generated by phorbol ester modulation of Na/K/Cl cotransport are necessary for growth stimulation of quiescent cells.

In contrast, Amsler et al. (1985) observed that, in Swiss 3T3 fibroblasts, although Na/K/Cl cotransport is rapidly activated upon stimulation with serum or mitogenic hormones, the cotransport system does not seem to play a major role in the regulation of intracellular K^+, nor in mitogen-stimulation of DNA synthesis. The same conclusions were reached by Paris and Pouyssegur (1986), who reported that complete inhibition of Na/K/Cl cotransport in hamster fibroblasts produced only a slight inhibition (28%) of thrombin-induced re-initiation of DNA synthesis. These authors postulated that Na/K/Cl cotransport might play a role in $[K^+]_i$ regulation equivalent to that of the Na^+/H^+ antiport in pH_i regulation. Namely, whereas Na^+/H^+ activity becomes of crucial importance for the mitogenic response below a threshold pH value (Pouyssegur et al. 1985), Na/K/Cl cotransport might play a major role only below a critical intracellular K^+ concentration.

Finally, it should be pointed out that there are several cell types where growth promoters have been shown to activate both Na^+/H^+ antiport and $Na^+/K^+/Cl^-$ cotransport (e.g., Swiss 3T3 fibroblasts, vascular smooth muscle cells, human fibroblasts and hamster fibroblasts). Panet et al. (1989) recently reported that amiloride or bumetanide, when added separately to BALB/c 3T3 fibroblasts, only partially inhibited DNA synthesis induced by serum. However, addition of both drugs together with serum was found to cause total inhibition of DNA synthesis. From these observations it can be suggested that Na^+/H^+ antiport and Na/K/Cl cotransport may play a synergistic role in mitogenic signal transduction.

As in the case of the Na^+/H^+ antiport, the effect of activation of Na/K/Cl cotransport on cell volume has been mostly studied in the context of volume regulation. Very few data on the consequences of modulation of Na/K/Cl cotransport by growth promoters on cell volume are available. Activation of Na/K/Cl cotransport is thought to produce a net uptake of salt accompanied by osmotically obliged water. Growth promoters are therefore likely to induce an increase in cell volume through stimulation of this pathway as well. Whether the mitogen-induced volume change is itself necessary for growth and proliferation still remains a matter of speculation.

4 Concluding Remarks

In conclusion, it has been convincingly demonstrated that a variety of growth promoters activate Na/H exchange and/or Na/K/Cl cotransport in many mammalian cell types. It is also clear that activation of these transport systems can induce substantial increases in cell volume. The latter is certainly the case for Na/H exchange, inasmuch as the combined chemical gradient of Na^+ plus H^+ is poised to drive Na^+ inwards. In contrast, volume gain will not necessarily occur when Na/K/Cl is activated, since this system can be at or near electrochemical equilibrium. Thus, the intra- and extracellular activities of the transported ions need to be defined before concluding that net transport, and therefore volume gain, will occur when cotransport is stimulated.

The mechanism of activation of the antiport and symport likely involves phosphorylation of the transport proteins themselves. Occurrence of direct phosphorylation of the Na/H antiport has been recently documented and available data are suggestive also of phosphorylation of the cotransporter. However, definitive molecular identification of the latter will be required to confirm this notion. Moreover, it remains to be shown that the functional changes undergone by the transporters are the consequence of phosphorylation. Site-directed mutagenesis experiments eliminating the putative phosphorylation sites will resolve this question.

Finally, it is important to stress that the requirement and significance of the activation of inorganic ion transport to the proliferative process remain ambiguous. It is presently unclear whether transport activation is necessary or even beneficial for cell growth. Volume gain may be important to accommodate newly synthesized macromolecules and organelles. Swelling may also transduce signals, possibly through mechanical distension of the cytoskeleton. Further work will be required to clarify these issues.

Acknowledgements. Research in the authors' laboratory is supported by the Canadian Cystic Fibrosis Foundation, the National Cancer Institute, the Medical Research Council and by an International Scholars Award of the Howard Hughes Medical Institute.

References

Altamirano AA, Breitwieser GE, Russell JM (1988) Vanadate and fluoride effects on Na^+-K^+-Cl^- cotransport in squid giant axon. Am J Physiol 254: C582–C586

Altamirano AA, Breitwieser GE, Russell JM (1991) Effects of okadaic acid and staurosporine on Na, K, Cl cotransport in squid giant axon. Biophys J 59: 331a

Amsler K, Donahue JJ, Slayman CW, Adelberg EA (1985) Stimulation of bumetanide-sensitive K^+ transport in Swiss 3T3 fibroblasts by serum and mitogenic hormones. J Cell Physiol 123: 257–263
Aronson PS (1985) Kinetic properties of the plasma membrane Na^+-H^+ exchanger. Annu Rev Physiol 47: 545–560
Aronson PS, Nee J, Suhm MA (1982) Modifier role of internal H^+ in activating the Na^+-H^+ exchanger in renal microvillus membrane vesicles. Nature 299: 161–163
Benos DJ (1988) Amiloride: chemistry, kinetics, and structure-activity relationships. In: Grinstein S (ed) Na^+/H^+ exchange. CRC Press, Boca Raton, pp 121–136
Benos DJ, Sapirstein VS (1983) Characteristics of an amiloride-sensitive sodium entry pathway in cultured rodent glial and neuroblastoma cells. J Cell Physiol 116: 213–220
Besterman JM, Cuatrecasas P (1984) Phorbol esters rapidly stimulate amiloride-sensitive Na^+/H^+ exchange in a human leukemic cell line. J Cell Biol 99: 340–343
Bialojan C, Takai A (1988) Inhibitory effect of a marine-sponge toxin, okadaic acid, on protein phosphatases. Biochem J 256: 283–290
Bianchini L, Woodside M, Sardet C, Pouyssegur J, Takai A, Grinstein S (1991) Okadaic acid, a phosphatase inhibitor, induces activation and phosphorylation of the Na^+/H^+ antiport. J Biol Chem 266: 15406–15413
Bierman AJ, Moolenaar WH (1987) Constitutive activation of Na^+/H^+ exchange in mouse embryonal carcinoma cells. Xth ISN Congress. Satellite Symp on Structure function and regulation of membrane transport proteins, stanssted, p 80
Bierman AJ, Tertoolen LGJ, de Laat SW, Moolenaar WH (1987) The Na^+/H^+ exchanger is constitutively activated in P19 embryonal carcinoma cells, but not in a differentiated derivative. J Biol Chem 262: 9621–9628
Bierman AJ, Cragoe EJ, de Laat SW, Moolenaar WH (1988) Bicarbonate determines cytoplasmic pH and suppresses mitogen-induced alkalinization in fibroblastic cells. J Biol Chem 263: 15253–15256
Boonstra K, Moolenaar WH, Harrison PH, Moed P, van der Saag PT, de Laat SW (1983) Ionic responses and growth stimulation induced by nerve growth factor and epidermal growth factor in rat pheochromocytoma (PC12) cells. J Cell Biol 97: 92–98
Boron WF, Boulpaep EL (1983) Intracellular pH regulation in the renal proximal tubule of the salamander. Na-H exchange. J Gen Physiol 81: 29–52
Brown CD, Murer H (1985) Characterization of a Na:K:2Cl cotransport system in the apical membrane of a renal epithelial cell line (LLC-PK_1), J Membr Biol 87: 131–139
Burnham CE, Kidd J, Palfrey HC (1990) Expression of the Na^+-K^+-$2Cl^-$ cotransporter in *Xenopus* oocytes. Am J Physiol 259: F383–F388
Burns CP, Rozengurt E (1983) Serum, platelet-derived growth factor, vasopressin and phorbol esters increase intracellular pH in Swiss 3T3 cells. Biochem Biophys Res Commun 116: 931–938
Cala PM (1980) Volume regulation by *Amphiuma* red blood cells – the membrane potential and its implications regarding the nature of the ion flux pathways. J Gen Physiol 76: 683–708
Cassel D, Rothenberg P, Zhuang YX, Deuel TF, Glaser L (1983) Platelet-derived growth factor stimulates Na^+/H^+ exchange and induces cytoplasmic alkalinization in NR6 cells. Proc Natl Acad Sci USA 80: 6224–6228
Cassel D, Zhuang YX, Glaser L (1984) Vanadate stimulates Na^+/H^+ exchange activity in A431 Cells. Biochem Biophys Res Commun 118: 675–681
Cassel D, Whiteley B, Zhuang YX, Glaser L (1985) Mitogen-independent activation of Na^+/H^+ exchange in human epidermoid carcinoma A431 cells: regulation by medium osmolarity. J Cell Physiol 122: 178–186
Dagher G, Brugnara C, Canessa M (1985) Effects of metabolic depletion on the furosemide-sensitive Na and K fluxes in human red cells. J Membr Biol 86: 145–155
Daniel TO, Ives HE (1987) Cyclosporin A inhibits kinase C-independent activation of the Na^+/H^+ exchanger by PDGF and vanadate. Biochem Biophys Res Commun 145: 111–117
Dixon SJ, Cohen S, Cragoe EJ, Grinstein S (1987) Estimation of the number and turnover rate of Na^+/H^+ exchangers in lymphocytes. J Biol Chem 262: 3626–3632

Doppler W, Maly K, Grunicke H (1986) Role of the Na^+/H^+ antiport in the regulation of internal pH of Ehrlich ascites tumor cells in culture. J Membr Biol 91: 147–155
Dunham PB, Jessen F, Hoffmann EK (1990) Inhibition of Na-K-Cl cotransport in Ehrlich ascites cells by antiserum against purified proteins of the cotransporter. Proc Natl Acad Sci USA 87: 6828–6832
Fehlman M, Freychet P (1981) Insulin and glucagon stimulation of (Na-K)-ATPase transport activity in isolated rat hepatocytes. J Biol Chem 256: 7449–7453
Feit PW, Hoffmann EK, Schiodt M, Kristensen P, Jessen F (1988) Purification of proteins of the Na/Cl cotransporter from membranes of Ehrlich ascites cells using a bumetanide-Sepharose affinity column. J Membr Biol 103: 135–147
Forbush B III, Palfrey HC (1983) [^{3}H]Bumetanide binding to membranes isolated from dog kidney outer medulla. J Biol Chem 19: 11787–11792
Franklin CC, Turner JT, Kim HD (1989) Regulation of $Na^+/K^+/Cl^-$ cotransport and [^{3}H] bumetanide binding site density by phorbol esters in HT29 cells. J Biol Chem 264: 6667–6673
Ganz MB, Boyarsky G, Sterzel RB, Boron WF (1989) Arginine vasopressin enhances pHi regulation in the presence of HCO_3^- by stimulating three acid-base transport systems. Nature 337: 648–651
Geck P, Pfeiffer B (1985) $Na^+/K^+/2Cl^-$ cotransport in animal cells and its volume regulation. Ann NY Acad Sci 456: 166–182
Geck P, Pietrzyk C, Burckhardt BC, Pfeiffer B, Heinz E (1980) Electrically silent cotransport of Na, K and Cl in Ehrlich cells. Biochim Biophys Acta 600: 432–447
Gelfand EW, Cheung RK, Grinstein S (1988) Calcium-dependent intracellular acidification dominates the pH response to mitogen in human T cells. J Immunol 140: 246–252
Giesen-Crouse EM, McRoberts JA (1987) Coordinate expression of piretanide receptors and Na, K, Cl cotransport activity in Madin-Darby canine kidney cell mutants. J Biol Chem 262: 17393–17397
Giesen-Crouse EM, Welsch C, Imbs JL, Schmidt M, Schwartz J (1985) Characterization of a high affinity piretanide receptor on kidney membranes, Eur J Pharmacol 114: 23–31
Gillies RJ, Martinez-Zaguilan R (1991) Regulation of intracellular pH in BALB/c 3T3 cells. J Biol Chem 266: 1551–1556
Greenberg SM, Simons ER (1984) Cation gradient dependence of the steps in thrombin stimulation of human platelets. J Biol Chem 259: 1502–1509
Greger R (1985) Ion transport mechanisms in thick ascending limb of Henle's loop of mammalian nephron. Physiol Rev 65: 760–797
Grinstein S, Foskett JK (1990) Ionic mechanisms of cell volume regulation in leukocytes. Annu Rev Physiol 52: 399–414
Grinstein S, Rothstein A (1986) Mechanisms of regulation of the Na^+/H^+ exchanger. J Membr Biol 90: 1–12
Grinstein S, Cohen S, Rothstein A (1984) Cytoplasmic pH regulation in thymic lymphocytes by an amiloride-sensitive Na^+/H^+ antiport. J Gen Physiol 83: 341–369
Grinstein S, Cohen S, Goetz JD, Rothstein A (1985a) Osmotic and phorbol ester-induced activation of Na^+/H^+ exchange: possible role of protein phosphorylation in lymphocyte volume regulation. J Cell Biol 101: 269–276
Grinstein S, Cohen S, Goetz JD, Rothstein A, Gelfand EW (1985b) Characterization of the activation of Na^+/H^+ exchange in lymphocytes by phorbol esters: change in cytoplasmic dependence of the antiport. Proc Natl Acad Sci USA 82: 1429–1433
Grinstein S, Elder B, Furuya W (1985c) Phorbol ester-induced changes of cytoplasmic pH in neutrophils: role of exocytosis in Na^+/H^+ exchange. Am J Physiol 248: C379–386
Grinstein S, Mack E, Mills GB (1986) Osmotic activation of the Na^+/H^+ antiport in protein kinase C-depleted lymphocytes. Biochem Biophys Res Commun 134: 8–13
Grinstein S, Rotin D, Mason MJ (1989) Na^+/H^+ exchange and growth factor-induced cytosolic pH changes. Role in cellular proliferation. Biochim Biophys Acta 988: 73–97
Gurich RW, Warnock DG (1986) Electrically neutral Na-H exchange in endosomes obtained from rabbit renal cortex. Am J Physiol 251: F702–F709

Haas M (1989) Properties and diversity of (Na-K-Cl) cotransporters. Annu Rev Physiol 51: 443–457
Haas M, Forbush B III (1986) [3H]Bumetanide binding to duck red cells. J Biol Chem 261: 8434–8441
Haas M, Forbush B III (1987) Photolabeling of a 150-kDa (Na-K-Cl) cotransport protein from dog kidney with a bumetanide analogue. Am J Physiol 253: C243–C252
Haas M, Forbush B III (1988) Photoaffinity labelling of a 150 kDa (Na-K-Cl) cotransport protein from duck red cells with an analog of bumetanide. Biochim Biophys Acta 939: 131–144
Haas M, McManus TJ (1985) Effect of norepinephrine on swelling-induced potassium transport in duck red cells. Evidence against a volume-regulatory decrease under physiological conditions. J Gen Physiol 85: 649–667
Haas M, Schmidt WF III, McManus TJ (1982) Catecholamine-stimulated ion transport in duck red cells. Gradient effects in electrically neutral (Na+K+2Cl) co-transport. J Gen Physiol 80: 125–147
Haas M, Dunham PB, Forbush B III (1991) [^{3}H]Bumetanide binding to mouse kidney membranes: identification of corresponding membrane proteins. Am J Physiol 260: C791–C804
Haggerty JG, Agarwal N, Amsler K, Slayman CW, Adelberg EA (1987) Stimulation by serum of the Na^+/H^+ antiporter in quiescent pig kidney epithelial (LLC-PK_1) cells and role of the antiporter in the reinitiation of DNA synthesis. J Cell Physiol 132: 173–177
Haggerty JG, Agarwal N, Cragoe EJ, Adelberg EA, Slayman CW (1988) LLC-PK_1 mutant with increased Na^+-H^+ exchange and decreased sensitivity to amiloride. Am J Physiol 255: C495–C501
Hall AC, Ellory JC (1985) Measurement and stoichiometry of bumetanide-sensitive (2Na:1K:3Cl) cotransport in ferret red cells. J Membr Biol 85: 205–213
Hannafin J, Kinne-Saffran E, Friendman D, Kinne R (1983) Presence of a sodium-potassium chloride cotransport system in the rectal gland of *Squalus acanthias*. J Membr Biol 75: 73–83
Haystead TAJ, Sim ATR, Carling D, Honnor RC, Tsukitani Y, Cohen P, Hardie DG (1989) Effects of the tumour promoter okadaic acid on intracellular protein phosphorylation and metabolism. Nature 337: 78–81
Hendey B, Mamrack MD, Putnam RW (1989) Thrombin induces a calcium transient that mediates an activation of the Na^+/H^+ exchanger in human fibroblasts. J Biol Chem 264: 19540–19547
Hesketh TR, Moore JP, Morris JDH, Taylor MV, Rogers J, Smith GA, Metcalfe JC (1985) A common sequence of calcium and pH signals in the mitogenic stimulation of eukaryotic cells. Nature 313: 481–484
Homma T, Burns KD, Harris RC (1990) Agonist stimulation of $Na^+/K^+/Cl^-$ cotransport in rat glomerular mesangial cells. J Biol Chem 265: 17613–17620
Ives HE, Yee VJ, Warnock DG (1983) Mixed type inhibition of the renal Na^+/H^+ antiporter by Li^+ and amiloride. J Biol Chem 258: 9710–9716
Kaji D (1991) Regulation of NaK2Cl cotransport in medullary thick ascending limb (mTAL) cells by phosphorylation and dephosphorylation. FASEB J 5: A739
Kakinuma Y, Sakamaki Y, Ito K, Cragoe EJ Jr, Igarashi K (1987) Relationship among activation of the Na^+/H^+ antiporter, ornithine decarboxylase induction and DNA synthesis. Arch Biochem Biophys 259: 171–178
Kinne R, Hannafin JA, Konig B (1985) Role of the NaCl-KCl cotransport system in active chloride absorption and secretion. Ann NY Acad Sci 456: 198–206
Kinsella JL, Aronson PS (1980) Properties of the Na^+/H^+ exchanger in renal microvillus membrane vesicles. Am J Physiol 238: F461–F469
Kinsella JL, Aronson PS (1981) Amiloride inhibition of the Na/H exchanger in renal microvillus membrane vesicles. Am J Physiol 241: F374–F379
Klip A, Ramlal T, Cragoe EJ Jr (1986) Insulin-induced cytoplasmic alkalinization and glucose transport in muscle cells. Am J Physiol 250: C720–C728
Kregenow FM (1977) Transport in avian red cells. In: Ellory JC, Lew LV (eds) Membrane transport in red cells. Academic Press, New York, pp 383–426

Kregenow FM (1981) Osmoregulatory salt transporting mechanisms: control of cell volume in anisotonic media. Annu Rev Physiol 43: 493–505

L'Allemain G, Paris S, Pouyssegur J (1984) Growth factor action and intracellular pH regulation in fibroblasts. Evidence for a major role of the Na^+/H^+ antiport. J Biol Chem 259: 5809–5815

L'Allemain G, Paris S, Pouyssegur J (1985) Role of Na^+-dependent Cl-$HCO3^-$ exchange in regulation of intracellular pH in fibroblasts. J Biol Chem 260: 4877–4883

Lauf PK (1986) K^+Cl^- cotransport: sulfhydryls, divalent cations and the mechanism of volume activation in a red cell. J Membr Biol 88: 1–13

Livne AA, Sardet C, Pouyssegur J (1991) The Na^+/H^+ exchanger is phosphorylated in human platelets in response to activating agents. FEBS Lett 284: 219–222

Lopez-Rivas A, Adelberg EA, Rozengurt E (1982) Intracellular K^+ and the mitogenic response of 3T3 cells to peptide factors in serum-free medium. Proc Natl Acad Sci USA 79: 6275–6279

Mahennsmith RL, Aronson PS (1985) The plasma membrane sodium-hydrogen exchanger and its role in physiological and pathophysiological processes. Circ Res 56: 773–788

McRoberts JA, Erlinger S, Rindler MJ, Saier MH Jr (1982) Furosemide-sensitive salt transport in the Madin-Darby canine kidney cell line: evidence for the cotransport of Na, K and Cl. J Biol Chem 257: 2260–2266

Mendoza SA, Schneider JA, Lopez-Rivas A, Sinnett-Smith JW, Rozengurt E (1986) Early events elicited by bombesin and structurally related peptides in quiescent Swiss 3T3 cells. II. Changes in Na^+ and Ca^{2+} fluxes, Na^+/K^+ pump activity and intracellular pH. J Cell Biol 102: 2223–2233

Meyer M, Maly K, Uberall F, Hoflacher J, Grunicke H (1991) Stimulation of K^+ transport systems by Ha-ras. J Biol Chem 266: 8230–8235

Mills GB, Cragoe EJ Jr, Gelfand EW, Grinstein S (1985) Interleukin 2 induces a rapid increase in intracellular pH through activation of a Na^+/H^+ antiport. J Biol Chem 260: 12500–12507

Moolenaar WH (1986) Effects of growth factors on intracellular pH regulation. Annu Rev Physiol 48: 363–376

Moolenaar WH, Mummery CL, van der Saag PT, de Laat SW (1981) Rapid ionic events and the initiation of growth in serum-stimulated neuroblastoma cells. Cell 23: 789–798

Molenaar WH, Tsien RY, van der Saag PT, de Laat SW (1983) Na^+/H^+ exchange and cytoplasmic pH in the action of growth factors in human fibroblasts. Nature 304: 645–648

Moolenaar WH, Tertoolen LGJ, de Laat SW (1984) The regulation of cytoplasmic pH in human fibroblasts. J Biol Chem 259: 7563–7569

Moolenaar WH, Bierman AJ, de Laat SW (1988) Effects of growth factors on Na^+/H^+ exchange. In: Grinstein S (ed) Na^+/H^+ exchange. CRC Press, Boca Raton, pp 227–234

Moran A, Biber J, Murer H (1986) A sodium-hydrogen exchange system in isolated apical membrane from LLC-PK epithelia. Am J Physiol 251: F1003–F1008

Moule SK, McGivan JD (1990) Epidermal growth factor and cyclic AMP stimulate Na^+/H^+ exchange in isolated rat hepatocytes. Eur J Biochem 187: 677–682

Muldoon LL, Dinerstein RJ, Villereal ML (1985) Intracellular pH in human fibroblasts: effect of mitogens, A23187, and phospholipase activation. Am J Physiol 249: C140–C148

Muldoon LL, Jamieson GA Jr, Kao AC, Dalfrey HC, Villereal ML (1987) Mitogen stimulation of Na^+-H^+ exchange: differential involvement of protein kinase C. Am J Physiol 253: C219–229

Murer H, Hopfer U, Kinne R (1976) Sodium/proton antiport in brush-border membrane vesicles isolated from rat small intestine and kidney. Biochem J 154: 597–604

O'Brien TG, Krzeminski K (1983) Phorbol ester inhibits furosemide-sensitive potassium transport in BALB/c 3T3 preadipose cells. Proc Natl Acad Sci USA 80: 4334–4338

O'Brien TG, Prettyman R (1987) Phorbol esters and mitogenesis: comparison of the proliferative response of parental and $Na^+K^+Cl^-$ cotransport-defective BALB/c 3T3 cells to 12-0-tetradecanoylphorbol-13-acetate. J Cell Physiol 130: 377–381

O'Brien TG, George K, Prettyman R (1988a) Protein kinase C and membrane transport: divergent responses of $Na^+/K^+/Cl^-$ cotransport and sugar transport to exogenous diacylglycerol. Biochim Biophys Acta 945: 41–50
O'Brien TG, Prettyman R, George KS, Herschman HR (1988b) A phorbol ester non-proliferative variant of Swiss 3T3 Cells is deficient in $Na^+K^+Cl^-$ cotransport activity. J Cell Physiol 134: 320–306
O'Donnell ME (1989a) Regulation of Na-K-Cl cotransport in endothelial cells by atrial natriuretic factor. Am J Physiol 257: C36–C44
O'Donnell ME (1989b) [^{3}H] Bumetanide binding in vascular endothelial cells. J Biol Chem 264: 20326–20330
O'Donnell ME, Owen NE (1988) Reduced Na-K-Cl cotransport in vascular smooth muscle cells from spontaneously hypertensive rats. Am J Physiol 255: C169–C180
O'Grady SM, Musch MW, Field M (1986) Stoichiometry and ion affinities of the Na-K-Cl cotransport. J Membr Biol 91: 33–41
O'Grady SM, Palfrey HC, Field M (1987a) Characteristics and functions of Na-K-Cl cotransport in epithelial tissues. Am J Physiol 253: C177–C192
O'Grady SM, Palfrey HC, Field M (1987b) Na-K-Cl cotransport in winter flounder intestine and bovine kidney outer medulla: [^{3}H] bumetanide binding and effects of furosemide analogues. J Membr Biol 96: 11–18
Owen NE (1984) Regulation of Na/K/Cl cotransport in vascular smooth muscle cells. Biochem Biophys Res Commun 125: 500–508
Owen NE (1985) Effect of TPA on ion fuxes and DNA synthesis in vascular smooth muscle cells. J Cell Biol 101: 454–459
Owen NE, Prastein ML (1985) Na/K/Cl cotransport in cultured human fibroblasts. J Biol Chem 260(3): 1445–1451
Palfrey HC, Greengard P (1981) Hormone-sensitive ion transport systems in erythrocytes as models for epithelial ion pathways. Annu NY Acad Sci 372: 291–308
Palfrey HC, Rao MC (1983) Na/K/Cl cotransport and its regulation. J Exp Biol 106: 43–54
Palfrey HC, Alper SL, Greengard P (1980a) Protein phosphorylation and the regulation of cation cotransport. J Exp Biol 89: 103–115
Palfrey HC, Feit PW, Greengard P (1980b) cAMP-stimulated cation cotransport in avian erythrocytes: inhibition by "loop" diuretics. Am J Physiol 238: C139–C148
Palfrey HC, Silva P, Epstein FH (1984) Sensitivity of cAMP-stimulated salt secretion in shark rectal gland to "loop" duiretics. Am J Physiol 246: C242–C246
Panet R (1985) Serum-induced net K^+ influx performed by the diuretic-sensitive transport system in quiescent NIH 3T3 mouse fibroblasts. Biochim Biophys Acta 813: 141–144
Panet R, Fromer I, Atlan H (1982) Differentiation between serum stimulation of ouabain-resistant and sensitive Rb^+ influx in quiescent NIH 3T3 cells. J Membr Biol 70: 165–169
Panet R, Amir I, Atlan H (1986) Fibroblast growth factor induces a transient net K^+ influx carried by the bumetanide-sensitive transporter in quiescent BALB/c 3T3 fibroblasts. Biochim Biophys Acta 859: 117–121
Panet R, Amir I, Snyder D, Zonenshein L, Atlan H, Laskov R, Panet A (1989) Effect of Na^+ flux inhibitors on induction of c-fos, c-myc, and ODC genes during cell cycle. J Cell Physiol 140: 161–168
Paris S, Pouyssegur J (1983) Biochemical characterization of the amiloride-sensitive Na^+/H^+ antiport in Chinese hamster lung fibroblasts. J Biol Chem 258: 3503–3508
Paris S, Pouyssegur J (1984) Growth factors activate the Na^+/H^+ antiporter in quiescent fibroblasts by increasing its affinity for intracellular H^+. J Biol Chem 259: 10989–10994
Paris S, Pouyssegur J (1986) Growth factors activate the bumetanide-sensitive $Na^+/K^+/Cl^-$cotransport in hamster fibroblasts. J Biol Chem 261: 6177–6183
Paris S, Chambard JC, Pouyssegur J (1987) Coupling between phosphoinositide breakdown and early mitogenic events in fibroblasts. J Biol Chem 262: 1977–1983
Pewitt EB, Hegde RS, Haas M, Palfrey HC (1990) The regulation of Na/K/2Cl cotransport and bumetanide binding in avian erythrocytes by protein phosphorylation and dephosphorylation. J Biol Chem 265: 20747–20756

Pouyssegur J, Sardet C, Franchi A, L'Allemain G, Paris S (1984) A specific mutation abolishing Na^+/H^+ antiport activity in hamster fibroblasts precludes growth at neutral and acidic pH. Proc Natl Acad Sci USA 81: 4833–4837
Pouyssegur J, Franchi A, L'Allemain G, Paris S (1985) Cytoplasmic pH, a key determinant of growth factor-induced DNA synthesis in quiescent fibroblasts. FEBS Lett 190: 115–119
Pouyssegur J, Franchi A, L'Allemain G, Magnaldo I, Paris S, Sardet C (1987) Genetic approach to the structure function and regulation of the Na^+/H^+ antiporter. Kidney Int 32: S144–S149
Prasad KVS, Severini A, Kaplan JG (1987) Sodium ion influx in proliferating lymphocytes: an early component of the mitogenic signal. Arch Biochem Biophys 252: 515–525
Putnam RW (1985) Effect of insulin on intracellular pH in frog skeletal muscle fibers. Am J Physiol 248: C330–C336
Rosoff PM, Cantley LC (1985) Lipopolysaccharide and phorbol esters induce differentiation but have opposite effects on phosphatidyl inositol turnover and Ca^{2+} mobilization in 70Z/3 pre-B lymphocytes. J Biol Chem 260: 9209–9215
Rosoff PM, Stein LF, Cantley LC (1984) Phorbol esters induce differentiation in a pre-B lymphocyte cell line by enhancing Na^+/H^+ exchange. J Biol Chem 259: 7056–7060
Rosskopf D, Barth C, Siffert W (1991) Estimation of carrier density and turnover rate of the Na^+/H^+ exchanger in human platelets using 5-(N-methyl-N-[^{3}H]isobutyl)-amiloride. Biochem Biophys Res Commun 176: 639–644
Rozengurt E (1980) Stimulation of DNA synthesis in quiescent cultured cells: exogenous agents, internal signals, and early events. Curr Top Cell Regul 17: 59–88
Rozengurt E (1981) Na^+ influx as a mitogenic signal for cultured fibroblastic cells. In: Moses LW, Schuttz J, Scotl WA, Werner R (eds) Miami Winter Symposium, vol 18. Academic Press, New York, pp 149–167
Rozengurt E, Heppel LA (1975) Serum rapidly stimulates ouabain-sensitive $^{86}Rb^+$ influx in quiescent 3T3 cells. Proc Natl Acad Sci USA 72: 4492–4495
Russell JM (1983) Cation coupled chloride influx in squid axon: role of potassium and stoichiometry of the transport process J Gen Physiol 81: 909–925
Sapirstein VS, Benos DJ (1984) Activation of amiloride-sensitive sodium transport in C6 glioma cells. J Neurochem 43: 1098–1105
Sardet C, Franchi A, Pouyssegur J (1989) Molecular cloning, primary structure and expression of the human growth factor-activatable Na^+/H^+ antiporter. Cell 56: 271–280
Sardet C, Counillon L, Franchi A, Pouyssegur J (1990) Growth factors induce phosphorylation of the Na^+/H^+ antipoter, a glycoprotein of 110 kD. Science 247: 723–726
Sardet C, Fafournoux P, Pouyssegur J (1991) α-Thrombin, epidermol growth Poctor, and okadaic acid activate the Na^+/H^+ exchanger, NHE-1, by phosphorylating a set of common sites. J Biol Chem 266: 19166–19171
Schlatter E, Greger R, Weidtke C (1983) Effect of "high ceiling" diuretics on acute salt transport in the cortical thick ascending limb of Henle's loop of rabbit kidney. Correlation of chemical structure and inhibitory potency. Pflügers Arch Eur J Physiol 396: 210–217
Schuldiner S, Rozengurt E (1982) Na^+/H^+ antiport in Swiss 3T3 cells: mitogenic stimulation leads to cytoplasmic alkalinization. Proc Natl Acad Sci USA 79: 7778–7782
Stewart DJ (1988) Sodium-proton exchanger in isolated hepatocytes exhibits a set point. Am J Physiol 255: G346–G351
Stokes JB (1984) Sodium chloride absorption by the urinary bladder of the winter flounder. A thiazide-sensitive, electrically neutral transport system. J Clin Invest 74: 7–16
Sussman I, O'Brien TG (1985) Characterization of a BALB/c 3T3 preadipose cell mutant with altered $Na^+K^+Cl^-$ cotransport activity. J Cell Physiol 124: 153–159
Swann K, Whitaker M (1985) Stimulation of the Na^+/H^+ exchanger of sea urchin eggs by phorbol ester. Nature 314: 274–277

Szwergold BS, Brown TR, Freed JJ (1988) Bicarbonate abolishes intracellular alkalinization in mitogen-stimulated NIH 3T3 cells. Ann NY Acad Sci 551: 277–279
Takasu N, Nagasawa Y, Komiya I, Shimizu Y, Yamada T (1989) Cytoplasmic pH in cultured porcine thyroid cells alkalinization by epidermal growth factor. Biochim Biophys Acta 1012: 117–120
Too CKL, Cragoe EJ, Friesen HG (1987) Amiloride-sensitive Na^+/H^+ exchange in rat NB2 node lymphoma cells. Stimulation by prolactin and other mitogens. Endocrinology 121: 1512–1520
Turner RJ, George NJ (1990) Solubilization and partial purification of the rabbit parotid Na/K/Cl dependent bumetanide binding site. J Membr Biol 113: 203–210
Ueberschar S, Bakker-Grunwald T (1985) Effects of ATP and cyclic AMP on the (Na^+ + K^+ + $2Cl^-$)-cotransport system in turkey erythrocytes. Biochim Biophys Acta 818: 260–266
Vairo G, Hamilton JA (1988) Activation and proliferation signals in murine macrophages: stimulation of Na/K-ATPase activity by hemopoietic growth factors and other agents. J Cell Physiol 134: 13–24
Vigne P, Frelin C, Audinot M, Borsotto M, Cragoe EJ, Lazdunski M (1984) [^{3}H] Ethylpropylamiloride, a radio labelled diuretic for the analysis of the Na^+/H^+ exchange system. Its use with kidney cell membranes. EMBO J 3: 2647–2651
Vigne P, Frelin C, Lazdunski M (1985) The Na^+/H^+ antiport is activated by serum and phorbol esters in proliferating myoblasts but not in differentiated myotubes. J Biol Chem 260: 8008–8013
Villereal ML (1981) Sodium fluxes in human fibroblasts. Effect of serum, Ca^{2+} and amiloride. J Cell Physiol 107: 359–369
Whiteley B, Cassel D, Zhuang YX, Glaser L (1984) Tumor promoter phorbol 12-myristate 13-acetate inhibits mitogen-stimulated Na^+-H^+ exchange in human epidermoid carcinoma A431 cells. J Cell Biol 99: 1162–1166
Wike-Hooley JL, Vandenberg AP, Vanderzee J, Reinhold HS (1985) Human tumour pH and its variation. Eur J Cancer Clin Oncol 21: 785–792

Chapter 9

The Regulation of Organelle Function Through Changes in Their Volume

A.P. Halestrap[1]

Contents

[1] Department of Biochemistry, School of Medical Sciences, University of Bristol, Bristol BS8 1TD, United Kingdom

1 Introduction

Although a search of the literature provides evidence for morphological changes in many subcellular organelles in response to a variety of stimuli, the only subcellular organelle whose volume regulation has been studied in detail is the mitochondrion. For this organelle there is evidence for an important physiological role for volume regulation, thus this chapter will focus on the regulation of mitochondrial volume and its metabolic consequences. I will first review the mechanisms involved in volume homeostasis and secondly describe how, in the liver at least, mitochondrial volume may also be regulated by extrinsic factors such as hormones. I will then describe how such volume regulation may have profound effects on mitochondrial metabolism that are important for the overall effects of the hormones on liver metabolism. I will suggest that similar effects may follow changes in cellular volume such as those described by Häussinger (this Vol.). The evidence for changes in the mitochondrial volume induced by physiological stimuli in other tissues will also be reviewed briefly. Finally, I will describe how some pathological states are accompanied by massive increases in mitochondrial volume which may cause irreversible damage both to the mitochondria and consequently to the cell. For other reviews on the regulation of mitochondrial volume and function, the reader is referred to Halestrap (1989) and McCormack et al. (1990) where some aspects of the material covered in the present review are considered in more detail.

2 The Regulation of Mitochondrial Volume

2.1 Mitochondria Are Perfect Osmometers

The morphology of isolated mitochondria and those in situ has been studied in great detail using electron microscopy. The effects of changes in diet, energy status and osmotic support have been investigated as well as differences observed between species and tissues (Munn 1974; Bereiter-Hahn 1990). It is well established that mammalian mitochondria are surrounded by two membranes, an outer membrane which is permeable to all molecules of about 6 kDa or less and an inner membrane which is impermeable to all but a limited number of metabolites and ions. The high permeability of the outer membrane is caused by the presence of porins or voltage-dependent anion-selective ion channels (VDACS), proteins which create a non-specific channel through which molecules can cross the membrane (Benz 1985; Colombini 1989). However, there is little evidence to suggest that the outer membrane exerts a permeability barrier to the entry of metabolites or ions into the intermembrane space under physiological conditions, or that the volume of this compartment is regulated (Halestrap 1989).

In contrast to the outer membrane, the permeability of the inner membrane is limited to small polar molecules such as polyols of C4 or less, lipid-soluble compounds, and metabolites for which specific transport processes exist (LaNoue and Schoolwerth 1984). Changing the osmolality of the osmotic support or entry of a permeant solute causes the matrix volume to change as water is taken up to equilibrate the osmotic pressure on either side of the membrane. Indeed, it was shown many years ago that the intramitochondrial volume of deenergized mitochondria responds exactly as would be expected for a well-behaved osmometer. Thus, the mitochondria expand and contract by the cristae folding and unfolding, the membrane supporting no apparent turgor pressure. When the physical limit of swelling is approached, the mitochondria become leaky and lose some of their osmotically active material (see Beavis et al. 1985; Halestrap 1989).

The limited permeability of the inner membrane is important for two interrelated aspects of mitochondrial function. Firstly, it allows the existence of a metabolic compartment within the cell quite distinct from the cytosol, although interconnected with it through the operation of specific transport mechanisms (LaNoue and Schoolwerth 1984). Such compartmentation of metabolism is an essential feature of metabolic control in higher organisms. Secondly, the limited permeability of the inner membrane is essential for proton pumping by the respiratory chain to generate a membrane potential and pH gradient across the membrane (see McCormack et al. 1990). These two gradients not only participate in the production of ATP (according to the chemiosmotic hypothesis) but also influence the activity of specific transport systems for ions and metabolites. This in turn allows the mitochondrial compartment to contain a different composition of permeant metabolites and ions to that of the cytosol. One such ion, K^+, is the major intramitochondrial cation and through the operation of specific influx and efflux pathways that are driven by the proton motive force, the mitochondrial content of K^+ can be regulated. If movement of a compensating anion such as phosphate also occurs, there will be an imbalance of osmotic pressure across the membrane and this will cause a change in the intramitochondrial volume. For more information on how such changes in mitochondrial volume can be measured in vitro and in vivo by isotopic or electron microscope techniques, the reader is directed elsewhere (Halestrap 1989).

2.2 Mitochondrial Volume Is Regulated by Movement of K^+ Across the Inner Mitochondrial Membrane

It is now well established that the mitochondrial inner membrane contains K^+/H^+ antiporter which pumps K^+ out of the mitochondria and an electrogenic uniporter which mediates K^+ entry into the mitochondria (Diwan 1987; Garlid 1988). It is thought that the operation of these two transporters, in conjunction with the parallel movement of permeant anions such as phosphate, is responsible for the regulation of the intramitochondrial volume (see Garlid 1988; Halestrap 1989). Increasing the matrix volume by hypoosmotic shock or addition of the K^+ ionophore valinomycin leads to an increase in the activity of the K^+/H^+ antiporter, loss of K^+ from the matrix with a compensating anion and thus prevention of further swelling. Garlid (1980) has proposed that Mg^{2+} acts as a natural inhibitor of the K^+/H^+ antiporter whose concentration in the matrix decreases as the matrix volume increases and so allows stimulation of the antiporter. Several laboratories have shown that treatment of mitochondria with A23187 and EDTA to deplete them of Mg^{2+} causes stimulation of K^+/H^+ fluxes (see Garlid 1980; Diwan 1987). However, it has been pointed out (Jung and Brierley 1986; Bernadi et al. 1989) that depletion of Mg^{2+} might cause non-specific effects on membrane permeability since it is known that bound Mg^{2+} is important for membrane stability. Furthermore, the physiological concentration of $[Mg^{2+}]$ within the matrix has now been measured using fluorescent probes and appears to be 0.5–1.5 mM (Jung et al. 1990; Rutter et al. 1990). Since the total concentration of Mg^{2+} within the matrix is 20–40 mM (Diwan 1987), it is clear that the majority of it must be bound to metabolites and phospholipids. Thus, the free $[Mg^{2+}]$ would not be expected to change greatly with a change in matrix volume (Brierley et al. 1987). Recent measurements of the sensitivity of the K^+/H^+ antiporter towards $[Mg^{2+}]$ suggest that half-maximal inhibition is observed at 0.05–0.15 mM at pH 7.6–7.8 in heart mitochondria and reconstituted antiporter respectively (Kakar et al. 1989; Brierley and Jung 1990). This implies that the antiporter must be very strongly inhibited at physiological $[Mg^{2+}]$ and thus unlikely to be regulated by $[Mg^{2+}]$ under normal physiological conditions (Jung and Brierley 1986; Brierley and Jung 1990). Furthermore, there is also substantial evidence for the K^+ uniporter being inhibited by both matrix and cytosolic $[Mg^{2+}]$ which would tend to negate any effect of $[Mg^{2+}]$ on the antiporter (see Diwan 1987; Bernadi et al. 1989; Halestrap and Davidson 1989). Thus, it is likely that another mechanism must operate to regulate the K^+/H^+ antiporter in response to matrix volume and Beavis and Garlid (1990) have suggested that this may be through activation of the antiporter by an increase in pH. However, as argued by Brierley and Jung (1990), the pH is unlikely to rise sufficiently high under physiological conditions for this to be an effective means of regulation. A more probable mechanism would be a volume-induced change in the folding of the cristae or some other perturbation of the membrane structure causing activation of the antiporter (Bernadi et al. 1982; Jung and Brierley 1986) as will be discussed below (Sect. 3.2).

The properties of the influx and efflux pathways for K^+ have been studied in some detail and are reviewed elsewhere (Diwan 1987; Brierley and Jung 1988; Halestrap 1989). The former has a K_m for K^+ of 6–12 mM and is activated by thiol reagents, phosphate and high pH. In addition, it shows activation by atractylate and adenine nucleotide depletion, and inhibition by ADP, which suggests some association with the adenine nucleotide transporter. The activity of the K^+/H^+ antiporter is inhibited by quinine analogues, Mg^{2+}, low pH and DCCD. It has been identified as an 82 kDa protein which has been purified and reconstituted into proteoliposomes (Li et al. 1990).

2.3 Hormones Increase the Mitochondrial Volume in Hepatocytes

Some 10 years ago we demonstrated that hormones which stimulated hepatic gluconeogenesis and respiration caused a decrease in the light scattering of hepatocyte suspensions (Quinlan et al. 1983). The use of light scattering to measure changes in the matrix volume of isolate mitochondria is well accepted (see Garlid and Beavis 1985; Halestrap 1989), and there were indications in the literature that this technique might be capable of detecting mitochondrial volume changes in situ (Hackenbrock et al. 1971). We were able to confirm that the changes in light scattering induced by various treatments of the hepatocyte correlated with measurements of mitochondrial volume but not with the whole cell volume (Quinlan et al. 1983; Halestrap 1989). To measure the mitochondrial volume in situ we devised two techniques (Quinlan et al. 1983). Firstly, we used a sonication technique to rapidly disrupt the liver cells in the presence of ^3H_2O and $[^{14}C]$-sucrose, followed by centrifugation and measurement of the ^3H and ^{14}C in the pellet.

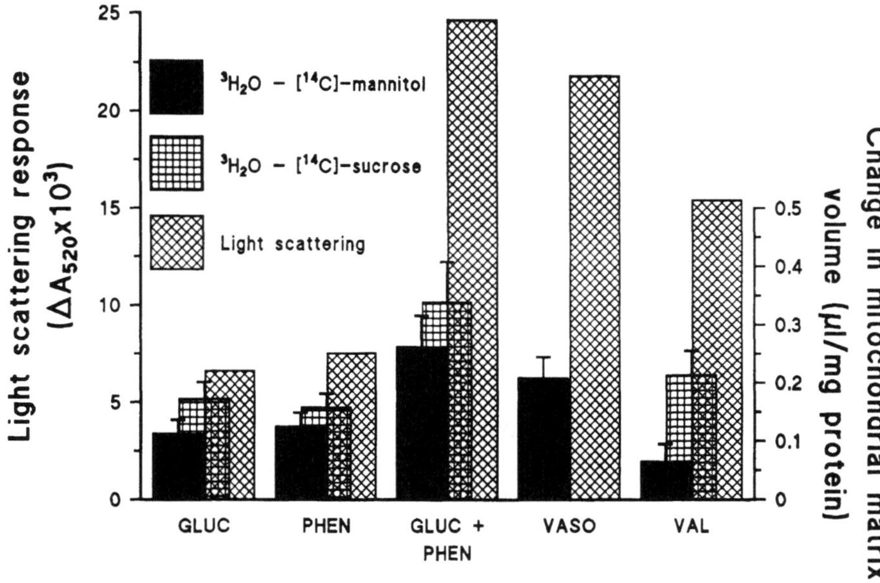

Fig. 1. Correlation between the hormonally induced changes in light scattering of hepatocytes and mitochondrial volumes measured in situ. Data are taken from the sources cited in Halestrap (1989). Where *error bars* are given they represent the SEM of 3–15 separate experiments

Secondly, we incubated cells in the presence of [^{14}C]-mannitol and ^3H_2O until the uptake of [^{14}C]-mannitol had reached equilibrium (45 min). The mannitol was assumed to equilibrate across the plasma membrane but not the inner mitochondrial membrane, and thus measurement of the ^{14}C and ^3H in the cell pellet after centrifugation should give an estimate of the mitochondrial volume. In our hands both the disruptive and non-disruptive techniques were able to detect an increase in matrix volume which correlated with the decrease in light scattering. This is summarized in Fig. 1. The largest effects were obtained with the hormones vasopressin or the α-adrenergic agonist phenylephrine in the presence of glucagon which increased the matrix volume by 20–30%, whilst the increase caused by phenylephrine or glucagon alone was only about 10%. The K^+ ionophore valinomycin produced similar results at low concentrations (1 nM) as might be expected, and this has been confirmed by Brown et al. (1990). However, these workers were unable to detect any hormonally induced changes, although the conditions used were not the same as those used in this laboratory. The use of [^{14}C]-mannitol and ^3H_2O has also been used by Lund and Wiggins (1987) to measure mitochondrial volumes in situ but these authors equilibrated with the mannitol for less time and thus obtained different results to our own (Halestrap 1989).

2.4 The Role of Ca^{2+} in the Hormonal Regulation of Mitochondrial Volume

For hormones to exert their effects on the intramitochondrial volume an intracellular signal must be generated to cause the increase in K^+ permeability of the inner mitochondrial membrane. In the case of the Ca^{2+}-mobilizing hormones such as phenylephrine and vasopressin an obvious candidate for this signal is Ca^{2+}. Evidence for an important role for Ca^{2+} came from the observation that vasopressin and phenylephrine failed to elicit a light-scattering response of any magnitude when extracellular Ca^{2+} was removed just before hormone addition (Quinlan et al. 1983; Halestrap et al. 1986). However, addition of Ca^{2+} back to the cells after vasopressin (but not glucagon) gave a large light-scattering response not observed in the absence of hormone (Halestrap et al. 1986). Furthermore, the addition of the Ca^{2+} ionophore A23187 to hepatocytes also produced a decrease in light scattering concomitant with an increase in gluconeogenesis (Halestrap et al. 1986).

Additional evidence for a role of Ca^{2+} came from studies of the effects of Ca^{2+} on isolated mitochondria (Halestrap et al. 1986). Mitochondria were incubated under energized conditions in the presence of albumin, Mg^{2+}, phosphate, ATP and EGTA, and Ca^{2+} added to give a free [Ca^{2+}] of 0.1–1.0 μM, similar to the range of cytosolic Ca^{2+} concentrations observed in hormone-stimulated hepatocytes. A decrease in light scattering occurred with a time course similar to that seen in hepatocytes following hormone addition. The half-maximal response was seen at about 0.3 μM Ca^{2+} and it was shown to be associated with an uptake of K^+ and an increase in matrix volume (Halestrap et al. 1986; Davidson and Halestrap 1987). Rb^+ could replace K^+, whilst Na^+, Li^+ and choline were progressively less effective. The presence of a respiratory chain inhibitor or Ruthenium Red to block the electrogenic influx pathway for Ca^{2+} prevented the swelling unless high concentrations of Ca^{2+} were added. Thus, it appears that Ca^{2+} must enter the mitochondria in order to exert its effect. Ba^{2+}, Mn^{2+} and Sr^{2+} were unable to substitute for Ca^{2+}. Swelling occurred without uncoupling or any non-specific permeability increases in the inner membrane and was not affected by inhibitors of phospholipase A_2. Neither were inhibitors of the known ion transport mechanisms of the plasma membrane able to inhibit swelling at concentrations that would cause maximal inhibition of plasma membrane ion fluxes, although quinine was slightly inhibitory at 100 μM as has been described for the K^+ influx mechanism for isolated mitochondria (Diwan 1987). Swelling was enhanced if ATP was omitted from the buffer or, in the presence of ATP, if carboxyatractyloside was added. Addition of carboxyatractyloside in the absence of Ca^{2+} had no effect on swelling. These data suggest some involvement of the adenine nucleotide translocator in the Ca^{2+}-stimulated entry of K^+ into mitochondria, and this is consistent with observations of others on the entry of K^+ into mitochondria (see Sect. 2.2).

Ca^{2+}-mobilizing hormones increase cytoplasmic [Ca^{2+}] through two mechanisms involving both intra- and extracellular pools (see Williamson and Monck 1989; Putney 1990). The initial and rapid release of Ca^{2+} from the IP_3-sensitive stores of the endoplasmic reticulum is only transient when extracellular Ca^{2+} is present to prevent loss of the released Ca^{2+} from the cell. A second phase of Ca^{2+} mobilization involves Ca^{2+} entry from outside and this leads to a net accumulation of Ca^{2+} by the cell and its mitochondria. Since the hormonal increase in mitochondrial volume requires extracellular Ca^{2+}, it must be the sustained increase in mitochondrial [Ca^{2+}] that is responsible for the increase in the mitochondrial volume rather than the rapid transient spikes. This conclusion is strengthened by the observed synergistic effect of glucagon (or cyclic AMP) on both the light-scattering response and mitochondrial accumulation of [Ca^{2+}] brought about by Ca^{2+}-mobilizing hormones (Quinlan et al. 1983; Altin and Bygrave 1986, 1987; Assimacopoulos-Jeannet et al. 1986; Davidson and Halestrap 1988).

The mechanism by which glucagon alone increases the mitochondrial volume is less clear. When added to suspensions of hepatocytes, glucagon produces a rapid and significant increase in cytoplasmic [Ca^{2+}] which can be mimicked by addition of cyclic AMP or forskolin (Sistare et al. 1985; Staddon and Hansford 1989). However, in single hepatocytes cyclic AMP appears to be unable to elicit an increase in cytosolic [Ca^{2+}] (Burgess et al. 1991; Schöfl et al. 1991) which implies that its action in cell suspensions may be through enhancing the response to endogenous adenine nucleotides that have leaked from the cells (Sistare et al. 1985, Kawanishi et al. 1989). Furthermore, the light-scattering response caused by glucagon is delayed by about 90 s, whereas the response to vasopressin or phenylephrine is almost immediate (Quinlan et al. 1983). Yet the initial rise in cytosolic [Ca^{2+}] in response to glucagon is rapid and is maximal within 30 s or so, not dissimilar from the response to vasopressin. In addition, net uptake of Ca^{2+} into the cell is not observed following glucagon treatment alone, whereas it is with phenylephrine or vasopressin (Altin and Bygrave 1986, 1987). Thus, it would seem that the rise in cytosolic and therefore mitochondrial [Ca^{2+}] caused by glucagon added alone is not sufficient to cause an increase in mitochondrial volume. Rather glucagon may have an alternative cyclic AMP-dependent mechanism by which it increases the matrix volume. Glucagon treatment is known to cause changes in liver mitochondrial function that are persistent through isolation of the mitochondria, and may involve some change in the membrane phospholipids (see Halestrap 1989). It is possible that such events may underlie the effects of glucagon on matrix volume.

2.5 An Increase in Matrix Pyrophosphate Concentration May Mediate the Ca^{2+}-Induced Increase in Matrix Volume

Incubation of either isolated liver mitochondria or hepatocytes with short chain fatty acids such as butyrate produces a light-scattering response similar to that induced by Ca^{2+} or hormones (Halestrap et al. 1985; Davidson and Halestrap 1987). The activation of butyrate to butyryl-CoA occurs within the mitochondrial matrix and produces AMP and PPi. Since PPi is a weak substrate for the adenine nucleotide translocator (Krämer 1985), and our own data had implicated this carrier in mitochondrial volume regulation, a role for PPi in this process seemed possible. In isolated mitochondria we were able to demonstrate a butyrate or Ca^{2+}-induced rise in PPi that correlated with the light-scattering response with respect to both its time course and magnitude. Both responses were half-maximal at 0.3 µM [Ca^{2+}] and only when matrix PPi was increased beyond about 500 pmol/mg protein (as in the presence of Ca^{2+} and butyrate) did the correlation breakdown (Davidson and Halestrap 1987). This is probably a consequence of the precipitation of CaPPi or MgPPi. Addition of PPi to energized mitochondria was also able to produce swelling provided that adenine nucleotides were not present (Davidson and Halestrap 1987).

Changes in mitochondrial PPi could also be demonstrated in freeze-clamped livers from rats treated with glucagon or adrenaline (Davidson and Halestrap 1987) and in isolated hepatocytes exposed to butyrate, hormones or other Ca^{2+}-mobilizing agents such as A23187 and ADP (Davidson and Halestrap 1988). Rapid subcellular fractionation showed this PPi to be almost exclusively mitochondrial (Davidson and Halestrap 1988). This might be expected in view of the importance of PPi hydrolysis to drive many cytosolic reactions which involve the production of AMP and PPi. The increase in PPi induced by a wide variety of hormone treatments correlated with the observed light-scattering response with respect to both its time course and magnitude. Furthermore, both processes were inhibited by omission of Ca^{2+} from the incubation medium (Davidson and Halestrap 1988). However, when PPi reached very high levels, such as when glucagon and vasopressin were added together, no further decrease in light scattering was observed. Indeed, when both butyrate and vasopressin were added together the PPi could reach levels as high as 800 pmol/mg cell protein which is equivalent to 2400 pmol/mg mitochondrial protein. This was accompanied by an inhibition of the light-scattering response much as was seen in isolated mitochondria incubated with both Ca^{2+} and butyrate. Such large increases in mitochondrial PPi have also been observed by others in livers perfused with acetate and Ca^{2+}-mobilizing hormones (Inoue et al. 1989). When glucagon alone was added to hepatocytes, the increase in matrix volume also appeared to correlate with an increase in matrix PPi but, as discussed above, it is unlikely that this is induced solely by a rise in mitochondrial [Ca^{2+}].

It should be noted that in addition to causing an increase in matrix volume, the elevated [PPi] can account for the increase in total adenine nucleotides seen in liver mitochondria isolated from hormone-treated animals or following Ca^{2+} treatment of isolated mitochondria (see Davidson and Halestrap 1987, 1988; Aprille 1988). The adenine nucleotide carrier is known to be capable of exchanging PPi for ADP (Krämer 1985; Halestrap and Davidson 1989) and thus as matrix PPi leaves the mitochondria it can stimulate net adenine nucleotide uptake in a carboxyatractyloside-sensitive manner. However, other workers have provided evidence for a carboxyatractyloside-insensitive uptake mechanism for MgATP (Aprille 1988).

2.6 The Mechanism by Which Ca^{2+} Regulates Mitochondrial [PPi]

The concentration of PPi within the mitochondrial matrix must represent a balance between its synthesis, degradation and transport out of the mitochondria. The effect of Ca^{2+} on the matrix [PPi] is a result of inhibition of the soluble matrix pyrophosphatase which has been studied in this laboratory and that of Baykov (Baykov et al. 1989a; Davidson and Halestrap 1989; Unguryte et al. 1989). The V_{max} activity of the enzyme is about 400 nmol PPi hydrolyzed per mg matrix protein at 37 °C, and the apparent K_m for MgPPi is about 5 μM. Ca^{2+} is a powerful inhibitor of the enzyme, the sensitivity to Ca^{2+} decreasing with increasing concentrations of Mg^{2+}, an essential cofactor for enzyme activity. This is because Ca^{2+} acts primarily through formation of CaPPi which is a very strong competitive inhibitor with respect to MgPPi ($K_i < 0.1$ μM). The effective K_i for Ca^{2+} inhibition of the enzyme at 0.5–1 mM [Mg^{2+}], the probable matrix concentration (Rutter et al. 1990), is 5–10 μM. In contrast, the $K_{0.5}$ for Ca^{2+}-mediated increases in matrix volume and PPi under physiological conditions is about 0.3 μM (Davidson and Halestrap 1987). This difference between the sensitivity of the isolated pyrophosphatase and the matrix PPi content to [Ca^{2+}] might be explained by the presence of an endogenous inhibitor in liver mitochondria. This would seem quite likely for two reasons. Firstly, the concentration of PPi within the matrix is high enough to saturate the pyrophosphatase unless most of the PPi were bound. Whilst it is likely that some of the PPi is bound to both the ATPase and the pyrophosphatase (Issartel et al. 1987; Baykov et al. 1989b), the free [PPi] is likely to be sufficient to allow near-maximal activity of the pyrophosphatase. This would require PPi to be involved in a fast and energetically wasteful, futile cycle for which there is no evidence. Indeed, the increase in matrix PPi following Ca^{2+} addition to isolated mitochondria takes some 2 min to complete (Halestrap and Davidson 1987) which would not be expected for a rapid turnover of PPi. Secondly, in isolated heart mitochondria, whose matrix contains about the same activity of pyrophosphatase as does liver mitochondria, there is very little effect of Ca^{2+} on matrix PPi unless Ca^{2+} is elevated to very high concentrations (E.J. Griffiths and A.P. Halestrap, 1993). This suggests that in these mitochondria the pyrophosphatase is working sufficiently fast to remove PPi as rapidly as it is formed even when substantially inhibited by Ca^{2+}. This is not the consequence of a different rate of synthesis of PPi since when this is enhanced by the addition of acetate, heart mitochondria still fail to show an increase in matrix PPi, unlike liver mitochondria. This is also true in the perfused heart where neither acetate, adrenaline or increased extracellular Ca^{2+}, together or alone, were found to increase tissue PPi (E.J. Griffiths and A.P. Halestrap, 1993).

The pathway of PPi synthesis within the matrix is less clear. There are two potential sources of PPi within the mitochondrial matrix. Firstly, it could be produced from the breakdown of ATP, perhaps as a consequence of the turnover of mitochondrial phospholipids by the operation of the Ca^{2+}-activated phospholipase A_2 and subsequent re-esterification of the released fatty acids. This involves activation of fatty acids to fatty acyl-CoA and consequently the production of both AMP and PPi (Beatrice et al. 1980). However, we have shown that the inhibitor of mitochondrial phospholipase A_2, tetracaine, does not influence the ability of Ca^{2+} to increase the matrix volume and [PPi] (Halestrap and Davidson 1987). Secondly, there are reports of the presence of a PPi-synthesizing, proton-translocating pyrophosphatase which is membrane-bound and driven by the proton-motive force (see Mansurova 1989). Such an enzyme is well documented in bacteria where it plays a part in energy conservation (Baltscheffsky and Nyren 1984) but less well studied in mammalian mitochondria. I have used the membrane potential-sensitive dye safranin to assess whether PPi added to de-energized liver or heart mitochondria is able to create a membrane potential. This would be expected if a proton-translocating pyrophosphatase were present. I was able to show clearly that PPi can energize mitochondria in a carboxyatractyloside-sensitive manner which confirms that PPi enters the mitochondria only on the adenine nucleotide translocase (see Halestrap and Davidson 1989). However, the effect of PPi was largely abolished by oligomycin and aurovertin, inhibitors of the proton-translocating ATPase, but not by fluoride which is known to inhibit pyrophosphatases. These data might suggest that PPi can act as a poor substrate for the ATPase, but available evidence suggests that it binds very tightly to a regulatory rather than a catalytic site (Issartel et al. 1987). It is more likely PPi is able to form a small amount of ATP from AMP and some metabolic intermediate such as acetyl-CoA or fatty acyl CoA within the matrix, and then the ATP would energize the mitochondria. In heart mitochondria no equivalent energization was seen with PPi, but this might reflect the much greater hydrolysis of PPi within the matrix of heart mitochondria as discussed above.

2.7 The Mechanism by Which PPi Increases the Mitochondrial Volume

We have demonstrated an effect of PPi on K^+ permeability directly by studying the swelling of de-energized mitochondria in isoosmotic KSCN (Davidson and Halestrap 1987; Halestrap and Davidson 1989). In this buffer swelling is limited by the entry of K^+ and can be greatly stimulated by the addition of valinomycin. Addition of ADP inhibited swelling and this effect was reversed by the addition of either carboxyatractyloside or bongkrekic acid. Swelling was stimulated by addition of PPi, 20 mM PPi producing about the same rate of swelling as 0.5 μM valinomycin; the concentration ratio for their comparative efficacy was 4×10^4. In energized mitochondria exposed to 1 μM Ca^{2+}, matrix [PPi] increased by about 300 μM and swelling of the same order was induced by 0.5 nM valinomycin; the concentration ratio was 6×10^4. The similarity between these concentration ratios lends additional support that the PPi-induced K^+ permeability increase is responsible for the Ca^{2+}-induced swelling. These data are consistent with a role for the adenine nucleotide translocase in mediating K^+ permeability and the ability of PPi to bind to this carrier makes this an attractive hypothesis. We have proposed a model whereby PPi and Pi can transiently displace adenine nucleotides from the carrier and expose a channel capable of translocating K^+ ions. This is shown in Fig. 2. Since the effect of externally added PPi was not inhibited by either carboxyatractyloside or bongkrekic acid, PPi must also be capable of acting by binding to an external substrate binding site as well as from within the matrix, and in the presence of inhibitors which lock the carrier in either the "m" or "c" conformation. However, we have shown (Halestrap and Davidson 1989) that the effects of externally added PPi can

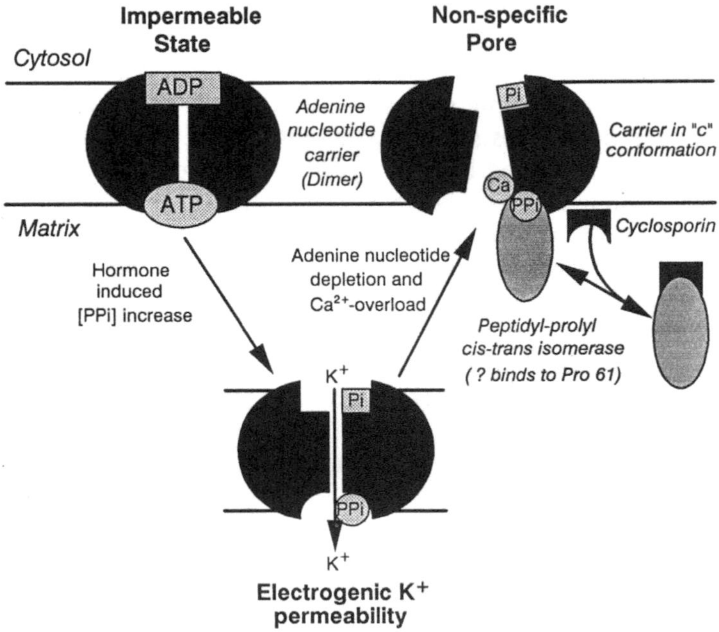

Fig. 2. A scheme illustrating how the adenine nucleotide translocase may be involved in both the PPi-induced increase in K^+ permeability of the inner mitochondrial membrane that is responsible for the hormonal effects on mitochondrial volume, and in the increase in non-specific permeability induced by Ca^{2+} overload of mitochondria during reperfusion following a period of ischaemia

be mimicked by N-(hydroxyethyl)ethylenediamine-triacetic acid (HEDTA), a metal chelator whose Mg complex has a K_d of 43 μM at pH 7.2 similar to that of MgPPi (55 μM). Thus, it is possible that PPi-induced swelling is merely due to chelation of Mg^{2+} which is known to have multiple effects on the ionic permeability of the inner mitochondrial membrane including a stimulation of electrogenic K^+ flux (see Bernadi et al. 1989; Nicolli et al. 1991).

It has been proposed that anionic monoazo dyes inhibit the adenine nucleotide translocase by chelating a Mg^{2+} ion within the carrier which is essential for activity (Boos and Schlimme 1983). If this were the case, then it may be that this Mg^{2+} is important in preventing non-specific cation movement through the pore of the gated carrier. Its chelation by PPi or HEDTA might allow the passage of K^+ to occur. We are currently developing techniques to test our hypothesis directly using the reconstituted adenine nucleotide translocase. However, the model presented in Fig. 2 receives some support from the observation that both the glutamate/aspartate carrier and the adenine nucleotide translocase can be converted into much less specific pores or channels by modification of specific thiol groups (Dierks et al. 1990a,b).

3 Changes in the Mitochondrial Matrix Volume as a Mechanism of Hormone Signal Transduction Across the Liver Mitochondrial Membrane

3.1 The Effects of Matrix Volume on Mitochondrial Metabolism

Adam and Haynes (1969) first demonstrated that incubation of mitochondria in hypoosmotic medium greatly stimulated their rate of pyruvate metabolism. These data were confirmed in this and other laboratories and it has been shown that a similar stimulation can be induced by addition of low concentrations (10^{-9} M) of valinomycin (see Halestrap 1989). It is now known that many other mitochondrial processes are stimulated by an increase in the matrix volume. These include the initial rate of citrulline synthesis (Armston et al. 1982), glutaminase (McGivan 1988), the rate of uncoupled and ADP-stimulated oxidation of all substrates entering the respiratory chain prior to ubiquinone (Armston et al. 1982; Halestrap 1982) and especially glycine oxidation (Halestrap et al. 1990) and the β-oxidation of fatty acids (Halestrap and Dunlop 1986; Halestrap 1987a). The relationship between the mitochondrial matrix volume and the activity of these processes is shown in Fig. 3. It is clear that all parameters are sensitive to small changes in the matrix volume between 0.9 and 1.4 μl/mg protein, which encompasses the range of

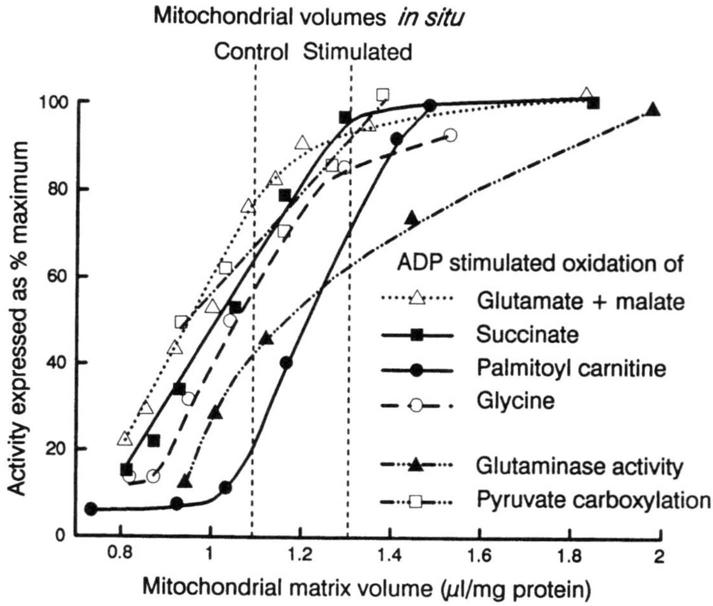

Fig. 3. The sensitivity of a variety of mitochondrial processes to changes in the matrix volume. Data are taken from the sources cited in the text

mitochondrial volumes we have determined in vivo. Furthermore, all these parameters are also stimulated in the perfused liver or isolated hepatocytes in response to glucagon, Ca^{2+}-mobilizing hormones and valinomycin as discussed below. Indeed, the ability of valinomycin to act like a glucogenic hormone on liver gluconeogenesis (see Thomas and Halestrap 1981) is best explained by its effect on mitochondrial volume. These considerations imply that hormonally induced increases in mitochondrial volume may be integral to the metabolic effects of the hormones. This will be discussed in Section 3.3.

3.2 Mechanisms by Which the Matrix Volume Influences Liver Mitochondrial Metabolism

Citrulline synthesis and pyruvate carboxylation both require a supply of ATP within the matrix and we have shown that the stimulation of these processes by an increase in matrix volume is secondary to the stimulation of the respiratory chain and rise in matrix ATP/ADP (Armston et al. 1982). Both these processes have been shown by others to be activated by increases in mitochondrial ATP/ADP ratio under appropriate conditions. Conversely, addition of low concentrations of a respiratory chain inhibitor such as amytal inhibited pyruvate carboxylation and citrulline synthesis by decreasing the ATP/ADP ratio (Halestrap and Armston 1984). We have used two techniques to identify the locus at which an increase in matrix volume stimulates electron flow through the respiratory chain (Armston et al. 1982; Halestrap 1982, 1987b; Halestrap and Dunlop 1986). Respiration in the presence of a variety of electron donors and acceptors that interact with different segments of the respiratory chain and spectrophotometric studies of the redox state of the cytochromes and flavoproteins both indicate that stimulation occurs on the reduced side of ubiquinone, leading to a regulation of electron flow into the ubiquinone pool. The oxidation of palmitoyl-carnitine is particularly sensitive to changes in the matrix volume (Halestrap and Dunlop 1986) and with this substrate it is possible to demonstrate that it is electron flow from the electron transferring flavoprotein (ETF) into the ubiquinone pool that is stimulated. This reaction is catalyzed by ETF dehydrogenase, an iron sulphur protein-containing enzyme (Ruzicka and Beinhert 1977) and it is of interest that electron transfer from NADH dehydrogenase and succinate dehydrogenase also involves iron sulphur proteins. Thus, some regulation of these proteins by matrix volume would seem likely, although the mechanism by which this is achieved is unknown.

It is quite possible that there is a volume-mediated effect on the fluidity of the inner mitochondrial membrane which is sensed by the respiratory chain, or unfolding of the cristae might alter the local environment around membrane-bound enzymes. In halotolerant bacteria changes in the osmotic strength of the external medium are known to influence membrane fluidity

(Curtain et al. 1983), whilst changing the fluidity of the inner mitochondrial membrane, e.g. by addition of benzyl alcohol, through the operation of phospholipase A_2 or by changes in diet, can all influence the activity of the respiratory chain significantly (Armston et al. 1982; Halestrap 1982; Narabayashi et al. 1982; McMurchie et al. 1983). Furthermore, glutaminase (McGivan 1988) and the glycine cleavage enzyme (Hampson et al. 1983) are both associated with the inner membrane and the properties of the former are known to be greatly influenced by that association. It is also possible that a change in membrane fluidity could affect the activity of the K^+/H^+ antiporter involved in osmoregulation as described above. In this context it is of interest that membrane stretching can activate a Ca^{2+} channel in the choroid plexus epithelium membrane (Christensen 1987; Bear 1990).

3.3 The Importance of Increases in the Mitochondrial Volume for the Hormonal Control of Liver Metabolism

3.3.1 Gluconeogenesis

Stimulation of gluconeogenesis by glucagon can be accounted for in part by an inhibition of pyruvate kinase and phosphofructokinase through cyclic AMP-dependent phosphorylation mechanisms (see Pilkis and El-Maghrabi 1988). However, there is also considerable evidence for a stimulation of pyruvate carboxylation following treatment of hepatocytes with glucagon and also Ca^{2+}-mobilizing hormones such as adrenaline and vasopressin. The latter exert little or no effect on phosphofructokinase and pyruvate kinase (see Thomas and Halestrap 1981; Patel and Olson 1986; Pryor et al. 1987), and under these conditions pyruvate carboxylase may be the major locus of the hormonal stimulation of gluconeogenesis. This enzyme is activated by acetyl-CoA and by increased mitochondrial ATP/ADP ratios and these have been thought to be the major mechanisms by which fatty acids stimulate gluconeogenesis from lactate and pyruvate in the liver (see Halestrap 1989). An increase in mitochondrial volume will stimulate both the respiratory chain and fatty acid β-oxidation and so increase the mitochondrial ATP/ADP ratio and acetyl-CoA concentration. This provides a mechanism by which an increase in matrix volume may regulate pyruvate carboxylase, and hence gluconeogenesis. Indeed, we have demonstrated that the respiratory chain exerts a strong influence on the rate of pyruvate carboxylation and gluconeogenesis in isolated hepatocytes (Quinlan and Halestrap 1986; Pryor et al. 1987; Halestrap and Owen 1991). Thus, the flux control coefficient of the respiratory chain for gluconeogenesis from L-lactate was found to be 0.61, indicating that it has a significant rate-controlling influence on gluconeogenesis. In the presence of glucagon or phenylephrine the flux control coefficient dropped to 0.39 and 0.25 respectively, confirming that the respiratory chain is stimulated by these hormones.

Measurement of the intermediates of the gluconeogenic pathway and of fluxes through pyruvate kinase has confirmed that the effect of mild respiratory chain inhibitors is mediated through a decrease in both cytosolic and mitochondrial ATP/ADP ratios. The former activates pyruvate kinase futile cycling, whilst the latter inhibits pyruvate carboxylation (Pryor et al. 1987; Halestrap and Owen 1991). At low pyruvate concentrations such as those found physiologically, an additional effect of the inhibitors is to increase the cytosolic $NADH/NAD^+$ ratio and so decrease cytosolic and mitochondrial pyruvate concentrations. This leads to further inhibition of pyruvate carboxylation. The biguanides, which have been widely used as hypoglycaemic agents in type 2 diabetes, are known to be mild respiratory chain inhibitors and probably exert their effects in a similar fashion (Halestrap and Owen 1991). The effects of such respiratory chain inhibitors on gluconeogenesis and the high flux control coefficient of respiration for gluconeogenesis argue strongly that volume-mediated stimulation of respiration plays an important role in the hormonal regulation of gluconeogenesis.

3.3.2 Fatty Acid Oxidation

In the fed state glucagon can activate fatty acid oxidation by cyclic AMP-dependent phosphorylation and inhibition of acetyl-CoA carboxylase which leads to a decrease in cytosolic [malonyl-CoA]. This metabolite acts as a powerful inhibitor of carnitine acyl transferase 1, which is thought to be the major rate-controlling step in the oxidation of fatty acids (McGarry and Foster 1983; Schulz 1991). However, there are reasons to believe that this is not the only site of regulation of fatty acid oxidation by glucagon (see Halestrap 1989 for detailed references). Thus, glucagon stimulates fatty acid oxidation in the starved state where carnitine acyl transferase 1 is relatively insensitive to inhibition by malonyl-CoA and acetyl-CoA carboxylase is largely inactive, and also in biotin-deficient rats where little if any acetyl-CoA carboxylase activity would be expected. Furthermore, Ca^{2+}-mobilizing hormones such as phenylephrine and vasopressin can activate fatty acid oxidation and yet these hormones either have no effect on acetyl-CoA carboxylase or activate the enzyme. Thus, carnitine acyl transferase 1 is unlikely to be the locus of action of these hormones and this is confirmed by their ability to stimulate the oxidation of octanoate which enters the mitochondria in a carnitine-independent fashion. An intramitochondrial activation of fatty acid oxidation would appear to be the most likely explanation of these observations and could be provided by the observed increase in mitochondrial matrix volume.

3.3.3 Nitrogen Metabolism

Hormones which stimulate gluconeogenesis from such substrates as lactate, alanine and glutamine also stimulate urea synthesis under appropriate conditions. It is generally accepted that a major rate-controlling step in this pathway is carbamoyl phosphate synthetase, an intramitochondrial enzyme (see Meijer and Hensgens 1982). Although it is thought by many workers that the primary stimulation of this pathway is mediated through increases in N-acetylglutamate concentrations (Meijer and Hensgens 1982), these changes occur much more slowly than the rapid increases in urea synthesis observed following phenylephrine treatment (Taylor et al. 1986). A volume-mediated increase in respiratory chain activity (see below) may be responsible for at least part of this activation by providing the additional ATP required for the stimulated rate of citrulline synthesis observed in the presence of hormones. Coincident with the stimulation of urea synthesis, the increase in matrix volume is also able to cause the observed activation of the glycine cleavage pathway (Jois et al. 1990) and glutaminase (McGivan 1988). The activation of the latter enzyme ensures that NH_3 produced by amino acid catabolism is usually excreted as urea rather than as NH_4^+ which is essential for pH homeostasis (Häussinger 1990).

3.4 Hormonal Activation of Respiration Is Achieved by an Activation of Ca^{2+}-Sensitive Dehydrogenases and a Volume-Mediated Stimulation of the Respiratory Chain

Hormonal stimulation of both gluconeogenesis and urea synthesis requires increased respiration to provide the additional ATP required. The conventional views of the regulation of respiration would imply that this must be achieved either by a decrease in proton-motive force (which should be reflected in a decrease in mitochondrial ATP/ADP*Pi) or by a rise in mitochondrial $NADH/NAD^+$. However, as reviewed elsewhere (Halestrap 1989; McCormack et al. 1990) what is actually observed when hormones stimulate respiration in the liver is an increase in mitochondrial ATP/ADP*Pi and possibly in the membrane potential. The $NADH/NAD^+$ ratio may increase transiently but returns to control levels or below even though respiration remains stimulated. These observations can be explained by activation of Ca^{2+}-sensitive dehydrogenases and a volume-mediated stimulation of the respiratory chain. The scheme shown in Figure 4 summarizes this, showing how hormones acting at the cell surface can stimulate mitochondrial respiration and hence metabolism.

Hormones such as vasopressin and phenylephrine cause a rise in cytoplasmic $[Ca^{2+}]$ which is relayed across the mitochondrial membrane to elevate mitochondrial $[Ca^{2+}]$. This rapidly activates the mitochondrial Ca^{2+}-sensitive dehydrogenases (McCormack and Denton 1989) causing a rise in the mitochondrial $NADH/NAD^+$ ratio, which can be detected by an increase in NAD(P)H fluorescence and an increase in the β-hydroxybutyrate/acetoacetate ratio, cytochrome c reduction and the rate of respiration (Quinlan and Halestrap 1986). A similar response is seen in isolated liver mitochondria exposed to submicromolar $[Ca^{2+}]$. A characteristic of the increase in NAD(P)H fluorescence is its transience. After the initial rapid

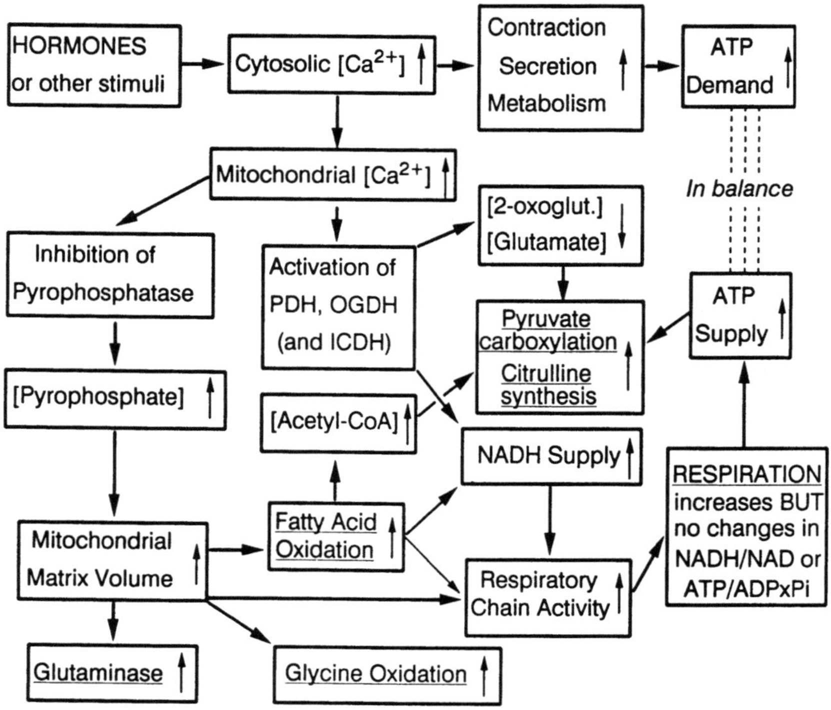

Fig. 4. A scheme illustrating how hormones acting at the cell surface may influence mitochondrial metabolism through increases in both matrix volume and Ca^{2+}-sensitive dehydrogenase activity. These allow ATP production to keep in step with ATP utilization without perturbation of ATP/ADP or $NADH/NAD^+$ ratios and also, in the liver at least, the hormonal stimulation of metabolic pathways involving mitochondrial steps (*underlined*)

rise the fluorescence decreases again with the same time course as the increase in mitochondrial PPi and matrix volume (as detected by the decrease in light scattering). Yet in all cases cytochrome c remains more reduced and the rate of respiration stimulated. This is a consequence of the increase in mitochondrial matrix volume stimulating electron flow from NADH to cytochrome c. The rise in NAD(P)H fluorescence caused by glucagon is more transient than with vasopressin or phenylephrine and may lead to the mitochondrial NADH becoming more oxidized than in the absence of hormones (Agius et al. 1986; Patel and Olson 1986; Quinlan and Halestrap 1986). This may reflect a sustained increase in mitochondrial volume that is not accompanied by a sustained increase in $[Ca^{2+}]$ and dehydrogenase activation. Evidence for such a mechanism is provided by the effects of valinomycin on intact cells or isolated mitochondria. Here, there is no rapid Ca^{2+}-induced increase in NADH and cytochrome c reduction, but rather a slow oxidation of NADH and reduction of cytochrome c whose time courses parallel that of the decrease in light scattering (Quinlan and Halestrap 1986). This is exactly what would be expected for a stimulation of electron flow between NADH and cytochrome c mediated by increases in mitochondrial matrix volume.

The hormonal stimulation of respiration by the perfused liver is also biphasic (see for example Buxton et al. 1982). There is an initial rapid increase in the rate of respiration which follows the time course of the increase in NAD(P)H fluorescence, then a slight decrease and finally another increase in rate. This latter phase follows the same time course as the increase in mitochondrial matrix volume, and thus may represent a volume-mediated stimulation of the respiratory chain. Such a two-phase mechanism for the stimulation of respiration offers distinct advantages for the cell. It allows the respiratory chain to synthesize ATP at a faster rate to fuel the increased rates of gluconeogenesis and citrulline synthesis, but it does so without decreasing the ATP/ADP or NAD^+/NADH ratios, both of which can be inhibitory for gluconeogenesis (see Sistare and Haynes 1985; Halestrap and Owen 1991). Indeed, mitochondrial ATP/ADP ratios often increase (see Halestrap 1989), which may be important for the activation of pyruvate carboxylation and citrulline synthesis as discussed above. Furthermore, the increase in matrix volume is able to stimulate fatty acid oxidation and hence increase production of acetyl-CoA to supply the stimulated citric acid cycle and to activate pyruvate carboxylase. There is therefore a concerted stimulation of both the production and utilization of NADH and ATP to supply the increased requirements of the cell with minimal disturbance of their concentration. In addition, the concentrations of glutamate and 2-oxoglutarate are decreased as a result of the stimulation of 2-oxoglutarate dehydrogenase (see McCormack et al. 1990), and this is desirable since glutamate acts as an inhibitor of pyruvate carboxylase (Scrutton and White 1974). Thus, a decrease in its concentration may be important in the hormonal stimulation of this enzyme.

3.5 Changes in Hepatocyte Volume May Also Stimulate Mitochondrial Metabolism Through an Increase in the Matrix Volume

The groups of Häussinger and Hue have drawn attention to the fact that amino acid uptake by the liver leads to 10–20% cell swelling. Such swelling, whether mediated by amino acids or hypotonic shock, is associated with a variety of metabolic responses. This is reviewed by Häussinger et al. (this Vol.). These responses include a stimulation of respiration, urea synthesis, glutaminase, glycine oxidation and a decrease in the mitochondrial NADH/NAD^+ ratio, which are all characteristic of a concomitant increase in mitochondrial volume. However, this has yet to be measured directly. It is unlikely that such a change in volume occurs as a result of a decrease in the osmolality of the cytosol, since it is an increase in osmolality, accompanying amino acid uptake, which drives the entry of water into the cell to cause swelling. Rather it is likely that swelling produces some other intracellular signal which can exert its effects on the volume regulatory mechanisms of the mitochondria. A rise in cytosolic inositol-1,4,5-trisphosphate and [Ca^{2+}] have been observed in response to hepatocyte swelling (Baquet et al. 1991), but the magnitude of the effect would seem too small to account for the observed changes in mitochondrial metabolism.

4 Regulation of the Mitochondrial Volume in Other Tissues

4.1 Brown Fat

Upon cold adaptation or stimulation with adrenaline the rate of oxygen consumption by brown fat increases dramatically. This is usually thought to be due to an increase in the concentration of the uncoupling protein (thermogenin) or its activation by fatty-acid displacement of guanine or adenine nucleotides (Nicholls and Locke 1984). However, it is also well documented from electron microscope studies of brown fat that activation of respiration is accompanied by increases in mitochondrial volume (see Nedergaard and Cannon 1987). More recently, evidence has been presented that a 1-h exposure to β-adrenergic agonists to the cold increased the mitochondrial matrix volume measured using ^3H_2O from almost undetectable levels to about 1 μl/mg protein. This was accompanied by an unmasking of existing thermogenin which Nedergaard and Cannon (1987) claimed could be mimicked in control mitochondria by swelling, although this has not been substantiated by others (Milner and Trayhurn 1988). It is also known that the oxidation of fatty acids and other substrates by brown fat mitochondria is negligible under conditions of matrix condensation (Nicholls and Lindberg 1972). Thus, the data suggest that increases in the mitochondrial volume of brown fat mitochondria may be an essential process in the stimulation of thermogenesis.

4.2 Secretory Tissues

A wide variety of exocrine glands, when stimulated to secrete, form watery vacuoles. Morphological studies on the rat parotid, lacrimal and sweat glands have shown that formation of these vacuoles is also accompanied by an increase in the size of the mitochondria which is dependent on the presence of extracellular [Ca^{2+}] (Sampson 1982; Leslie and Putney 1983). It is not known whether this process involves a rise in PPi or whether it is important for secretion, but secretion is usually accompanied by an increase in respiration.

4.3 The Heart

The volume of isolated rat heart mitochondria appears to be relatively insensitive to [Ca^{2+}] within the physiological range as is matrix [PPi]. However, increases in volume induced by hypoosmotic treatment or valinomycin are able to stimulate the respiratory chain of heart mitochondria and especially the rate of β-oxidation of fatty acids in a similar manner to that observed in liver (Halestrap 1987a). There is no available data for the mitochondrial volume in the heart nor any studies on whether it might be regulated under physiological conditions. Nevertheless, under conditions of incubation thought to mimic the situation in vivo the mitochondrial volume of isolated heart mitochondria is in the range over which fatty acid oxidation and respiration are extremely sensitive. There are reasons to believe that increases in mitochondrial volume may be an important response of the heart to stimulatory hormones such as adrenaline or to increased workload (see McCormack et al. 1990). Such a mechanism would account for how stimulation of respiration can occur without a significant decrease in ATP/ADP*Pi, which would be undesirable for conditions of increased muscle contraction, or a large increase in $NADH/NAD^+$, which would have the undesirable effect of increasing lactate output by the heart. When fatty acids rather than glucose are the respiratory substrate, an increase in matrix volume might play an additional role. Stimulation of flux through the citric acid cycle cannot be achieved by Ca^{2+} activation of 2-oxoglutarate and isocitrate dehydrogenases alone, since an increased supply of acetyl-CoA is also required. When glucose is the respiratory substrate, this is achieved by Ca^{2+} activation of pyruvate dehydrogenase. However, when fatty acids are the main respiratory fuel, an alternative mechanism must be available. It would seem appropriate that this should also be a Ca^{2+}-dependent mechanism, and a Ca^{2+}-mediated increase in the matrix volume leading to stimulation of β-oxidation of fatty acids would provide such a mechanism.

5 Swelling of Mitochondria Occurs in Pathological States

5.1 The Liver in Disease

Electron microscope analysis of the liver of humans and animals in various diseased or drug-induced states has revealed the existence of giant mitochondria (see Tandler and Hoppel 1986). Whilst such mitochondria do not necessarily have increased matrix volumes per milligram mitochondrial protein, and indeed no measurements of this parameter have been made, it would seem quite probable that this is so. Two situations are of particular interest; firstly, alcohol-induced damage to liver (Taraschi et al. 1986). It is known that alcohol can cause a rise in cytosolic [Ca^{2+}] which might be expected to increase mitochondrial [PPi] and matrix volumes in its own right. However, in addition to this, a major product of alcohol metabolism is acetate (Lundquist et al. 1962) and this can be metabolized by liver mitochondria with the generation of PPi as discussed earlier. Since we know that butyrate and Ca^{2+} can produce massive increases in PPi in isolated mitochondria, it is possible that alcohol abuse might do the same and that this might account for the large mitochondria observed. Secondly, such giant mitochondria are sometimes observed in diabetes or prolonged starvation (Tandler and Hoppel 1986). Under these conditions ketosis occurs and acetate is produced by hydrolysis of acetyl-CoA (Seufert et al. 1974). This might also be expected to yield high levels of PPi and it is therefore of interest that Cohen (1987) has observed that about 25% of the livers of diabetic rats show a large PPi NMR signal; we have confirmed this by enzymatic assay (Halestrap 1989).

5.2 Cell Death and Reperfusion Injury – The Opening of a Non-Specific Pore in the Mitochondrial Inner Membrane

When cells are insulted with hypoxia or chemicals that deplete the cell of ATP, cellular [Ca^{2+}] increases as the ATP-dependent pumps that normally expel excess Ca^{2+} from the cell are unable to operate. At these higher [Ca^{2+}] concentrations mitochondria become overloaded with Ca^{2+}, especially if they are energized as may occur under conditions of reperfusion following a period of ischaemia. This leads to massive mitochondrial swelling and considerable ultrastructural changes which are irreversible and a prelude to cell death (Post et al. 1985; Orrenius et al. 1989; Crompton 1990; Herman et al. 1990). The massive swelling is a consequence of the opening of a non-specific pore for small molecular weight (<1500 daltons) solutes which become freely permeant (Crompton 1990). The molecular nature of this pore remains to be established, but several features of it are widely accepted (Crompton 1990; Gunter and Pfeiffer 1990; Halestrap and Davidson 1990; Griffiths and Halestrap 1991). It requires the presence of Ca^{2+} to open, is blocked by the presence of the immunosuppressant cyclosporin A and is associated in some way with both matrix peptidyl-prolyl *cis-trans* isomerase (PPIase) and the adenine nucleotide translocase. These observations have led us to present a hypothetical model of the mechanism of pore opening which is summarized in Fig. 2. We propose that in the presence of Ca^{2+} and under conditions of adenine nucleotide depletion the adenine nucleotide translocase takes up the "c" conformation and binds the isomerase to a proline residue exposed on the matrix face of the carrier. The resulting conformational change causes the opening of the pore. We have demonstrated that pore opening can be prevented by lowering the pH to 6.5 or below (Halestrap 1991) and it may be significant that lowering of the perfusion pH enhances recovery during re-oxygenation following a period of ischaemia (see Herman et al. 1990).

5.2.1 Pore Opening May Also Have a Physiological Role

Under normal physiological conditions the opening of the non-specific pore must be extremely limited, otherwise massive swelling and damage to mitochondria would result. Nevertheless, there is evidence that it does occur. Firstly, hormonally induced light-scattering changes in hepatocytes are modulated by cyclosporin, suggesting that at the cytosolic $[Ca^{2+}]$ present in stimulated cells some pore opening is occurring (Davidson and Halestrap 1990). Secondly, $[^{14}C]$-sucrose introduced into hepatocytes by electropermeabilization enters the mitochondria over a period of hours (Tolleshaug and Seglen 1985). Thirdly, cells stained with Rhodamine G, a dye which is concentrated within mitochondria in response to their membrane potential, show oscillations in the Rhodamine fluorescence, indicating that the mitochondria may undergo cycles of energization and de-energization (Bereiter-Hahn et al. 1983). Fourthly, contact sites between inner and outer membranes are enriched in the adenine nucleotide translocase (Brdiczka et al. 1990; Bucheler et al. 1991) and are involved in protein import which requires pore formation (Pfanner et al. 1990). In addition, it may be significant that prolonged cyclosporin therapy can lead to severe liver and kidney damage which is associated with changes in mitochondrial morphology. Damage is less severe with cyclosporin G than cyclosporin A, whilst FK-506 does not suffer from these side effects. This correlates with the potency of these immunosuppressants as inhibitors of pore formation (Griffiths and Halestrap 1991). Thus, it is possible that occasional pore opening is a normal part of mitochondrial function, perhaps for protein import or for removal of waste products for which specific transporters do not exist. Total blockage would then be damaging to the mitochondria and lead to the observed side effects.

Acknowledgements. Research in the author's laboratory is funded by grants from The Medical Research Council and British Diabetic Association. I am grateful to the many colleagues who have contributed towards the work presented here, and especially to Annie E. Armston, Andrew P. Thomas, Paul T. Quinlan, Anne M. Davidson and Elinor J. Griffiths, and also to Prof. R.M. Denton, Dr. J.G. McCormack and Dr. J.D. McGivan for many helpful discussions.

References

Adam PAJ, Haynes RC (1969) Control of mitochondrial CO_2 fixation by glucagon, epinephrine and cortisol. J Biol Chem 224: 6444–6450

Agius L, Chowdhury MH, Alberti KGMM (1986) Regulation of ketogenesis, gluconeogenesis and the mitochondrial redox state by dexamethasone in hepatocyte monolayer cultures. Biochem J 239: 593–601

Altin JG, Bygrave FL (1986) Synergistic stimulation of Ca^{2+} uptake by Ca^{2+}-mobilizing hormones in the perfused rat liver. A role for mitochondria in long term Ca^{2+} homeostasis. Biochem J 238: 653–661
Altin JG, Bygrave FL (1987) Ca^{2+} uptake stimulated by the synergistic action of glucagon and Ca^{2+}-mobilizing agents in the perfused rat liver occurs through the activation of a unidirectional Ca^{2+} influx pathway. Biochem Biophys Res Commun 142: 745–753
Aprille JR (1988) Regulation of the mitochondrial adenine nucleotide pool size in liver, mechanism and metabolic role. FASEB J 2: 2547–2556
Armston AE, Halestrap AP, Scott RD (1982) The nature of the changes in liver mitochondrial function induced by glucagon treatment of rats. The effects of intramitochondrial volume, aging and benzyl alcohol. Biochim Biophys Acta 681: 429–439
Assimacopoulos-Jeannet F, McCormack JG, Jeanrenaud B (1986) Vasopressin and/or glucagon rapidly increases mitochondrial calcium and oxidative enzyme activities in the perfused rat liver. J Biol Chem 261: 8799–8804
Baltscheffsky M, Nyren P (1984) The synthesis and utilization of inorganic pyrophosphate. N Compr Biochem 9: 187–197
Baquet A, Meijer AJ, Hue L (1991) Hepatocyte swelling increases inositol 1,4,5-trisphosphate, calcium and cyclic AMP concentration but antagonizes phosphorylase activation by Ca^{2+}-dependent hormones. FEBS Lett 278: 103–106
Baykov AA, Volk SE, Unguryte A (1989a) Inhibition of inorganic pyrophosphatase of animal mitochondria by calcium. Arch Biochem Biophys 273: 287–291
Baykov AA, Pavlov AR, Kasho VN, Avaeva SM (1989b) Allosteric regulation of yeast inorganic pyrophosphatase by substrate. Arch Biochem Biophys 273: 301–308
Bear CE (1990) A nonselective cation channel in rat liver cells is activated by membrane stretch. Am J Physiol 258: C421–C428
Beatrice MC, Palmer JW, Pfeiffer DR (1980) The relationship between mitochondrial membrane permeability, membrane potential and the retention of Ca^{2+} by mitochondria. J Biol Chem 255: 8663–8671
Beavis AD, Garlid KD (1990) Evidence for the allosteric regulation of the mitochondrial H^+/K^+ antiporter by matrix protons. J Biol Chem 265(5): 2538–2545
Beavis AD, Brannan RD, Garlid KD (1985) Swelling and contraction of the mitochondrial matrix. 1. A structural interpretation of the relationship between light scattering and mitochondrial volume. J Biol Chem 260: 13424–13433
Benz R (1985) Porin from bacterial and mitochondrial outer membrane. CRC Crit Rev Biochem 19: 145–190
Bereiter-Hahn J (1990) Behaviour of mitochondria in the living cell. Int Rev Cytol 122: 1–63
Bereiter-Hahn J, Seipel K-H, Voth M, Ploem JS (1983) Fluorimetry of mitochondria in cells vitally stained with DASPMI or Rhodamine 6 GO. Cell Biochem Funct 1: 147–155
Bernadi P, Pozzan M, Azzone GF (1982) Mitochondrial oscillation and activation of H^+/cation exchange. J Bioenerg Biomembr 14: 387–403
Bernadi P, Angrilli A, Ambrosin V, Azzone FA (1989) Activation of latent K^+ uniport in mitochondria treated with the ionophore A23187. J Biol Chem 264: 18902–18906
Boos K-S, Schlimme E (1983) A model for metallo-protein catalysed ADP,ATP transport in mitochondria. FEBS Lett 160: 11–15
Brdiczka D, Bucheler K, Kottke M, Adams V, Nalam VK (1990) Characterization and metabolic function of mitochondrial contact sites. Biochim Biophys Acta 1018(2–3): 234–238
Brierley GP, Jung DW (1988) K^+/H^+ antiport in mitochondria. J Bioenerg Biomembr 20: 193–209
Brierley GP, Jung DW (1990) Kinetic properties of the K^+/H^+ antiport of heart mitochondria. Biochemistry 29: 408–415
Brierley GP, Davis M, Jung DW (1987) Respiration-dependent uptake and extrusion of Mg^{2+} by isolated heart mitochondria. Arch Biochem Biophys 253: 322–332

Brown GC, Lakinthomas PL, Brand MD (1990) Control of respiration and oxidative phosphorylation in isolated rat liver cells. Eur J Biochem 192: 355–362

Bucheler K, Adams V, Brdiczka D (1991) Localization of the ATP/ADP translocator in the inner membrane and regulation of contact sites between mitochondrial envelope membranes by ADP – a study on freeze-fractured isolated liver mitochondria. Biochim Biophys Acta 1056: 233–242

Burgess GM, Bird GSJ, Obie JF, Putney JW (1991) The mechanism for synergism between phospholipase-C-linked and adenylylcyclase-linked hormones in liver-cyclic AMP-dependent kinase augments inositol trisphosphate-mediated Ca^{2+} mobilization without increasing the cellular levels of inositol polyphosphates. J Biol Chem 266: 4772–4781

Buxton D, Barron LL, Olson MS (1982) The effects of α-adrenergic agonists on the regulation of the branched chain α-ketoacid oxidation in the perfused rat liver. J Biol Chem 257: 14318–14323

Christensen O (1987) Mediation of cell volume regulation by Ca^{2+} influx through stretch-activated channels. Nature 330: 66–68

Cohen SM (1987) ^{13}C and ^{31}P NMR study of gluconeogenesis: utilization of ^{13}C-labeled substrates by perfused liver from streptozotocin-diabetic and untreated rats. Biochemistry 26: 563–572

Colombini M (1989) Voltage gating in the mitochondrial channel, VDAC. J Membr Biol 111: 103–111

Crompton M (1990) The role of Ca^{2+} in the function and dysfunction of heart mitochondria. In: Langer GA (ed) Calcium and the heart. Raven Press, New York, pp 167–198

Curtain CC, Looney FD, Regan DL, Ivancic NM (1983) Changes in the ordering of lipids in the membrane of *Dunaliella* in response to osmotic-pressure changes. An e.s.r. study. Biochem J 213: 131–136

Davidson AM, Halestrap AP (1987) Liver mitochondrial pyrophosphate concentration is increased by Ca^{2+} and regulates the intramitochondrial volume and adenine nucleotide content. Biochem J 246: 715–723

Davidson AP, Halestrap AP (1988) Pyrophosphate in the hepatocyte is located primarily in the mitochondria and is elevated by gluconeogenic hormones and A23187. Biochem J 254: 379–384

Davidson AM, Halestrap AP (1989) Inhibition of mitochondrial matrix inorganic pyrophosphatase by physiological [Ca^{2+}], and its role in the hormonal regulation of mitochondrial matrix volume. Biochem J 258: 817–821

Davidson AM, Halestrap AP (1990) Partial inhibition by cyclosporin A of the swelling of liver mitochondria in vivo and in vitro induced by sub-micromolar [Ca^{2+}] but not by butyrate. Evidence for two distinct swelling mechanisms. Biochem J 268: 147–152

Dierks T, Salentin A, Heberger C, Kramer R (1990a) The mitochondrial aspartate/glutamate and ADP/ATP carrier switch from obligate counterexchange to unidirectional transport after modification by SH-reagents. Biochim Biophys Acta 1028: 268–280

Dierks T, Salentin A, Kramer R (1990b) Pore-like and carrier-like properties of the mitochondrial aspartate/glutamate carrier after modification by SH-reagents – evidence for a preformed channel as a structural requirement of carrier-mediated transport. Biochim Biophys Acta 1028: 281–288

Diwan J (1987) Mitochondrial transport of K^+ and Mg^{2+}. Biochim Biophys Acta 895: 155–165

Garlid KD (1980) On the mechanism of regulation of the mitochondrial K+/H+ exchanger. J Biol Chem 255: 11273–11279

Garlid KD (1988) Mitochondrial volume control. In: Lemasters JJ, Hackenbrock CR, Thurman G, Westerhoff HV (eds) Integration of mitochondrial function. Plenum, New York, pp 257–276

Garlid KD, Beavis AD (1985) Swelling and contraction of the mitochondrial matrix. II. Quantitatve application of the light scattering technique to solute transport across the inner membrane. J Biol Chem 25: 13434–13441

Griffiths EJ, Halestrap AP (1991) Further evidence that cyclosporin A protects mitochondria from calcium overload by inhibiting a matrix peptidyl-prolyl cis-trans isomerase. Implications for the immunosuppressive and toxic effects of cyclosporin. Biochem J 274: 611–614

Griffiths EJ, Halestrap AP (1993) Pyrophosphate metabolism is the perfused heart and isolated heart mitochondria and its role in regulation of mitochondrial function by calcium. Biochem J (in press)

Gunter TE, Pfeiffer DR (1990) Mechanisms by which mitochondria transport calcium. Am J Physiol 258: C755–C786

Hackenbrock CR, Rehn TG, Weinbach EC, Lemasters JJ (1971) Oxidative phosphorylation and ultrastructural transformations in mitochondria in the intact ascites tumour cell. J Cell Biol 51: 123–157

Halestrap AP (1982) The nature of the stimulation of the respiratory chain of rat liver mitochondria by glucagon pretreatment of animals. Biochem J 204: 37–47

Halestrap AP (1987a) The regulation of the oxidation of fatty acids and other substrates in rat heart mitochondria by changes in matrix volume induced by osmotic strength, valinomycin and Ca^{2+}. Biochem J 244: 159–164

Halestrap AP (1987b) Glucagon treatment of rats activates the respiratory chain of liver mitochondria at more than one site. Biochim Biophys Acta 927: 280–290

Halestrap AP (1989) The regulation of the matrix volume of mammalian mitochondria in vivo and in vitro, and its role in the control of mitochondrial metabolism. Biochim Biophys Acta 973: 355–382

Halestrap AP (1991) Calcium dependent opening of a non-specific pore in the mitochondrial inner membrane is inhibited by pHs below 7. Implications for the protective effect of low pH against chemical and hypoxic cell damage. Biochem J 278: 715–719

Halestrap AP, Armston AE (1984) Re-evaluation of the role of mitochondrial pyruvate transport in the hormonal control of rat liver mitochondrial pyruvate metabolism. Biochem J 223: 677–685

Halestrap AP, Davidson AM (1989) The role of the adenine nucleotide transporter in the regulation ofintramitochondrial volume. In: Azzi A, Fonyo A, Nalecz MJ, Vignais PV, Wojtcak L (eds) Anion carriers of mitochondrial membranes. Springer, Berlin Heidelberg New York, pp 337–348

Halestrap AP, Davidson AM (1990) Inhibition of Ca^{2+}-induced large amplitude swelling of liver and heart mitochondria by cyclosporin A is probably caused by the inhibitor binding to mitochondrial matrix peptidyl-prolyl cis-trans isomerase and preventing it interacting with the adenine nucleotide translocase. Biochem J 268: 153–160

Halestrap AP, Dunlop JL (1986) Intramitochondrial regulation of fatty acid β-oxidation occurs between flavoprotein and ubiquinone. A role for changes in matrix volume. Biochem J 239: 559–565

Halestrap AP, Owen MR (1991) Hormonal control of respiration in the liver. In: Grunnet N, Quistorff B (eds) Regulation of hepatic function. Metabolic and structural interactions. Munksgaard, Copenhagen, pp 198–207

Halestrap AP, Quinlan PT, Armston AE, Whipps DE (1985) Mechanisms involved in hormone signal transduction across the mitochondrial membrane. Biochem Soc Trans 13: 659–663

Halestrap AP, Quinlan PT, Whipps DE, Armston AE (1986) Regulation of the mitochondrial matrix volume in vivo and in vitro. The role of calcium. Biochem J 236: 779–787

Halestrap AP, Davidson AM, Potter WD (1990) Mechanisms involved in the hormonal regulation of mitochondrial function through changes in the matrix volume. Biochim Biophys Acta 1018: 278–281

Hampson RK, Barron LL, Olson MS (1983) Regulation of the glycine cleavage system in isolated rat liver mitochondria. J Biol Chem 258: 2993–2999

Häussinger D (1990) Nitrogen metabolism in liver – structural and functional organization and physiological relevance. Biochem J 267: 281–290

Herman B, Gores GJ, Nieminen AL, Kawanishi T, Harman A, Lemasters JJ (1990) Calcium and pH in anoxic and toxic injury. Crit Rev Toxicol 21: 127–148
Inoue T, Yamada T, Furuya E, Tagawa K (1989) Ca^{2+}-induced accumulation of pyrophosphate in mitochondria during acetate metabolism. Biochem J 262: 965–970
Issartel J-P, Favre-Bulle O, Lunardi J, Vignais PV (1987) Is pyrophosphate an analogue of adenosine diphosphate for beef heart mitochondrial F1-ATPase. J Biol Chem 262: 13538–13544
Jois M, Hall B, Brosnan JT (1990) Stimulation of glycine catabolism in isolated perfused rat liver by calcium mobilizing hormones and in isolated rat liver mitochondria by submicromolar concentrations of calcium. J Biol Chem 265: 1246–1248
Jung DW, Brierley GP (1986) Matrix magnesium and the permeability of heart mitochondria to potassium ions. J Biol Chem 261: 6408–6415
Jung DW, Apel L, Brierley GP (1990) Matrix free $Mg\text{-}^{2+}$ changes with metabolic state in isolated heart mitochondria. Biochemistry 29: 4121–4128
Kakar SS, Mahdi F, Li X, Garlid KD (1989) Reconstitution of the mitochondrial non-selective $Na^+/H^+:K^+/H^+$ antiporter into proteoliposomes. J Biol Chem 264: 5846–5851
Kawanishi T, Blank LM, Harootunian AT, Smith MT, Tsien RY (1989) Ca^{2+} oscillations induced by hormonal stimulation of individual Fura-2-loaded hepatocytes. J Biol Chem 264: 12859–12866
Krämer R (1985) Characterisation of pyrophosphate exchange by the reconstituted adenine nucleotide translocator from mitochondria. Biochem Biophys Res Commun 127: 129–135
LaNoue KF, Schoolwerth AC (1984) Metabolite transport in mammalian mitochondria. In: Ernster L (ed) New comprehensive biochemistry, vol 9. Elsevier, Amsterdam, pp 221–268
Leslie BA, Putney JW (1983) Ionic mechanisms in secretogogue-induced morphological changes in rat parotid gland. J Cell Biol 97: 1119–1130
Li XQ, Hegazy MG, Mahdi F, Jezek P, Lane RD, Garlid KD (1990) Purification of a reconstitutively active K^+/H^+ antiporter from rat liver mitochondria. J Biol Chem 265: 15316–15322
Lund P, Wiggins D (1987) The matrix water space of mitochondria in situ in isolated hepatocytes. Biosci Rep 7: 59–66
Lundquist FT, Tygstrup N, Winkler K, Mellemgaard K, Munck-Petersen S (1962) Ethanol metabolism and the production of free acetate in the human liver. J Clin Invest 41: 955–961
Mansurova SE (1989) Inorganic pyrophosphate in mitochondrial metabolism. Biochim Biophys Acta 977: 237–247
McCormack JG, Denton RM (1989) The role of Ca^{2+} ions in the regulation of intramitochondrial metabolism and energy production in rat heart. Mol Cell Biochem 89: 121–125
McCormack JG, Halestrap AP, Denton RM (1990) The role of calcium ions in the regulation of mammalian intramitochondrial metabolism. Physiol Rev 70: 391–425
McGarry JD, Foster DW (1983) Glucagon and ketogenesis. Glucagon I. Handb Exp Pharmacol 66: 383–398
McGivan JD (1988) Liver glutamine metabolism: regulation and physiological significance. In: Kvamme E (ed) Glutamine and glutamate metabolism in mammals, vol 1. CRC Press, Boca Raton, pp 183–202
McMurchie EJ, Gibson RA, Abeywardena MY, Charnock JS (1983) Dietary lipid modulation of rat liver mitochondrial succinate, cytochrome c reductase. Biochim Biophys Acta 727: 163–169
Meijer AJ, Hensgens HESJ (1982) Ureogenesis. In: Sies H (ed) Metabolic compartmentation. Academic Press, New York, pp 259–286
Milner RE, Trayhurn P (1988) Evidence that the acute unmasking of GDP-binding sites in brown adipose tissue mitochondria is not dependent on mitochondrial swelling. Biochem Cell Biol 66: 1226–1230
Munn EA (1974) The structure of mitochondria. Academic Press, London

Narabayashi H, Takeshige K, Minakami S (1982) Alteration of inner-membrane components and damage to electron-transfer activities of bovine heart submitochondrial particles induced by NADPH-dependent lipid peroxidation. Biochem J 202: 97–103
Nedergaard J, Cannon B (1987) Apparent unmasking of [^{3}H]-GDP binding and thermogenin antigen in rat brown fat is due to mitochondrial swelling. Eur J Biochem 164: 681–686
Nicholls DG, Lindberg O (1972) Inhibited respiration and ATPase activity of rat liver mitochondria under conditions of matrix condensation. FEBS Lett 25: 61–64
Nicholls DG, Locke RM (1984) Thermogenic mechanisms in brown fat. Physiol Rev 64: 1–64
Nicolli A, Redetti A, Bernardi P (1991) The K^+ conductance of the inner mitochondrial membrane – a study of the inducible uniport for monovalent cations. J Biol Chem 266: 9465–9470
Orrenius S, Mcconkey DJ, Bellomo G, Nicotera P (1989) Role of Ca^{2+} in toxic cell killing. Trends Pharmacol Sci 10: 281–285
Patel TB, Olson MS (1986) Regulation of gluconeogenesis from pyruvate and lactate in the isolated perfused rat liver. Biochim Biophys Acta 888: 315–324
Pfanner N, Rassow J, Wienhues U, Hergersberg C, Sollner T, Becker K, Neupert W (1990) Contact sites between inner and outer membranes – structure and role in protein translocation into the mitochondria. Biochim Biophys Acta 1018: 239–242
Pilkis SJ, El-Maghrabi MR (1988) Hormonal regulation of hepatic gluconeogenesis and glycolysis. Annu Rev Biochem 57: 755–783
Post JA, Leunissen-Bijvelt J, Ruigrok TJC, Verkleij AJ (1985) Ultrastructural changes of sarcolemma and mitochondria in the isolated rabbit heart during ischaemia and reperfusion. Biochim Biophys Acta 845: 119–123
Pryor HJ, Smyth JE, Quinlan PT, Halestrap AP (1987) Evidence that the flux control coefficient of the respiratory chain is high during gluconeogenesis from lactate in hepatocytes from starved rats. Biochem J 247: 449–457
Putney JW (1990) Receptor-regulated calcium entry. Pharmacol Ther 48: 427–434
Quinlan PT, Halestrap AP (1986) The mechanism of the hormonal activation of respiration in isolated rat heptocytes and its importance in the regulation of gluconeogenesis. Biochem J 236: 789–800
Quinlan PT, Thomas AP, Armston AE, Halestrap AP (1983) Measurement of the intramitochondrial volume in hepatocytes without cell disruption and its elevation by hormones and valinomycin. Biochem J 214: 395–404
Rutter GA, Osbaldeston NJ, McCormack JG, Denton RM (1990) Measurement of matrix free Mg^{2+} concentration in rat heart mitochondria by using entrapped fluorescent probes. Biochem J 271: 627–634
Ruzicka FJ, Beinhert H (1977) A new iron-sulphur protein of the respiratory chain – a component of the fatty acid β-oxidation pathway. J Biol Chem 252: 8440–8443
Sampson HW (1982) Increased mitochondrial size during cholinergic induced exocrine secretion. Cell Biol Int Rep 6: 981
Schöfl C, Sanchezbueno A, Brabant G, Cobbold PH, Cuthbertson KSR (1991) Frequency and amplitude enhancement of calcium transients by cyclic AMP in hepatocytes. Biochem J 273: 799–802
Schulz H (1991) Beta-oxidation of fatty acids. Biochim Biophys Acta 1081(2): 109–120
Scrutton MC, White MD (1974) Pyruvate carboxylase. Inhibition of the mammalian and avian liver enzymes by α-ketoglutarate and L-glutamate. J Biol Chem 249: 5405–5415
Seufert CD, Graf M, Janson G, Kuhn A, Soling HD (1974) Formation of free acetate by isolated perfused livers from normal, starved and diabetic rats. Biochem Biophys Res Commun 57: 901–909
Sistare FD, Haynes RC (1985) The interaction between the cytosolic pyridine nucleotide redox potential and gluconeogenesis from lactate/pyruvate in isolated rat hepatocytes. Implications for investigations of hormone action. J Biol Chem 260: 12748–12753

Sistare FD, Picking RA, Haynes RC (1985) Sensitivity of the response of cytosolic calcium in quin-2-loaded rat hepatocytes to glucagon, adenine nucleosides and adenine nucleotides. J Biol Chem 260: 12744–12747

Staddon JM, Hansford RG (1989) Evidence indicating that the glucagon-induced increase in cytoplasmic free Ca^{2+} concentration in hepatocytes is mediated by an increase in cyclic AMP concentration. Eur J Biochem 179: 47–52

Tandler B, Hoppel CL (1986) Studies on giant mitochondria. Ann NY Acad Sci 488: 127–139

Taraschi TF, Thayer WS, Ellingson JS, Rubin E (1986) Effects of ethanol on the structure and function of rat liver mitochondrial and microsomal membranes. Ann NY Acad Sci 488: 127–139

Taylor WM, Van De Pol E, Bygrave FL (1986) On the stimulation of respiration, by α-adrenergic agonists in perfused rat liver. Eur J Biochem 155: 319–322

Thomas AP, Halestrap AP (1981) The role of mitochondrial pyruvate transport in the stimulation by glucagon and phenylephrine of gluconeogenesis from L-lactate in isolated rat hepatocytes. Biochem J 198: 551–564

Tolleshaug H, Seglen PO (1985) Autophagic-lysosomal and mitochondrial sequestration of [^{14}C]-sucrose. Density gradient distribution of sequestered radioactivity. Eur J Biochem 153: 223–229

Unguryte A, Smirnova IN, Baykov AA (1989) Kinetic models for the action of cytosolic and mitochondrial inorganic pyrophosphatases of rat liver. Arch Biochem Biophys 273: 292–300

Williamson JR, Monck JR (1989) Hormone effects on cellular Ca^{2+} fluxes. Annu Rev Physiol 51: 107–124

Subject Index